JAMES CLERK MAXWELL

Faith, Church and Physics

This gripping book works at three levels. It is an intellectual and spiritual biography of a pioneering and world-class scientist, whose theologically informed faith and empirically based scientific knowledge grew in tandem. Second, it has an 'encyclopaedic' aspect, making all sorts of connections with people and movements and events of Maxwell's time. Third, it is truly a treatment of issues to do with created life and purpose seen from a believing viewpoint (inter alia evolution, energy and entropy, the soul and human freedom, purpose), which is essential for understanding anything we might wish to call Natural Theology in the 21st century. Ritchie's book is erudite, crystal clear in explanatory force as well as winsome and warm in tone and a most welcome major contribution to the field.

Mark W. Elliott, FRSE
Professor of Biblical and Historical Theology, UHI
and Professorial Fellow, Wycliffe College, University of Toronto

When it comes to science and religion in 19th-century Britain, historians inevitably point to James Clerk Maxwell as the foremost scientist who persevered in traditional Christian faith and belief. Yet, surprisingly, until now there has been no comprehensive historical study devoted to Maxwell's faith. Ritchie's monograph more than fills this gap, and in the process throws much needed light on Maxwell's activities, relationships, and intellectual development in every phase of life from his Scottish upbringing to his years at Cambridge. Ritchie writes with an impressive command of primary sources and a nuanced appreciation of social, philosophical, and religious contexts. As a result, this study is essential reading for anyone interested in either science or religion in 19th-century England or Scotland, as well as, of course, Maxwell himself. As a pure delight to read, with brilliant insight in all the hard places, it is a most constructive contribution to scholarship. I couldn't put it down.

Kerry V. Magruder
The John and Drusa Cable Chair of the History of Science
Curator, History of Science Collections
Department of the History of Science, Technology, and Medicine
University of Oklahoma

Through careful research and insightful deduction, Ritchie has produced a wonderful record of Maxwell's life. Ritchie gives remarkably clear explanations of the massively complicated questions and debates of the nineteenth century world of science. As he approaches it from the position of Maxwell's evolution in thinking Ritchie does not assume prior knowledge and therefore gently introduces readers to the beginning of the arguments before taking the reader through the various contemporary debates, which makes this book accessible to beginners and experts alike. He carefully analyses the available sources and draws his own conclusions about Maxwell. He is careful to locate Maxwell within the culture and debate of the nineteenth century and does not try to relocate his thinking into our modern-day world.

Ritchie paints an honest picture of the man. Maxwell is revealed as a wonderful, kind and gentle man. He did not shy away from scientific debate and argument yet maintained deep and precious friendships even in the midst of disagreement. He was a man of deep faith with a preference for simple Christianity, eschewing 'mental gymnastics' required by some preachers of his day. His faith was practically based, deeply influenced by the love and kindness he experienced as a young man when he was deeply in need. It also becomes evident that he was not a brilliant lecturer, but a wonderful mentor.

Many years after my time as a DPhil studies in physics I am undertaking biblical research from the position of faith. As I grapple again with questions of science and faith again I find myself indebted to Maxwell. As I read this book I constantly found myself agreeing wholeheartedly with the faith of Maxwell that underpinned the science I had previously loved. For as Ritchie carefully argues, for Maxwell it was never a question of science being compatible with faith. Rather they were completely inextricable, and that neither was possible without the other. This is a brilliant book for anyone who wants to understand and think deeply about the faith and life of a man on whose work much of modern science is built, and I highly recommend it.

Revd Dr Liz Grier
Curate and Associate Formation Tutor, St Mellitus College

Bruce Ritchie is well-placed to author this splendid volume on one of Scotland's most eminent scientists, James Clerk Maxwell. Ritchie is himself a Maths and Science graduate who went on to study Theology under T.F. Torrance, in whose class he had a 'Wow!' moment on hearing that Maxwell's theology had enabled him to think radically in his physics. Indeed, while Ritchie covers much other ground in this volume, it is to this core element that he returns in his conclusion, that it was Maxwell's 'humble Christian faith' that 'illuminated and advanced his scientific thinking,' giving him 'a foundation of optimism' – not the kind of thing the new atheists would like you to know!

Ritchie's aptness for this task has been further enhanced by his continuing interest in, and grasp of, both maths and physics, and by his academic teaching of Systematic Theology in Zomba, Malawi, and his teaching of Scottish Church History at Highland Theological College UHI; not to mention a lifetime of service in a geographical spread of congregations of the Church of Scotland, from Galloway to the Highlands. And then there is also his recently published volume on T. F. Torrance.

There is much in this present volume to delight the reader – and not just those interested in the science-faith debate. Along the way we learn about what appear to be Maxwell's evangelical beliefs and spirituality, often outworked in very practical expressions of compassionate care, but also alongside an appropriate ecumenical spirit. We learn much too about 19th century Scottish life and Church History, and also, through his letter writing, we get some insights into Maxwell's romance that led him into marriage to Katherine Dewar. We learn too of the way in which he and Katherine coped with life's afflictions and sufferings. While some of the more scientific passages may stretch your mind a little further than you may have been used to for some time – as they did mine – Ritchie writes in a very clear and winsome way and with a style that is easy to follow.

Hector Morrison
Principal HTC, University of the Highlands & Islands

G J Stodart

JAMES CLERK MAXWELL

Faith, Church and Physics

Bruce Ritchie

British Library Cataloguing in Publication Data: a catalogue record for this publication is available from the British Library

ISBN 978-1-912052-85-1

Typeset in 11pt Minion Pro at Haddington, Scotland

Printed by Bell and Bain, Glasgow

Cover images reproduced with the permission of the Master and Fellows of Trinity College, Cambridge

Frontispiece courtesy of the History of Science Collections, University of Oklahoma Libraries

The Publishers acknowledge with thanks grants from the Hope Trust, and from the Drummond Trust, 3 Pitt Terrace, Stirling

Almighty God, who hast created man in thine own image, and made him a living soul that he might seek after thee and have dominion over thy creatures; teach us to study the works of thy hands that we may subdue the earth to our use, and strengthen our reason for thy service; and so to receive thy blessed Word, that we may believe on him whom thou hast sent to give us the knowledge of salvation and the remission of our sins. All which we ask in the name of the same Jesus Christ our Lord. [**JCM,** *C&G1*, 323]

Lord, our Lord, how excellent is thy name in all the earth, who hast set thy glory above the heavens, and out of the mouths of babes and sucklings hast perfected praise. When we consider thy heavens, the work of thy fingers, the moon and the stars which thou hast ordained, teach us to know that thou art mindful of us, and visitest us, making us rulers over the works of thy hands, showing us the wisdom of thy laws, and crowning us with honour and glory in our earthly life.

Looking higher than the heavens, may we see Jesus, made a little lower than the angels for the suffering of death, crowned with glory and honour, that he, by the grace of God, should taste death for every man. Lord, fulfil thy promise, and put all things in subjection under his feet. Let sin be rooted out of the earth and let the wicked be no more. Bless thou the Lord, my soul, praise the Lord. [**JCM**, *C&G1*, 323]

Contents

Acknowledgements

Thanks are owed to many: Professor Kerry Magruder, who read an early draft and supplied invaluable insights from the history and philosophy of science; Professor Donald Meek with his knowledge of nineteenth-century Scottish evangelicism; Elizabeth McHugh of UHI who accessed numerous documents on my behalf; Catherine Booth and David Kerridges of the *James Clerk Maxwell Foundation* in Edinburgh whose kindness made my visit to 14 India Street so profitable; the staff of Highland Archives in Inverness, and the staff of the National Library of Scotland, who were obliging as always; Jock Stein and Handsel Press who have given immense assistance; Nicolas Bell of Trinity College, Cambridge, plus the Master and Fellows of Trinity College for permission to use the images on the cover. Thanks also to my wife Grace and daughter Alison who read and improved the text. For any remaining deficiencies or errors, I alone am responsible.

Bruce Ritchie,
Inverness, August 2023

Abbreviations

AV	Authorised Version (Bible)
BA(AS)	British Association (for the Advancement of Science)
BLOR	British Library Online Reader
C&G1	Campbell & Garnett, The Life of JCM, First Edition
C&G2	Campbell & Garnett, The Life of JCM, Second Edition.
CPS	The Cambridge Philosophical Society
DCP	Darwin Correspondence Project (www.darwinproject.ac.uk)
DNB	Dictionary of National Biography
DSCHT	Dictionary of Scottish Church History and Theology
FES	Fasti Ecclesiae Scoticanae
FRSE	Fellow of the Royal Society of Edinburgh
HPGN	History of Physics Group, Newsletter
JCM	James Clerk Maxwell
JCMF	James Clerk Maxwell Foundation
JCMFL	James Clerk Maxwell Foundation Library (14 India Street, Edinburgh)
JCMFN	James Clerk Maxwell Foundation Newsletter
NSAS	The New Statistical Account of Scotland
NRS	National Records of Scotland
NVN1	Niven (ed.), The Scientific Papers of James Clerk Maxwell, Vol. 1
NVN2	Niven (ed.), The Scientific Papers of James Clerk Maxwell, Vol. 2
PM	The London, Edinburgh, and Dublin Philosophical Magazine and Journal of Science
PRSE	Proceedings of the Royal Society of Edinburgh
RSE	Royal Society of Edinburgh
RSSA	Royal Scottish Society of Arts
SBET	Scottish Bulletin of Evangelical Theology
TCPS	Transactions of the Cambridge Philosophical Society
TGSI	Transactions of the Gaelic Society of Inverness
TRSE	Transactions of the Royal Society of Edinburgh

For Mary and Abraham,

that they may delight in learning about

the world and the One who made it

Prologue

'It is a universal condition of the enjoyable that the mind must believe in the existence of a law, and yet have a mystery to move about in.' [JCM, C&G1, 231]

When I was thirteen one of my Christmas presents that year was *The Illustrated Encyclopaedia of Science,* published by Purnell and Sons. I had spotted it on the shelves of a newsagent's shop in Jedburgh, and my cousin Graham and his wife Elsie gave it to me as a gift. The book was a treasure trove for a teenager interested in science, and it still has a place in my bookcase. As the years passed, its sections on the history of science, which highlighted prominent figures through the centuries, interested me more and more. Each scientist was given an artist's illustration plus a concise pen-portrait, and the list included great names: Archimedes, Bohr, Curie, Dalton, Darwin, Davy, Edison, Einstein, Galileo, Hertz, Mendeleev, Newton, Planck, Röntgen, Rutherford, Simpson, etc.

Tucked away in a secondary list, and without the prestige of an illustration, was a brief, twenty-six-word reference to James Clerk Maxwell: *Physicist. Studied heat properties of gases. Predicted existence of radio waves before they were discovered, and completely changed chemical theory with his electromagnetic theory of light.*[1] Despite this fleeting mention, Maxwell was of such stature that the American physicist Richard Feynman predicted that, in the long view of history, the most significant event of the nineteenth century would be Maxwell's discovery of the laws of electrodynamics. Similarly, Steven Weinberg argued that without Maxwell's famous equations contemporary physics would be unimaginable. Concerning the same formulae, Ludwig Boltzmann exclaimed: 'Was it a God who wrote these signs!' The equations are carved into the wall of Warsaw's University Library.

---

1 Carter etc. (eds.), *The Illustrated Encyclopaedia of Science,* 152.

Albert Einstein said that the four men who laid the foundations of physics on which he constructed his theories were Galileo, Newton, Maxwell, and Lorentz.[2] He also stated that one scientific epoch ended and another began with James Clerk Maxwell.[3] When Einstein made his first visit to the United Kingdom, he was asked if he stood on the shoulders of Newton. He replied: 'That statement is not quite right; I stood on Maxwell's shoulders.' Einstein esteemed Maxwell so highly that an image of the Scottish scientist adorned the wall of his study at Princeton. Commenting on Maxwell's influence, Einstein stated in 1931 that people used to conceive physical reality as material points, whereas after Maxwell they conceived physical reality as represented by continuous fields, which were not explicable mechanically but which were subject to partial differential equations.[4] Einstein regarded this transformation in the conception of reality as the most profound and fruitful change for physics since Newton.[5] For Einstein, the special theory of relativity owed its origin to Maxwell's equations of the electromagnetic field; he also held that Maxwell's notion of the electromagnetic field could only be grasped properly through his theory of relativity.[6]

In fairness, Purnell's *Illustrated Encyclopaedia of Science* did have another reference to Maxwell. Near the end, in a section dealing with historic events in science, it identified 1873 as the year when Maxwell put forward his electromagnetic theory of light. The *Encyclopaedia* was a young person's volume with no pretensions to being a comprehensive almagest of science, and the very fact that it mentioned Maxwell at all is to its credit. Moreover, its statement

---

2 Calaprice (ed.), *The Ultimate Quotable Einstein,* 373: ref. *New York Times,* 4 Apr 1921.

3 Forfar, 'Maxwell, Einstein, Newton and Faraday,' *JCMFN* (Spring 2012).

4 In his 1871 Cambridge Inaugural Lecture, Maxwell explained that differential calculus was the proper vehicle for dealing with the relations of a continuous quantity: JCM, *Introductory Lecture on Experimental Physics,* 20.

5 Einstein, 'Maxwell and Physical Reality,' *James Clerk Maxwell: A Commemoration Volume 1831–1931,* 66–73.

6 Einstein, *Autobiographical Notes,* 59. Einstein knew that Maxwell's equations had limitations. Keswani, 'Clerk Maxwell Centenary,' *Nature,* Vol. 283/5745 (24 Jan 1980), 330, argued that Maxwell may have been the first to use the term 'relativity' with reference to the doctrine of relativity of motion, not only for mechanical but also for electromagnetic phenomena.

that Maxwell completely changed chemical theory with his electromagnetic theory of light, should have been enough to alert any careful reader to science of immense consequence in his work. But at the time, and with my limited knowledge, the information on Maxwell failed to spark a 'Wow!' moment. He seemed less important than the familiar big names. Since then, Maxwell's significance has increased in the consciousness of society, with a stream of books and the occasional television programme highlighting his critical contribution to understanding the fundamentals of reality.

I read mathematics at university, and although we looked at equations associated with Einstein's relativity theory we did so as an exercise in mathematics rather than as a reflection on the nature of the universe. Nonetheless, like other students of the time, I purchased James Coleman's *Relativity for the Layman* and Banesh Hoffman's *The Strange Story of the Quantum*, read them, and tried to get my head round the concepts involved.

Unexpectedly, it was when I studied theology that Maxwell became more significant. Our professor, Thomas F. Torrance, was working on the relationship between methodology in the natural sciences and methodology in theology. This involved reflecting on how to think objectively in each discipline, plus consideration of the disruption imposed on human thought by the type of concepts which our minds attempt to comprehend. Our ability or inability to 'picture' an idea can profoundly affect how we understand it. Torrance focussed on what he termed 'the deep level conceptual interface' between theology and science, rather than on the old questions of science and religion such as whether science proved or disproved God.[7] Important to him was how Maxwell enabled physics to move beyond Newton, and how Maxwell explored the source of the rationality embedded in the created order. Torrance pointed out to us how Maxwell envisioned fields of force, instead of picturing individual objects acting on one another. He also stressed that our visual imagination is so accustomed to thinking in terms of objects hitting objects—since that is what we see around us—that we limit what we consider to be possible. Maxwell himself eventually moved beyond a visual representation of reality to a mathematical one. Though his deep-rooted instinct was to start

7 Torrance, 'The Significance of John Philoponos as a Forerunner of James Clerk Maxwell', *HPGN* (Spring 2000), 48.

with the empirical and the geometrical and to source everything in these, he came to realise that he needed to work differently, by-passing constraints imposed by the visual imagination. This radical new way of thinking inhibited early acceptance of his ideas. It was only after his death that Heinrich Hertz (1857–1894) and Oliver Lodge (1851–1940) validated his theories empirically.

Torrance taught us that theology needed to break away from similar limitations. He insisted that theology had to reconceptualise how things are: not from a visual construct, but from an auditory construct (analogous to Maxwell's mathematical modelling) and proceed from there. As students we found this massively stimulating. And Torrance went further. He argued that Maxwell was able to exchange a Newtonian view of reality for a more relational understanding—as required in Field Theory—precisely *because of* his Christian faith and *because of* the relational way of thinking which the doctrines of the Trinity and Incarnation required. That did bring a 'Wow!' moment!

Further Maxwell-consciousness came when I was Church of Scotland minister in the parishes of Borgue, Tongland, and Twynholm in the rural south-west of Scotland. A few miles to the north-east were the parishes of Corsock and Kirkpatrick-Durham, whose minister, Andrew Mackay, took a keen interest in Maxwell, after discovering that one of the stained-glass windows in his church was dedicated to the scientist.

In the mid-nineteenth century, Maxwell had been parishioner, churchgoer, and elder of Corsock Kirk, and was buried in nearby Parton churchyard. The Maxwell family home of Glenlair was also in the parish. This stimulated Mackay to find out more. He contacted Torrance and, encouraged by him, he deepened his research. Torrance then made at least two visits to the parishes. He preached in Corsock at the 1979 James Clerk Maxwell Memorial Service, and he preached in Parton Kirk in 1989 on 'Christ the Light of the World with reference to James Clerk Maxwell.'[8] The fruit of Mackay's work

8 Copies of the following are in the James Clerk Maxwell archive at 14 India Street, Edinburgh: (i) Torrance, 'Sermon: James Clerk Maxwell Memorial Service', Corsock Parish Church, 4 Nov 1979, 6pp typescript. (ii) Torrance, 'Christ the Light of the World with reference to James Clerk Maxwell', Notes of a sermon at Parton Kirk, 11 Jun 1989, 3pp typescript.

was shared with local clergy when we met for our monthly Ministers' Fraternal. Along with local historian Sam Callander, he subsequently published a short monograph.[9]

The Corsock Kirk memorial window, that drew Andrew Mackay's attention, is in the south wall of the church, gathering the sunlight whose electromagnetic characteristics were at the heart of some of Maxwell's greatest discoveries. The stained-glass depicts the three wise men following the star of Bethlehem and has the inscription (in Greek): '*All good things come from above.*' Adjacent to the window is a memorial plaque stating that Maxwell was the creator and first director of the Cavendish Laboratory at Cambridge. Another Maxwell plaque is on the north wall to the right of the entrance.

In the year 2000, a Millennium poll voted Maxwell the third greatest physicist of all time, behind Newton and Einstein. In December 2015, Professor Ian Stewart, geologist and broadcaster, presented a BBC documentary on Maxwell which had the title 'Scotland's Einstein'.[10] Stewart emphasised that if we evaluate Maxwell's contribution only in terms of the technological advances that his work made possible—such as radio, television, and telecommunications—then we undervalue his achievements. Maxwell not only enabled new things to be invented, but changed the way we understand the nature of reality. In the programme, Sir Peter Knight, professor of quantum optics at Imperial College in London, summed up Maxwell's genius, stating that, before Maxwell, the world was understood as a machine-like entity, but Maxwell made us understand the interaction between material bodies through the concept of a *field* penetrating space, with the *field* governing the way the world behaves. In this way of thinking all the forces in the universe are seen as interacting through *fields* rather than in direct physical contact as traditionally thought. Moreover, objects and forces and their interactions are not things which happen *in* space or time, but are part of the continuum *of* space and time itself. Maxwell

---

9 Mackay & Callander, *James Clerk Maxwell: Pathfinder of Modern Science*, (1979).

10 Ian Stewart, *James Clerk Maxwell: The Man who made the Modern World*. BBC Scotland, 2 Dec 2015.

did not just rearrange previously known theories. He fostered a radical change of conceptualisation which moved beyond Newton's view of physics, and which influenced all physics after him. It was this which Einstein had acknowledged.

What of Maxwell as a Christian believer? J. Oliver Rendell wrote that behind the name of James Clerk Maxwell lay one of the rarest, most original, and most loveable personalities.[11] But was he just a conventional theist of his time? Was he a brilliant scientist *despite* his faith, or *because of* his faith?

From an early age he was interested in church and religious issues. Throughout his life he discussed a wide range of ecclesiastical and theological matters with his friends. The aim of this book is to examine that faith and church life in relation to his science. Church was important to him and we have sought to place him within that context, since understanding his engagement with it is key to unlocking his mind, attitude, and vocation. Maxwell saw himself as discovering laws which God imprinted in the universe. He wrote: 'I think Christians whose minds are scientific are bound to study science, that their view of the glory of God may be as extensive as their being is capable of.'[12]

Though Maxwell rarely mentioned God in his scientific papers he saw the totality of the universe, including its origin, order, and sustained existence, as contingent upon the creative hand of the Almighty. At the same time, he held that a scientific analysis of the physical universe should never call on God as a quick-fix *Deus ex Machina* to fill a gap of understanding. Maxwell never had a faith based on a 'God of the gaps', so he was untroubled when apparent 'gaps' were closed or if natural explanations were able to be offered for phenomena previously thought to be mysterious. He believed that, in the exploration of nature, the scientist could always find yet another physical cause lying behind the present state of knowledge of reality. For Maxwell, that capability was rooted in the divine perfection of God's creation. At the same time, the ultimate logic of physical reality—and not only the enigma of its existence—was inexplicable without God. Maxwell's approach to issues of faith and science prompted the scholar B.F. Westcott (1825–1901) to state

11 Rendell, *Clerk Maxwell of Glenlair,* 5.

12 Maxwell, *Letter to Secretary of the Victoria Institute,* March 1885: C&G1, 404ff.

that for Maxwell there was an immensely close connection between physics and theology.[13]

Notable publications on Maxwell's faith include the following. Campbell and Garnet's *Life of James Clerk Maxwell* (1882) comments on his beliefs throughout their book; Campbell, however, was more theologically liberal than Maxwell, and this may have affected his choice of material from Maxwell's letters, plus shaping the occasional remark which he made on his friend's piety. Torrance's 1980 essay, 'Christian Faith and Physical Science in the Thought of James Clerk Maxwell' in his book *Transformation and Convergence in the Frame of Knowledge*, is important, as is his 1996 'Introduction' to his republication of Maxwell's *A Dynamical Theory of the Electromagnetic Field*. Torrance was especially interested in Maxwell in relation to a non-Newtonian view of time and space, and was fascinated by Maxwell using a methodology which, in Torrance's view, anticipated his own emphasis on intuitive or tacit knowledge.[14] Ivan Tolstoy's *James Clerk Maxwell: A Biography* (1981), is an essential read. Paul Theerman's article 'James Clerk Maxwell and Religion' appeared in the *American Journal of Physics* in 1983; and the Hungarian scholar Botond Gaal, who studied under Torrance and was heavily influenced by him, published *The Faith of a Scientist: James Clerk Maxwell,* in 2003. A key emphasis made by Gaal is how Maxwell sought to grasp the *whole* in the light of which he interpreted the *parts,* rather than vice versa.

Jerrold L. McNatt discusses 'James Clerk Maxwell's refusal to join the Victoria Institute' in *Perspectives on Science and Christian Faith: Journal of the American Scientific Affiliation* (2004). More recently, Andy Colebrooke has produced *Science and Religion: The Spirituality of James Clerk Maxwell* (2020), in which he—controversially in our opinion—states that Maxwell would be uncomfortable with the notion of Intelligent Design. Another significant contribution is Philip Marston's 'Maxwell, Faith, and Physics', in *James Clerk Maxwell: Perspectives on his Life and Work.*

---

13 Westcott, 'Cambridge Speech': C&G1, 358.

14 Torrance, 'Review of Ivan Tolstoy's *James Clerk Maxwell: A Biography*', *Life and Work,* December 1981, 42, refers to Maxwell's 'distinctive, intuitive way of thinking'.

Concerning primary sources, nearly all of Maxwell's extant papers are reproduced in one of the following: Campbell and Garnett's *Life of James Clerk Maxwell*; W.D. Niven's *The Scientific Papers of James Clerk Maxwell*, volumes I and II; or P.M. Harman's monumental *The Scientific Letters and Papers of James Clerk Maxwell,* volumes I, II, and III (printed in five books). Harman considers Maxwell's greatest work to be in field physics and statistical molecular theory.[15] Harman also gives a comprehensive account of which primary manuscripts are still extant, why others are missing—many lost in a house fire at Glenlair—and where surviving versions can be found. The reader is directed to Harman's introduction for detail on source materials.[16]

Maxwell's articles for the ninth edition of the *Encyclopaedia Britannica* are all available online, as are his various contributions to *Nature* and *The Philosophical Magazine*. Other letters lie in libraries including Edinburgh, Cambridge, Durham, and Basel. Maxwell also wrote anonymously on subjects of a scientific nature, some of which articles can now be attributed to him.

Marston offers the tantalising possibility that a few anonymous writings on religion may have been authored by Maxwell. Regrettably, we think not. The 'young man' with whom Professor George Wilson of Edinburgh corresponded in the 1840s and 1850s was Daniel MacMillan, not Maxwell. Similarly, the author of an 1859 document on Revivalism in Aberdeen was almost certainly not Maxwell but an unknown. These are discussed in the appropriate chapters. We have also chosen not to rely overmuch on Maxwell's two prayers which are reproduced in our opening pages. Though Maxwell would only compose a prayer whose sentiments he agreed with in both mind and heart, the words and phrases he used were liturgically conventional for the time. We have sought to explore his faith through his more direct statements.

15 Harman, II/1, xi.

16 Harman, I, xvii–xxv.

Galloway and Edinburgh
(1831–1850)

Chapter 1
Morning Light

On the fifteenth of June 1831, the twice-weekly *Scotsman* newspaper reported riots in Scotland, South Wales, and the Forest of Dean. These, and other disturbances taking place across the United Kingdom, were linked to growing pressure for parliamentary reform which would culminate in the ground-breaking 1832 Reform Act and its Scottish counterpart. That same day, the *Scotsman* also reported the tragic death in custody, in Tain in the north of Scotland, of a woman who had concealed her pregnancy; printed an indignant letter criticising the massive Henry Dundas statue in Edinburgh; supplied notes on naval appointments; and published economic and business news. But one of its most keenly anticipated features was the list of births, marriages, and deaths on page four. That morning the attentive reader would note: 'At No. 14, India Street, on the 13th inst. Mrs Clark Maxwell of a son.' This is the first reference in print to James Clerk Maxwell. A few weeks later an entry was made in the local Church of Scotland Birth and Baptismal register:

> John Clerk Maxwell of Middlebie esq., Advocate, and Mrs Francis Cay his spouse, St Stephen's Parish, a son born 13th June 1831, named James. Baptised 29th July 1831 by the Rev Dr David Ritchie.

The *Scotsman* spelt the middle name Clark whereas the baptismal entry had Clerk. The surname had long been Clerk without the Maxwell until the family inherited the property of Middlebie in south-west Scotland. Linked to Middlebie were 1,500 acres by the river Urr, seven miles from the market town of Castle Douglas, comprising Nether Corsock and the moorland of Little Mochrum. It was there, in the region of Scotland known as Galloway, that John Clerk Maxwell built the family house of Glenlair on land added to the original estate.[1] When he died in 1856, the parish register referred to him simply as John Clerk. The family rarely hyphenated the two names, and the modern practice is to refer to his son James as simply Maxwell.

1 Rendell, *Clerk Maxwell of Glenlair*, 9.

John Clerk Maxwell (1790–1856)

The Parish entry mentioned three significant figures: John Clerk Maxwell, Frances Cay, and the Revd Dr David Ritchie. It described John Clerk Maxwell as an Advocate, which in the Scottish legal system is like a Barrister in England and Wales. John married Francis Cay in 1826, and their Edinburgh home was the substantial terraced house of 14 India Street, set on a broad, sloping, cobbled thoroughfare in Edinburgh's prestigious and elegant New Town district. The land had been purchased in 1807 by the architect and local entrepreneur William Wallace, and was sold to John Clerk Maxwell in 1820 when the house was built. He was its first owner.[2]

In December 1829 the couple lost an infant daughter, Elizabeth, who lived for only a day. Following James' birth in 1831, the family stayed at India Street for two years, after which Glenlair was their main home but with India Street still available. In due course the house was rented out. Ownership transferred to James in 1856 on the death of his father, and when he died in 1879 the house was bequeathed to his wife, Katherine. After passing out of the Maxwell family it was acquired by the James Clerk Maxwell Foundation in 1993, and now houses the Foundation's James Clerk Maxwell Museum. Its rooms are furnished with period items associated with James and his contemporaries, including equipment, portraits, manuscripts, medals, and books.

When John Clerk Maxwell examined his copy of the *Scotsman* to check his son's birth notice, the reports of Reform Bill disturbances would temper paternal delight. He was uneasy about the proposed legislation, and the riots did nothing to alleviate his concerns. Three months before James' birth, he had added his name to a petition asking Parliament to consider the issues more carefully.[3] Along with other petitioners, he was of the view that the Reform Bill was ill-conceived and that it would fail to produce a reform which was moderate and gradual. In his opinion it would destroy the delicate balance which 'the stability of the British Constitution' depended upon, and would threaten social foundations essential for the security of persons and property. These worries increased as 1831 progressed, especially when any parliamentary vote unfavourable to reform was followed

2 Iain MacIvor, 'The Story of 14 India Street, Edinburgh' (JCMFL: ED06).

3 *Scotsman*, 16 Mar 1831.

by trouble and unrest. Consequently, in November 1831 he was one of several well-known Edinburgh gentlemen who called a public meeting with the aim of submitting yet another address to Parliament. They were disturbed because a recent rejection of the current form of the Reform Bill by the House of Lords—a temporary setback as it turned out—had 'been made a pretext for inciting the ignorant and unwary to acts of violence.'[4]

The easy-going John Clerk Maxwell was not a natural politician. Nor was he a high-flying legal eagle. His career in law had been modest and unspectacular. By 1831 he had abandoned the profession, preferring to run his estate and take pride in his herd of Galloway cattle. As a man of leisure he pursued amateur scientific interests, as well as exploring the Scottish Highlands and north of England.[5] At Glenlair he occupied a minor political post as chair of the Parochial Board for the district. In 1853, for example, he had to meet with representatives of the other local Parochial Boards to discuss implementation of the recent and contentious Poor Law (Scotland) Act.[6] Each Board was responsible for Poor House provision for all destitute persons in their parishes. The meeting decided that the Poor House in Kirkcudbright could serve all the local area.

In 1821 John Clerk Maxwell was elected a Fellow of the Royal Society of Edinburgh, and his zest for inventions and new discoveries meant he was a regular attender of RSE lectures, passing this enthusiasm on to his son. Several forebears had been members of Edinburgh's numerous intellectual societies, with James born into an enquiring family.[7] As an amateur inventor, one of John Clerk Maxwell's projects involved producing square-toed shoes which were shaped to give more room and comfort to the wearer. Less successfully, he designed clothes for his young son. Boy's outfits which were unremarkable in rural Galloway stood out as downright peculiar in Edinburgh!

John Clerk Maxwell had two siblings, George and Isabella. His elder brother was the politician Sir George Clerk (1787–1867), Sixth

---

4 *Scotsman,* 26 Nov 1831.

5 *Dumfries & Galloway Standard,* 12 Nov 1845.

6 *Ibid.,* 20 Apr 1853.

7 Arthur, *Brilliant Lives: The Clerk Maxwells and the Scottish Enlightenment,* (2016).

Baronet of Penicuik, who shared John's dislike of the Reform Bill.[8] Sir George later claimed to have changed his stance: 'The Reform Bill is now the law of the land, and there is no man in the multitude I see before me who will be more ready than myself to defend and maintain it.'[9] Sir George insisted that his original objection to the Act had been the great and sudden change he thought it would produce. To his credit, in 1826 Sir George had presented a petition to Parliament from the University of St Andrews against 'negro slavery'.[10] When the Slavery Abolition Act was finally passed in 1833, Sir George was temporarily out of Parliament having lost his seat in 1832, before being re-elected in 1835.

In the House of Commons, Sir George was well-known for raising matters on behalf of the Church of Scotland. He presented two petitions: one for enforcing the due observance of the Lord's Day, and the other against the sale of spirits on any portion of the Sabbath.[11] As a 'Peelite' he was a close friend of Sir Robert Peel (1788–1850) who was Prime Minister on two occasions: first briefly in the 1830s, and then for five years in the 1840s. It was through Sir George that his niece Jemima Wedderburn (1823–1909) was invited to a viewing of Peel's private collection of Dutch paintings. She commented that the pictures were very fine, but she did not care for Dutch pictures which, though beautifully done, were of ugly subjects! She considered the Prime Minister to be a better judge of politics than of art![12] Sir George, like his brother John, stayed within the Church of Scotland when the Disruption took place in 1843, but he was not obstructionist towards the new Free Church as were some. He ensured that road blockages, preventing easy access to the new Free Church building in Morningside, were cleared.[13]

John Clerk Maxwell's sister, Isabella Clerk (1789–1865), was known in her youth as the *Daisy of Pentland*, referencing the beauty of the Pentland Hills which lie south of Edinburgh. In 1813 she married the Scottish judge James Wedderburn (1782–1822) who came from a

8 *Scotsman*, 28 Sep 1831.

9 *Caledonian Mercury*, 20 Dec 1832; also 12 Jul 1832.

10 *Times*, 11 Apr 1826. (page 2).

11 *Scotsman*, 20 June 1840.

12 Hurd, *Robert Peel*, 86. Cf. Fairley, *Jemima: The Paintings and Memoirs of a Victorian Lady*.

13 Smith, *Historic South Edinburgh*, Vol. I, 114.

well-known Scottish family. In the late eighteenth century one of the Wedderburn ancestors challenged the claim of former slave Joseph Knight, that the Wedderburn family no longer 'owned' him. In a historic judgement the Scottish courts ruled that slavery was illegal in Scotland, and therefore Knight was a free man who was subject to no limitations on habitation or employment. Happily, by the time Isabella became one of the family, that unfortunate episode was in the past. One of Isabella's sons, Andrew Wedderburn (1821–1896), inherited the Middlebie and Glenlair estate when James Clerk Maxwell died in 1879 leaving no heir. At that point Andrew took the name Wedderburn Maxwell. It was through the Wedderburns that links were made between the Clerk Maxwells and the Mackenzie family from Peebles whom James knew well, especially Charles Mackenzie (1825–1862) who became an Anglican missionary in southern Africa.

Aunt Isabella's husband, James Wedderburn, was Solicitor General for Scotland, but died suddenly in November 1822 whilst in Kirkcudbright on a visit to his sister, the Countess of Selkirk. That was six months before their daughter Jemima was born. After his death, Aunt Isabella stayed on at 31 Heriot Row, which was also in the Edinburgh New Town district. This was a large and spacious house of five stories which became James' base when he later attended Edinburgh Academy and Edinburgh University. Though Jemima was nine years older than James, the cousins were close companions, and she participated in some of James' projects, supplying artwork for his devices.

Father to Sir George, Isabella, and John, was James' paternal grandfather, Sir James Clerk (d. 1793). He had been a naval captain with the East India Company, and had a talent for playing the bagpipes. A favourite family legend told of how, when his ship was wrecked near where the Hugli River enters the Bay of Bengal, Sir James swam ashore using the bag of his pipes as a float! On gaining land, he played a vigorous tune, not only cheering the survivors but frightening away any tigers in the area![14] Whatever truth or exaggeration was in the story, James loved to tell the tale and show the bagpipes to visitors at Glenlair.

14 C&G1, 4.

Frances Clerk Maxwell (1792–1839)

Under traditional Scottish law a married woman retains her birth name throughout life, hence the baptismal register refers to James' mother as Mrs Frances Cay rather than as Mrs Frances Clerk Maxwell. The Cays were from Charlton Hall of North Charlton in Northumberland, though by the early nineteenth century Edinburgh was the family base. Her father was Robert Hodshon Cay (1758–1810) who rose to be Principal Judge in the High Court of the Admiralty in Scotland, overseeing naval trials. The post gave him the grand title of 'Judge Admiral of Scotland'. Religiously, Cay was Episcopalian and a trustee of the Episcopal Charlotte Chapel in Rose Street, Edinburgh. That was before the congregation moved to the new-build St John's Episcopal Church at the west end of Princes Street.[15] The Charlotte Chapel building was then sold, and in time became the home of one of Edinburgh's foremost Baptist congregations. The Scottish Episcopal Church is part of the world-wide Anglican communion, though governmentally independent of the Church of England.

When John Clerk Maxwell and Frances Cay married on the fourth of October 1826, the ceremony took place in St John's, with Bishop Daniel Sandford officiating. Their union meant that, through his maternal line, James was brought up with an appreciation of the Episcopalian tradition which complemented his father's Presbyterianism. Further exposure to Anglican liturgy and theology came at school in Edinburgh and at Cambridge University. He bestrode both traditions with ease and was comfortable within the worship-life of each. Nor was he limited to these denominations. During his years in London, especially, he was no stranger to nonconformist preachers.

Frances Clerk Maxwell was a gentle, cultured, and deeply religious woman, and in her late thirties when James was born. The *Life of James Clerk Maxwell,* authored by Lewis Campbell (1830–1908) and William Garnett (1850–1932), portrays her as able, intelligent, musical, practical, and gentle, with kindness a notable part of her personality as it would be for her son. On one occasion, workers were seriously hurt during blasting at a quarry near Glenlair.[16] Frances

15 Cf. Balfour, *Revival in Rose Street. (2007)*

16 C&G1, 12. Cf. *Perthshire Courier*, 10 Sep 1835.

responded immediately, looking after the injured before medical help arrived. In the *Life,* her quality of instinctive kind-heartedness is reflected in the practical nature of James' own Christian faith, with James frequently tending friends and family.

Frances enjoyed life at Glenlair. When an informal local archery group was formed, it was Frances who organised the ladies of Nether Corsock to make a silver belt with tassel, as the prize for a joint male/female competition.[17] On another occasion, she wrote to her sister, telling her that at a local church reopening she had loaned a baby's hat to Jane Crosbie, the minister's wife.[18] The Crosbies had been married for just over a year, and their first son, Robert, was only a few months old at the grand event.[19] Frances had an instinctive generosity. She was full of life, love, and laughter. Her death in December 1839 was a massive blow to her husband and eight-year-old son.

Frances had seven siblings, including Robert Dundas Cay (1806–1888), sometime Registrar of the Supreme Court of Hong Kong, who became James' lawyer. But the most significant member of Frances' family was her younger sister, Jane Cay (1797–1876). During James' school years, Aunt Jane lived at 6 Great Stuart Street which was near Aunt Isabella's house, and she was a major spiritual influence in James' life. Like the Maxwells, the Cays took faith seriously, and the *Life* refers to two 1841 letters from Aunt Jane to her nephew, which, even at James' young age, involved religious topics.[20] James had been with her in Edinburgh earlier that summer and was due to return in November to start school at Edinburgh Academy. The letters touched on spiritual issues though, regrettably, that content is not reproduced in the *Life.* Campbell and Garnett commented that Aunt Jane wrote of 'theological and other matters' and assumed—rightly or wrongly—that these would 'not generally be thought interesting to a boy of ten.' In the same correspondence, Aunt Jane thanked James for his thoughtfulness in gathering ferns for her, and expressed delight in his enjoyment in carrying out experiments and adventures. The Cay

17 *Perthshire Courier,* 24 Sep 1835 (from *Dumfries Advertiser*).

18 John Clerk Maxwell, *Letter to Jane Cay,* 25 Apr 1834 (Frances' insert): C&G1, 28.

19 Cf. *FES,* II,420–421.

20 C&G1, 44, fn. 1, refers to letters to JCM from Jane Cay, dated 1 Sep 1841 & 21 Oct 1841.

sisters also had an artistic bent, inherited from their mother, Elizabeth Liddell (1770–1831), which James shared. Lewis Campbell recalled James helping Aunt Jane with patterns and arrangement of colours, and one of his earliest applications of geometry was to correct the perspective of Aunt Jane's picture of the interior of Roslin Chapel, south of Edinburgh.

Aunt Jane died on the first of January 1876, less than four years before James' own death.[21] Their lively correspondence started during James' schooldays and lasted for most of his life. In an early letter from June 1845, when Aunt Jane was out of Edinburgh, James told her that he and his father went to her house to water the plants, and that he had 'got the lend of the whole of Home's Introduction to the *Knowledge of the Scriptures*, and Humphrey Prideaux's *Connection of the Old and New Testaments*, and Townshend's *Harmony*, which are of great use.'[22] These religious books were most likely borrowed from Aunt Jane's own library, and James knew she was interested in his spiritual education. She took increasing charge of James when Aunt Isabella became less able, with John Clerk Maxwell's diary—now lost—recording that James was 'continually' at his aunt's house.

Revd Dr David Ritchie (1763–1844)

The third name in the Parish Register is that of the Revd Dr David Ritchie. When the New Town area of Edinburgh was developed, its rapid population growth necessitated new parishes to be created and new churches to be built. One of the most architecturally admired of these was St Andrew's with its stylish elliptical classical design, which opened on George Street in 1784. John and Frances' home at 14 India Street was technically in St Stephen's parish, but they attended St Andrew's. In 1800, when the Revd William Moodie (1759–1812) was minister of the congregation, it was decided by the Town Council that St Andrew's should become collegiate. Though Edinburgh was a city, and Scotland's capital, its ruling body was termed the Town Council.

21 NRS has Jane Cay's death as 1 Jan 1876, as does the *Scotsman*, 4 Jan 1876. C&G1, 396 & Harman, III/1, 421, print: JCM, *Letter to Campbell*, 25 Dec 1876: 'We intended to have gone round by Edinburgh and pay Aunt Jane a visit; but we both had such bad colds that we came home to nurse them.' Campbell and Garnett misread the year date; Harman copied them.

22 C&G1, 71–72.

Soon afterwards, the Revd David Ritchie was called to be Moodie's colleague. Around the same time the congregation established a preaching station in nearby Elder Street, which had seating for over five hundred.[23]

David Ritchie was a Clerk Maxwell family friend and not just their minister. As a young man he had been employed as a tutor by the Clerks of Penicuik and had taught James' father (John) and his uncle (Sir George) when they were boys, plus assisting in the parish when the minister was ill.[24] After studying at the University of St Andrews he served as schoolmaster at Trinity-Gask in Perthshire before training for the ministry of the Church of Scotland. In 1799 he was inducted into his first parish in Kilmarnock. However, before going to Kilmarnock he had been assistant to William Moodie at St Andrew's church in Edinburgh. During that period, he had also been chaplain to the Royal Edinburgh Volunteers city militia, and a sermon on Psalm 127:1, '*Except the Lord build the house, they labour in vain that build it: except the Lord keep the city, the watchman waketh but in vain*', which he delivered in that capacity, went down well with the authorities.[25] They marked him as a man of promise. Consequently, in 1801 when the Town Council, as patron of St Andrew's parish, decided a second minister was required for the congregation, Ritchie was the man they wanted, and they persuaded him to return to the capital.

Ritchie's academic abilities led to his appointment in 1808 as professor of logic and metaphysics at the University of Edinburgh, a post which he held in conjunction with his parish duties. As at St Andrew's Church, where the Council had patronage powers, the university chair was in the gift of the Council which controlled all university posts in Edinburgh until the system was reformed in 1858. In 1811, Ritchie became a Fellow of the Royal Society of Edinburgh. As a keen sportsman he loved curling during the winter months, with some reckoning he was more illustrious on the ice than in the professor's chair![26] After retiring from the university in 1836 he continued as a minister at St Andrew's until his death in 1844.[27] His

23 Christie, *The Story of St Andrew's Edinburgh*, 49-50.

24 Ramsay, *Reminiscences of Scottish Life and Character*, 322.

25 *Scots Magazine*, 1 Jan 1796 & 1 May 1797

26 Watt, *The Book of Edinburgh Anecdotes*, 63.

27 *FES*, I, 90–91.

university successor was Sir William Hamilton (1790–1865), who became one of the most influential of James' teachers. Sir William was also a friend of James' uncle, John Cay (1790–1865).

What of Ritchie the preacher? Not every listener in the pew agrees with every point of doctrine from the pulpit, but John Clerk Maxwell was a discerning man who took faith seriously, and he must have been comfortable with what he heard. Ritchie's theological emphasis was Reformed and evangelical. Definitions can be elusive, but, within a mid-nineteenth-century context, to be evangelical involved emphasising doctrines such as: the centrality and authority of the Bible; the fall and depravity of man; justification by faith in Christ; spiritual rebirth; sanctification by the Holy Spirit; and the heart moved by a love of Christ. An evangelical preacher, whether Calvinist or non-Calvinist, would also stress the urgent importance of becoming right with God through the atoning work of Christ.[28]

In his published writings, Ritchie affirmed the mainstream doctrines of the Reformed faith. His main work was a two-volume theological analysis of Paul's Epistle to the Romans, titled *Lectures Explanatory and Practical on the Doctrinal Part of the Epistle to the Romans* (1831). Ritchie had set out to expound Reformed orthodoxy and was shocked when he was accused of heresy at the 1834 General Assembly of the Church of Scotland. The charge was made by the radical politician Andrew Johnston (1798–1862), who was a well-known firebrand in Scottish politics and church life. Johnston was Member of Parliament for St Andrews' Burghs in Fife, and a lay member of the Assembly that year. He moved that a committee be set up to inquire into alleged heretical doctrine in Ritchie's book. This proposal, which discomfited such a respected and distinguished figure as Dr Ritchie, was resisted on two grounds. First, Johnston had raised the matter without seeking clarification from Ritchie himself. Second, he should have taken his concern to a lower church court before bringing it to the Assembly.

28 Bebbington, *Evangelicalism in Modern Britain*, (1989), offers a modern definition: evangelicals emphasise (i) a high view of the Bible as ultimate authority; (ii) Christ's work of salvation on the cross; (iii) the necessity of conscious personal faith commitments and changed lives; (iv) a commitment to activism in person-to-person evangelism, missions, and moral reform.

Notwithstanding the procedural irregularities the allegations were serious, particularly for a minister of one of Edinburgh's most prominent churches and a professor at the university. What Johnston thought he detected in Ritchie's book was a weakening of the Reformed doctrine of justification by faith alone. He homed in on Ritchie's treatment of the Apostle's phrase, '*Not the hearers of the law only shall be justified, but the doers of it*' (Romans 2:13). He argued that Ritchie's interpretation of the text allowed the possibility that, to some degree, a person was put right with God by what he or she did, rather than relying solely, totally, and entirely on God's grace received through faith. When called to speak in his defence, Ritchie knew he needed to clarify his position on this important doctrine. He conceded that he may have been ambiguous in his wording, but stated to the General Assembly:

> I do, in the most explicit manner, disclaim all heresy, and maintain in the most decided manner, my adherence to the doctrine justification by faith alone without works; and if there be any passage in any of my writings which is at variance with this statement, I most solemnly disclaim that it was my intention to convey such a meaning.[29]

Stung and hurt by Johnston's accusations, Ritchie reminded the General Assembly that it had always given a clergyman an opportunity to refute alleged wrong doctrine, and that a man should be allowed to explain the meaning of his own words. The Assembly was satisfied with the 'solemn statement which the reverend Doctor had made', and it was agreed that Ritchie would furnish the Assembly's clerk with a written statement for the Assembly's minutes. To widespread relief the matter was dropped. Given that the General Assembly met each year in Dr Ritchie's own church, the accusation of heresy would be embarrassing. The principle of justification by grace alone, through faith alone, is pivotal in evangelical Reformed theology, and it was a major theme for clergy who exerted substantial influence on James Clerk Maxwell's faith in the years to come.

St Andrew's was a collegiate charge, and its evangelical credentials were strengthened in 1837 when the Revd John Bruce (1794–1880) became Ritchie's ministerial colleague. Bruce was

29 *Inverness Courier*, 4 June 1834; *Morning Chronicle*, 27 May 1834.

a vigorous preacher of evangelical doctrine. He was a son of the manse, with his early ministry at Guthrie, a small rural parish in Forfarshire. When Bruce went to Guthrie his style at first bemused the congregation, but was then loved by his increasingly devoted parishioners. In 1831 Bruce transferred from Guthrie to the New North Kirk in Edinburgh, which was part of the St Giles' complex on the capital's Royal Mile. But his Guthrie people resisted the loss of their beloved pastor.

At that time the New North Kirk appointment was in the gift of the Edinburgh Town Council. The Council had a great interest in procuring able and popular ministers to fill the pulpits of the city churches, and if a vacancy occurred they often sent a deputation to some country parish whose minister had been drawn to their attention. When the deputation came to Guthrie, they came upon a workman breaking stones at the roadside. The stone-breaker eyed the party with suspicion, before saying in his broad Scots, 'Ye'll nae doubt be some o' thae men that's come to take away oor minister!' He continued with considerable emotion, 'We didna care for him at the first, for we didna understand him; but when once we got ower the curiosity o' him then we thought, and we still think, there's naebody like him. Ye'll better no try tae take him, for I can assure ye, we had him twa year here in Guthrie afore we understood him; and if ye take him tae Edinburgh ye'll no understand him at aa!'[30]

Despite the reluctance of Guthrie to lose its minister, John Bruce did move to Edinburgh where, especially during his years of ministry at the New North Kirk, he profoundly influenced several notable young men who went on to become prominent evangelicals in both the Church of Scotland and the future Free Kirk. His regular hearers included Alexander Moody Stuart (1809–1898), minister, religious author, and Free Kirk historian.[31] There was also William C. Burns (1815–1868), one of the preachers at the Kilsyth revival of 1839 where he ministered alongside Daniel Dewar, Maxwell's future father-in-law. Burns, who was also at the centre of spiritual revival in St Peter's Dundee during his interim ministry there before becoming a missionary to China, attributed his own religious conversion

30 *Brechin Advertiser*, 10 Aug 1880.

31 Kenneth Moody Stuart, *Alexander Moody Stuart: A Memoir*, 440.

and 'change of heart' to Bruce.[32] Yet another moulded by Bruce's ministry was the short-lived Robert Murray McCheyne (1813–1843) whose memoirs are still hallowed in Scottish evangelical circles. McCheyne's diary for 1832 states: 'Jan. 29, Sabbath: Afternoon heard Mr Bruce on Malachi 1:1–6. A most noble discourse.'[33] From that date McCheyne sat almost entirely under John Bruce's ministry, taking copious notes of his lectures and sermons.

In his preaching John Bruce was said to have had 'an unction, a fervour, a rich flavour of evangelism, which made his ministrations as acceptable to the humblest members of his flock as it was attractive to the most intelligent and refined.'[34] This was the man whom St Andrew's congregation sought in 1837. The congregation was aware of Bruce's intense type of ministry at the New North Kirk and wanted it for themselves. With Ritchie and Bruce as colleagues, St Andrew's—as the Edinburgh spiritual home for John Clerk Maxwell and his family—had a clear evangelical emphasis.

---

32 Cf. W.C. Burns, *Letter to his Sister,* (20 Feb 1832): 'I had another conversation with Mr. Bruce about a week ago; I was as much as on the former occasion delighted with him, and I trust edified . . . They were both very much suited to my state, and I trust I was much benefited by them.' Also, Burns, *Journal,* (17 Nov 1841): 'Yesterday morning I breakfasted with Mr. Bruce [who] promised to give me the dates of several sermons which I was benefited by when preached. The means by which my change of heart was brought about were these, I think—Mr. Bruce's preaching, which engaged me much, and the fear of sudden death from the approach of cholera.'. Both documents cited by Islay Burns, *Memoir of the Rev. William C. Burns, M.A.,* pages 28 and 26.

33 Cited by Andrew Bonar, *Memoir and Remains of the Rev. Robert Murray McCheyne,* (Edinburgh: Oliphant, 1894), Ch. 1

34 *The Daily Review,* 6 Aug 1880.

Chapter 2
Galloway Childhood

A heavy shadow fell over Glenlair when James' mother died of cancer in 1839. He later wrote to his friend William Huggins (1824–1910) saying, though he had but eight years with his mother, and though his father became his companion in all things, he felt her loss for many years.[1] Despite this, his childhood at Glenlair had an idyllic quality, with happiness and contentment marking his Galloway upbringing. When he died in 1879, the local newspaper described his early years as having been spent in an atmosphere of lavish love and old-fashioned piety.[2] There was an affectionate home. There was fun with local children. There was paddling in the burn. There was sailing on the pond. There were picnics. There were games. This environment was a powerful factor in producing the person he became. Serenity and joy were as potent stimuli as the sadness he had experienced.

Cousin Jemima recalled James telling her that one of his earliest memories was of lying on the grass in front of his father's house, looking at the sun, and *wondering*.[3] Maggie Murdoch, the Glenlair housekeeper, remembered the two-and-a-half-year-old James being given a tin-plate toy to play with. It was a bright sunny day. He held it to the sun, and the reflection went around the room. He said to her: 'Do look, Maggie! Go for papa and mamma!' As they came in, he sent the reflection across their faces. His father was delighted: 'What is this you are about, my boy?' To which James replied: 'It is the sun, papa! I got it in with the tin plate!'[4]

He needed to know how and why things worked. Famously, his repeated question was: 'What's the go o' that? What does it do?' If the answer did not go deep enough, his supplementary question

1 JCM, *Letter to W. Huggins,* 13 Oct 1868, (Harman, II/1, 451).

2 *Galloway News & Kirkcudbrightshire Advertiser,* 10 Nov 1882.

3 C&G1, 28.

4 Margaret Murdoch, *Letter to Campbell,* undated: C&G1, 29.

was, 'But what's the particular go of it?'[5] His father encouraged his enquiring mind, and one of the scientific toys which he bought James was the thaumatrope or phenakistoscope, which was essentially a stroboscope.[6] When he looked through apertures near the rim of the revolving disc, a series of drawn figures gave the impression of continuous movement. James was entranced by his 'Magic Disc'.

As he grew older there were excursions: to New Abbey's Sweetheart Abbey, famously associated with the thirteenth-century John Balliol and Devorgilla; to the dramatic moated castle of Caerlaverock; and into the wild and rugged hill-country around the granite massif of Cairnsmore of Fleet.[7]

Another of the other local landowning families were the Fletchers of Corsock House, a few miles from Glenlair.[8] They were also part of local church life, and Major Fletcher and John Clerk Maxwell were both involved in financing the building of a new place of worship in the vicinity. Major Fletcher first married in 1830 and he and his wife had two sons, Henry (b. 1833) and Philip (b. 1834), who were James' regular playmates. Mrs Fletcher died in 1835. The Major remarried in 1838 and a daughter, Margaret Isabella, was added to the family. Lewis Campbell recalled the summer of 1845 when James spent much of the time with the Fletcher boys, as well as learning estate-management from his father. That same summer there was an extended jaunt to Newcastle, Durham, and Carlisle, suggested by Aunt Jane. This introduced James to the cathedral architecture of the Church of England.[9] The following year saw a tour of the Scottish Borders, with visits to Walter Scott's Abbotsford and nearby Melrose Abbey.

---

5 C&G1, 28.

6 Ibid., 36.

7 Ibid., 65.

8 Colonel Edward Charles Fletcher became Major Fletcher by 1844, before succeeding his father to a Baronetcy in 1846. M'Kerly, *History of the Lands and their Owners in Galloway*, Vol. 5, 46f, details Corsock House ownership: Fletcher sold it to John Murray of Edinburgh around 1846; Murray died in 1849 leaving an only child, Eliza Esther; she married the formidable Free Church lawyer Alexander Dunlop who added the name Murray, becoming Alexander Colquhoun Stirling Murray-Dunlop.

9 C&G1, 72.

Covenanting Heritage

A favourite local spot was the Covenanters' Pool which formed part of the stream above Upper Corsock.[10] During the seventeenth century, the south-west of Scotland was a cauldron of Covenanting activity. The Covenanters took their name from the National Covenant (1637) which was a Scottish declaration of resistance to Charles I's dictatorial policies in both Church and State, and which sprang from the fear that the King was determined to undermine the independence of the Scottish (Presbyterian) Church and the authority of the Scottish Parliament. After the civil wars of the 1640s, Charles I was executed by the English Parliament in 1649. Oliver Cromwell's Commonwealth then took power, before Charles II was restored to the throne in 1660.

Though Charles II had earlier signed the National Covenant along with the Solemn League and Covenant, he soon imposed Episcopalian Church government in Scotland. In 1662 an Act was forced through the Scottish Parliament which abolished Presbyterianism and ejected all ministers from their parishes if they had been inducted after 1649, unless they acknowledged the legitimacy of Episcopalian Church government, and unless they applied to the local bishop and patron for the appropriate licence. This was unacceptable to staunch Presbyterians. To their mind, Episcopalianism was rule by bishops, and bishops were regarded as puppets of a totalitarian monarchy with Roman Catholic leanings. Many of the ejected Presbyterian ministers initiated open-air worship meetings called conventicles. These became immensely popular, especially in Scotland's Galloway region in the south-west. Then the government clamped down. Conventicles were declared illegal. Those who led or attended conventicles were dealt with ruthlessly.

The Covenanters' Pool of Upper Corsock marked the site of one of these conventicle meeting-places, scores of which are scattered across Galloway. Corsock estate also had the remains of an old castle, the erstwhile residence of Robert Nelson of Corsock, who was involved in events of that time.[11] Though James' spiritual upbringing made him appreciate both Presbyterianism and Episcopalianism, he was proud of the covenanting traditions of his home country.

10 To the north of Balmaclellan is Holy Linn, a small waterfall and pool in the Garpol burn used by the Covenanters as a place of worship and where children were sometimes baptised. (Sturrock, *Corsock Parish Church,* 22).

11 Crosbie, 'Parish of Parton', *NSAS* (1845), Vol. IV, 283.

This led him, whilst a student at Cambridge, to read a paper on the Covenanting movement for a college declamation.[12] His father was intrigued, wanting to know more:

> What sort of thing is a 'College Declamation'? You say you have chosen the Scottish Covenanters. Do you take the part of Advocate or Apologist for them? Or do you try the impartial historian? It would be difficult to give the Scots Prelates their due without offending some of the Order. During the Persecutions both the civil and ecclesiastical government of Scotland was getting ripe for the Revolution.[13]

James' paper sparked his father into further reflection on the Covenanters, and he wrote another letter to James in which he mused on a 'What might have been?' scenario.[14] What if, during the tumultuous seventeenth century, someone like the godly and irenic Bishop Robert Leighton of Dunblane had been Bishop of Galloway? Or if Leighton had been Archbishop of St Andrews instead of the ill-respected James Sharp? Might Episcopacy have become acceptable? Might it have been secured in Scotland? James' father was aware of the subtle nuances of seventeenth-century history. He knew there had been Episcopalian bishops in Scotland whose example might have paved the way for a more tranquil and peaceful era. This reveals John Clerk Maxwell as a reflective Presbyterian. Though he was committed to the Kirk—as the established Church of Scotland is still known—that commitment was never a narrow one. A similar ecumenicity of spirit was handed on to James.

Parton Kirk

Throughout James' childhood years the family attended Parton Kirk, six miles from Glenlair. This was the Maxwell's local church before Corsock Kirk was built and was where John Clerk Maxwell served as an elder in the congregation. In the Church of Scotland, the leadership of each congregation comprised the parish minister

12 C&G1, 161.

13 John Clerk Maxwell, *Letter to JCM*, 9 Nov 1852: C&G1, 182. 'Revolution' refers to the 'Glorious Revolution' of 1688 when the Protestant William and Mary took the throne from the Roman Catholic James VII & II.

14 John Clerk Maxwell, *Letter to JCM*, 11 Nov 1852: C&G1, 182.

plus elders, who together formed the Kirk Session. A person did not become an elder by any hereditary right but through election by the existing elders or congregation, and in the mid-nineteenth century, the Kirk Session of a country congregation such as Parton was responsible not only for the spiritual and pastoral oversight of church members but also for wider concerns including education and care of the poor.

John Clerk Maxwell was also a heritor. In the parish system, especially in rural areas, the heritors were the wealthier landowning parishioners. They had the legal responsibility for providing funds for the upkeep of the church, the minister's manse, and the minister's stipend. This responsibility was inherited with a property, irrespective of the property-owner's personal church allegiance.[15] James himself became a heritor immediately he became owner of Glenlair in 1856 on the death of his father, though he did not automatically become an elder. In Parton parish there were twelve landowners in all. The principal ones were Miss Glendonwyn of Parton, Lady Gordon of Letterfourie, John Clerk Maxwell of Middlebie, and Edward Fletcher of Corsock.[16]

As moneyed people, the heritors were heavily involved in the fiscal arrangements for the parish. The tiny cash offerings from the congregation were used mainly for charitable purposes within the parish, with the fund boosted by donations from the heritors. In 1844, for example, the Parton minister recorded that for the preceding three years, an average of twenty-four pounds and eight shillings per annum came from church collections, with forty pounds and six shillings added voluntarily by the heritors. A few years earlier, the *Collections for the Poor of the Parish of Parton* (1837) recorded an unusual donation from John Clerk Maxwell of one pound, nine shillings and sixpence, in lieu of 'the price of two strayed sheep on his property' which were unclaimed. There were also donations of one pound each from Messrs Barbour and Hannay; a donation of five pounds from Mr Fletcher of Corsock; a donation of three pounds and two shillings from Miss Glendonwyn;

15 The Church of Scotland (Property and Endowments) Act, 1925, transferred ownership of church buildings from heritors to the Church of Scotland General Trustees. (Weatherhead (ed.), *The Constitution and Laws of the Church of Scotland,* 192).

16 W.G. Crosbie, 'Parish of Parton', *NSAS,* Vol. IV (1845), 283.

plus the weekly offerings which were typically a few shillings.[17] In the early 1840s seventeen persons in Parton parish were classified as poor, with the average amount allotted to each being three pounds and fifteen shillings annually. Other expenditure included: payments to James W. George for precenting (leading singing of the psalms) at Merkland; payments to Robert Bell for precenting at Parton; reimbursements to the same Robert Bell for providing lodgings for a poor family; a donation to widow Grieve; a donation to Abraham Dickson for making clogs for poor boys; and contributions towards medical costs for various persons. That was how the rural parish system worked.

As well as elders and heritors there was also the patron of the parish. This was often the highest-ranking local aristocrat or foremost landowner. It was the patron who had the right to nominate the next minister of the parish when a vacancy occurred. In some urban situations the patron might be a corporate body, such as the town or city council. In a few parishes the patron was the monarch. During many of the Maxwell years at Parton the patroness was Miss Xaveria Glendonwyn (1787–1858), owner of Parton House and estate, whose family had held the property since 1458. As with the legal position of a heritor, the right of patronage was a heritable commodity in law, passed down within a family from generation to generation even if the patron belonged to a different denomination. This was the case in Parton. Miss Glendonwyn was from an old Catholic family and remained a practising Catholic all her life.

This type of circumstance, with the patron rarely living in the parish nor belonging to its church, was one of the factors which created unrest with the patronage system in Scotland, and which resulted in almost one third of the Church of Scotland's ministers and members seceding in 1843 to form the Free Church of Scotland. Miss Glendonwyn held the parish church patronage privileges despite not even being a Protestant. A wise patron took soundings before nominating a new minister, though this was not required. In Parton, with Miss Glendonwyn mostly absent, her factor, Alexander Fergusson of Kilquhanity, acted on her behalf, with the heritors such as John Clerk Maxwell (and James after him) also involved in such

17 *Collections for the Poor of the Parish of Parton,* 1837: Parton Kirk Session Accounts:1832–1852, (NRS/CH2/1473/3/64).

decisions. During James' lifetime, there were four Church of Scotland parish ministers at Parton:

Maxwell's Parton Ministers	
1830–1845	William Glendonwyn Crosbie (1806–1845)
1845–1856	Thomas Leckie (1812–1856)
1857–1858	John Young (1819–1859)
1859–1886	Henry Alexander Pattullo (1832–1886)

William Crosbie was born within Parton parish at Merkland in 1806. He authored the 1844 entry 'Parish of Parton' for the *New Statistical Account of Scotland,* and he is mentioned in Campbell and Garnett's *Life* on the occasion in 1834 when the new Parton church building was opened and Frances Clerk Maxwell loaned a baby's bonnet to William's wife, Jane Crosbie.

The historical record has little about Thomas Leckie, apart from a few details published at his death. A country minister completed his stipend by working the kirk glebe as a small farm, with the Parton glebe comprising 16 acres. The stipend for Parton in the mid-1840s was 'eight chalders of barley, and eight chalders of meal.'[18] When Leckie died in 1856, his farm possessions for sale included: a black mare and foal; a brown yearling filly; two Ayrshire cows; five ewes and seven lambs; one cart; one iron plough and swingletree; one drill harrow; barn fanners; one weighing machine and weights; one phaeton carriage and harness; two acres of growing oats; plus grass fields available for let until November.[19] The sale took place at Parton Manse on Saturday the ninth of August, conducted by Thomas Wallets, auctioneers from Castle Douglas. The glebe was also farmed by Leckie's successors, Young and Pattullo, who reappear in later chapters.

18 Crosbie, 'Parish of Parton', *NSAS*, Vol. IV, 285. A chalder varied between one and one and a half gallons.

19 *Dumfries & Galloway Standard,* 6 Aug 1856.

Corsock Kirk

As a heritor, John Clerk Maxwell helped to pay for the rebuilding of Parton Kirk in 1834. The heritors were aware that the kirk's traditional site was at the extremity of the parish, and John Clerk Maxwell of Glenlair and Edward Fletcher of Corsock House argued that it should not be rebuilt at the same place but located more centrally.[20] However, they could not persuade their fellow heritors to take the step. Accordingly, all the heritors, including the two dissidents, Clerk Maxwell and Fletcher, financed the new Parton Kirk on its original site. Despite this expenditure, Clerk Maxwell and Fletcher were still keen to provide a place of worship in their part of the parish, nearer to parishioners who lived in and around the village of Corsock. The other heritors left them to it. They regarded the burden of financing one new church building as enough, without paying for another! Having discussed the possibility of relocation and rejecting it, the other heritors saw their responsibilities as fully discharged.[21]

John Clerk Maxwell and Edward Fletcher pressed ahead. Informal religious services were already taking place at nearby Drumhumphry Schoolroom, with Lady Frances Fletcher taking a keen interest.[22] Not only did she finance the work at Drumhumphry, but she founded a Dorcas Society which focused on practical Christianity.[23] In May 1838, a public meeting was held at which John Clerk Maxwell was elected to chair a committee whose aim was to provide a permanent place of worship in the locality.[24] By the autumn of 1839 the building was complete. The 1845 *New Statistical Account* notes that a chapel on the banks of the Urr had been built, and that a preacher had been supplying regular services, with the preacher's stipend paid for partly by seat rents and partly by the liberality of Major Fletcher of Corsock. Though the new church was within Parton parish the Parton heritors had no financial responsibility for it or for its minister; their responsibilities only concerned the official parish kirk and its incumbent. The appointee

---

20 Crosbie, 'Parish of Parton', *NSAS,* Vol. IV, 285.

21 *NSAS,* Vol. IV (1845), 285.

22 Services also took place in Monybuie School (Balmaclellan), and Merkland School (Parton) (Sturrock, *Corsock Parish Church,* 5).

23 *Dumfries & Galloway Standard,* 14 Oct 1939.

24 *Corsock Church Records,* 25 May 1838, (NRS/CH2/446/1/1).

at Corsock depended on the pew-rents and local liberality, though in terms of Presbyterian governance it was under the Parton Kirk Session.

Corsock Kirk became the regular place of worship for the Maxwell family, though Parton still played an important part in their lives. This was because Corsock was not yet itself a parish, but was categorised simply as a preaching station. Its ministers were described as 'missionaries at Corsock' and were not necessarily fully ordained ministers. Hence, at each Communion Season the Maxwell family returned to Parton Kirk for the sacrament. Moreover, until Corsock became a full charge in the 1860s, John Clerk Maxwell, and James after him, still had heritor responsibilities for Parton. This explains James' involvement in 1856 in the appointment of a minister for Parton when Parton became vacant on the death of Thomas Leckie. Also, Corsock did not yet have a burial ground, and so it was in Parton churchyard that John and Frances Clerk Maxwell were laid to rest. James and Katherine were also buried there.

The new Corsock church was a boon to the local populace. Earlier in the century there had been considerable increase in the population of the parish, chiefly because of the subdivision of farms which resulted in the increased cultivation of land on the Corsock estate.[25] Edward Fletcher and John Clerk Maxwell believed that the enlarged population needed easier access to a local church, and this was why they were determined to provide one. In an article celebrating Corsock Kirk's centenary in 1939, the *Dumfries & Galloway Standard* described how, in the first half of the nineteenth century, some of the outlying roads were rutted to the wheel-axles, with other routes being little more than vague moorland tracks. Churchgoing for isolated families, some of whom lived ten miles from the nearest church, was a rare event apart from communion seasons. Their only contact with the church was the annual visit of the missionary or minister when he catechised young and old, and administered 'an admonitory word in season.'[26] The new church changed all that.

The Corsock Kirk building was completed on the twentieth of September 1839, and opened for public worship on the sixth of October at a service led by the Revd Dr Henry Duncan of Ruthwell,

25 Crosbie, 'Parish of Parton', *NSAS* (1845), Vol. IV, 284.

26 *Dumfries & Galloway Standard*, 14 Oct 1939; Sturrock, *Corsock Parish Church*, 5.

who was Moderator of the Church of Scotland General Assembly that year. The celebrated Dr Duncan was also founder of the Savings Bank movement, a scheme which he had initiated in his own parish. That day, John Clerk Maxwell stood bareheaded at the church entrance gate, welcoming everyone, with young James by his side. Frances Clerk Maxwell was already suffering from the abdominal cancer which killed her two months later.

During Sunday worship, young James sometimes collected the congregation's offering, using the 'spoon' or 'offering-ladle', which consisted of a small wooden box at the end of a long-handled wooden pole. In his youth he taught the Corsock Church Sabbath School.[27] During his lifetime Corsock had six ministers, each of whom he knew well:

Maxwell's Corsock Ministers	
1839–1844	William Corson (1809–1887)
1845–1855	John Milligan (1810–1875)
1855–1858	John Thomson (?–?)
1858–1861	John Allan Hunter Paton (1830–1911)
1861–1861	Joseph Burnett (?–1892)
1861–1902	George Sturrock (1832–1902)

William Corson was a local man, born at the nearby farmstead of Craigenputtock, which belonged to the writers Thomas and Jane Carlyle. When Corson was appointed to Corsock in 1839 he was an officially licensed preacher but unordained. He had been assistant minister at nearby St John's Town of Dalry, and had also helped the Drumhumphry mission before the new church was built. He was eventually ordained in 1848, four years after he left Corsock. As a licensee of the Church, Corson could preach and pastor but not conduct baptisms or Holy Communion. Sacramental occasions were covered by the Parton minister who was still the official incumbent of the parish. During his Corsock ministry, Corson lived in the Carlyle's house at Craigenputtock by permission of Mrs Carlyle.[28]

27 *Galloway News & Kirkcudbrightshire Advertiser*, 29 May 1885.

28 *Ayr Observer*, 20 Dec 1887.

Corson was popular and successful. This makes surprising the resistance which arose in the town of Girvan in 1847 when he was presented as minister by Queen Victoria who was patroness of the parish. Between Corsock and Girvan he had been assistant minister at Gilmerton, Edinburgh, where his ministry was also highly regarded.[29] However, when his Girvan nomination became public knowledge, local malcontents circulated a petition with complaints about his ability to speak, his intellectual capacity, and his hard doctrine. In face of this petition the local Presbytery hesitated about sustaining the appointment.

At that point Corson called on John Clerk Maxwell, whom he knew from his Corsock days, to testify on his behalf, which Maxwell was willing to do. Procedural wrangles about whether evidence from persons outside Girvan was permissible resulted in the matter going to the General Assembly, which ruled in Corson's favour. He then served forty years as minister of one of the largest congregations in Girvan until his death in 1887.[30] The reasons behind the agitation in Girvan in 1847 may have been rooted in a general discontent stirred by the patronage debate, rather than because of Corson in particular.

In 1845 the Revd John Milligan, 'a man of considerable antiquarian taste and knowledge', succeeded Corson.[31] Like his predecessor, Milligan was licensed but unordained, only being so when he moved to Twynholm parish in 1855. The stipend was increased for Milligan, with John Clerk Maxwell contributing an extra £10 per annum, and Major Fletcher £30.[32] Following Milligan, the Revd John Thomson took the post in 1855, before emigrating to New Zealand in 1858 where he became a member of the Legislature.

Between 1858 and 1861 the Corsock minister was John A.H. Paton, another unordained probationer.[33] His period of service was followed by the exceptionally brief 1861 ministry of Joseph Burnett, who emigrated to Ceylon (Sri Lanka). Then the Revd George Sturrock arrived in 1861 and remained minister at Corsock until 1902.

---

29 *Caledonian Mercury*, 6 Mar 1845.

30 'William Corson: Obituary', *Ayr Observer*, 20 Dec 1887.

31 Sturrock, *Corsock Parish Church*, 9.

32 John Clerk Maxwell, *Letter to Major Fletcher*, 1845; Sturrock, *Corsock Parish Church*, 8f.

33 *Dumfries & Galloway Standard*, 3 Aug 1859.

The three brief ministries which preceded Sturrock had prompted James to write from London, where he was a professor at King's College, with the advice that, as the situation was but that of an unordained preacher, the appointment should be made only for a limited time. This, he argued, would give full opportunity of judging whether the appointee satisfied the people and whether he could keep the congregation together.[34] His hopes for a minister in tune with the parish were amply satisfied in Sturrock.

When James succeeded his father as Laird of Glenlair, he continued his father's campaign to have Corsock Kirk raised to full status as a parish church. That milestone was reached in 1863 when a new Corsock parish was created, comprising outlying parts of the older parishes of Kirkpatrick Durham, Parton, and Balmaclellan, and including Nether Corsock.[35] Today, the church in Corsock which houses the James Clerk Maxwell memorial window is not the 1839 building which James' father helped to build and which the family attended. The current church was erected in 1852 for the Free Church, though it now belongs to the Church of Scotland. The James Clerk Maxwell window was transferred to it from what had been James' own church. The original 1839 Corsock Kirk is now a house.

Education: Sacred and Secular

At Glenlair, daily family worship and Christian education were central, with Frances encouraging James to 'look up through nature to nature's God.'[36] She ensured he knew his Bible, a quality noted later by almost everyone who knew him. It was claimed that, at eight years old, he could repeat all one hundred and seventy-six verses of Psalm 119, the Old Testament's longest psalm. Along with Scripture, lengthy passages of poetry were also part of his education. James was gifted with a mind which could absorb data and could recall it for instant use.

Frances Clerk Maxwell took responsibility for her son's general education, tutoring James at home. This left a mammoth gap when she died. A sixteen-year-old tutor was hired but this proved unsuccessful.

---

34 JCM, *Letter to Heritors,* 1861; Sturrock, *Corsock Parish Church,* 10.

35 Sturrock, *Corsock Parish Church,* 5.

36 C&G1, 19.

He appears to have treated James harshly. But by November 1841 a tutor was no longer required. John Clerk Maxwell decided that James should be educated in Edinburgh, with Aunt Jane heavily involved in persuading him to make the decision when he did. Even if Frances had lived, it is likely that James would have been sent off to school at some point. He was to attend Edinburgh Academy and during term-time was to stay with his father's sister, Isabella Wedderburn, at 31 Heriot Row, close to Aunt Jane's house in Great Stuart Street.[37] All of the Wedderburn boys had been educated at Edinburgh Academy.

In preparation for James' move to Edinburgh, several visits were made to the Wedderburns during the summer of 1841. Although the new school term started on the first of October that year, James' transfer to Edinburgh was delayed until November because Aunt Isabella, along with cousins Andrew, George, and Jemima, were holidaying on the Continent during the early autumn. After the tour George and Andrew stayed in Europe, with Andrew travelling on to India to take up a post in Madras, whilst Jemima and her mother made their way back home, visiting Glenlair *en route*. Final arrangements were then made.

On the evening of the eighteenth of November 1841, the Clerk Maxwell retinue arrived at 31 Heriot Row, which became James' Edinburgh home for the next nine years, apart from occasional breaks at Aunt Jane's house. His father returned to Glenlair to run the estate but came up to the city on a regular basis, especially during the winter months. After 14 India Street was rented out, he stayed at his sister Isabella's house in Heriot Row.

37 C&G1, 46. Rendell, *Clerk Maxwell of Glenlair*, 15, suggests that the original intention was to have James educated at home until he was thirteen, then enter him as a student at Edinburgh University.

Chapter 3
Edinburgh Schoolboy

Aunt Isabella had seven children, all boys apart from Jemima. When James arrived aged ten, she was eighteen. The Wedderburns were Episcopalian, though Aunt Isabella retained her loyalties to the Church of Scotland. As Jemima recorded in her *Journal*: 'My mother's family were Presbyterians, my father's Episcopalians. I found the former dull and chose to frequent the latter.'[1]

In the library of 31 Heriot Row, Bishop William Paley's works were on the bookshelves, and by the age of ten Jemima had read Paley's *Natural Theology*. Paley's proofs for the existence of God—including the Watchmaker Argument—impressed her intellectually if not devotionally: 'I was not religious and believed in nothing. I thought religion to be all got up for women and children as men did not go to church often unless taken there by women.'[2] Jemima's observations on male church-going habits were supported by Dean Ramsay (1793–1872) of St John's Church, who commented that the late Bishop Sandford had told him, when he first came to Edinburgh, that few gentlemen attended church.[3] Jemima added: 'My mother told me always to say my prayers at night, and I did so in a perfunctory manner because I did not like to break my word.'[4] In later years she wondered what effect Darwin would have had on Paley's thinking.[5]

Jemima admired Dean Ramsay. He was: 'An excellent, most lovable man, a good preacher, and the most beautiful reader of the prayers, much beloved by his congregation at St John's and the Sunday School children.'[6] She was also extremely fond of James' father, describing him as a great favourite. It was through his influence that

1 Fairley, *Jemima*, 15.

2 Ibid., 3.

3 Ramsay, *Reminiscences of Scottish Life and Character*, 57.

4 Fairley, *Jemima*, 98.

5 Ibid., 101.

6 Ibid., 106.

she gained access to the Surgeons' Museum in Edinburgh where she made drawings from the anatomical models, including sketches of horses in various poses with muscles coloured to show the action. Concerning James, she wrote: 'He was truly religious, and had the sweetest temper of any man I knew. I do not think I ever saw him angry or heard him say a word against anyone.'

The Wedderburn library was more extensive than the one at Glenlair, and James made good use of it.[7] He read Swift, Dryden, Butler, and Hobbes. Like Jemima, he was introduced to Paley's works. As well as reading he made practical experiments, with Jemima assisting him by painting some of his devices.[8] One of these, when spun, gave the illusion of a circus performer leaping from a horse's back through a hoop. They made other toys together, with James designing and Jemima decorating.

Heriot Row's spacious attic became home for James' experiments.[9] Years later, Lewis Campbell recalled preparations of jelly and gutta-percha, which may have been used to explore the compression of solids or try out ideas on the theory of glaciers. The attic also housed apparatus linked to James' developing interest in light, with experiments sometimes occupying whole days. When at Glenlair he spent countless hours making apparatus for even more investigations. His father encouraged all of this, getting the Corsock village blacksmith to produce steel bars which James could make into magnets.[10] This practical experimentation from an early age resulted in James developing a firm conviction that scientific theories were derived from, and could truly represent, physical reality. Quite apart from his later philosophical thinking on the nature of scientific enquiry, his early grounding in physical experimentation fed his lifelong conviction that science could reveal reality.

In June 1844, James sent a letter to his father which touched on his growing interest in solid geometry. After describing some touring Virginian Minstrels,[11] and after asking about the pets at

7 C&G1, 52.

8 Fairley, *Jemima,* 14.

9 C&G1, 80.

10 Ibid., 85.

11 For the Virginia Minstrels, see: *Greenock Advertiser,* 21 Jun 1844.

Glenlair, he wrote that he had made a tetrahedron, a dodecahedron, and two other hedrons 'whose names I don't know.'[12] In future years, geometrical analogy played a significant role in shaping his physics, especially in the application of projective geometry.[13] He instinctively thought geometrically. For example, whilst he was an undergraduate at Cambridge, one of the lecturers covered the blackboard three times trying to unravel a problem concerning three dimensions. The *Life* reports that James suggested it would be simpler if the tutor applied geometrical methods. With a diagram and a few lines of reasoning, the problem was solved![14]

Electromagnetic Amazement!

When John Clerk Maxwell came up to Edinburgh to visit his son they took regular walking expeditions, examining many of the historical sites which abound in and around Scotland's capital. They inspected the preparations for the new Granton railway. They visited Leith Fort, which had been built during the American War of Independence to resist a possible attack from John Paul Jones—who had been a native of Kirkbean, only a score or so miles from Glenlair.[15] They went with Jemima to Edinburgh's Botanical Gardens where she sketched the palm trees.[16] There were excursions to Salisbury Crags to examine rock stratifications. There was a trip to Blackford Hill to view evidence of glacial action, recently discovered by Louis Agassiz (1807–1873). A metal plaque placed in front of the great rock at the foot of the hillside in Blackford Glen still marks the spot where, in 1840, Charles Maclaren (1782–1866) showed this scratched and polished stone to Agassiz who exclaimed: 'This is the work of the ice!'[17] Maclaren was

---

12 JCM, *Letter to his Father,* June 1844 (C&G1, 56).

13 Harman, II/1, 21.

14 W. N. Lawson, *Letter to Campbell,* 6 Jan 1882 (C&G1, 175).

15 On 15 Sep 1779, John Paul Jones took a squadron of French ships, sailing under the American flag, up the Forth estuary, intending to burn the shipping at Leith but bad weather hampered him from reaching the shore. An alarmed James Boswell (1740–1795)—of Boswell and Johnson fame—recorded the event in his *Journal* entry for 16 Sep 1779. (Reed & Pootle, *Boswell: Laird of Auchinleck 1778–1782,* [Edinburgh: EUP, 1993 edition], 136.)

16 John Clerk Maxwell, *Diary,* Dec 1845, (cited by Fairley, *Jemima,* 18).

17 Smith, *Historic South Edinburgh,* Vol. 4, 214f.

co-founder and editor of the *Scotsman,* as well as a later President of Edinburgh Geological Society. All this fed James' awareness of geological theories.

In February 1842 there was a special treat. James and his father went to see Robert Davidson's demonstration of electromagnetic machines. Davidson, from Aberdeen, had built the first known electric locomotive in 1837. Seven months after James' visit to his 1842 demonstration, Davidson's '*Galvani*' rail-vehicle, powered by zinc-acid batteries, was tested on the Edinburgh-Glasgow line, reaching the speed of four miles per hour. Like several of Davidson's innovations, his electric locomotive was ahead of its time, and nothing came of it until years later.

The February 1842 demonstration was almost certainly not James' first encounter with Davidson's use of electromagnetism. In the summer of 1841, shortly before James transferred to Edinburgh for his schooling—but when we know that he and his father were in the city preparing for his move—Davidson had set up an earlier Electro-Magnetic Exhibition in St Andrew's Square, no more than a ten-minute walk from the Wedderburn home in Heriot Row. Given John Clerk Maxwell's enthusiasm for new inventions, and given his eagerness to introduce James to everything of interest, it is almost certain that father and son would have visited Davidson's exhibition where, for a one-shilling admission fee, visitors could view several of his inventions which employed 'magnetic influence as a moving power.'[18] Davidson was convinced that electromagnetic power would soon supersede steam-power both in safety and effectiveness, and his Electro-Magnetic Exhibition featured a range of equipment powered by electromagnetism. The star attraction was a locomotive carriage running on a circular railway. There was also a turning-lathe; a printing-machine; a sawmill; and—for those who dared try it—a machine giving an electromagnetic shock! A giant 500-pound electromagnetic magnet was on show: it was claimed that it could suspend a weight of several tons by magnetic attraction alone. Davidson also demonstrated 'combustion of metals by electromagnetism'. These 1841 and 1842 exhibitions would stimulate James' scientific imagination and excite his interest, kick-starting his future fascination with electromagnetic phenomena.[19]

18 *The Witness,* 17 July 1841.

19 Cf. C&G1, 84-5, with its note on James' interest and experimentations in magnetism, polarization of light, and prisms.

Optics were another enthralling area. In April 1847, his uncle, John Cay, took him to see William Nicol (c. 1768–1851) who invented the Nicol prism. This was the first device for reliably producing plane-polarized light. The visit to Nicol's private laboratory was a significant event in James' life, inspiring him to construct his own polariscope from wood and cardboard with glass reflectors and lenses.[20] When he sent water-colour drawings of his results to Nicol, the delighted man gifted him a pair of his own prisms enabling James to make more apparatus and immerse himself in further observations on polarized light. James treasured the gift. After the prisms survived an over-zealous housekeeper who was about to throw them out, they were given to Cambridge University three weeks before James' death, and placed in one of the showcases of the Cavendish Laboratory.[21]

The Royal Society of Edinburgh

John Clerk Maxwell enjoyed novel ideas, innovative discoveries, and the latest ground-breaking inventions. He was a regular attender at Royal Society of Edinburgh lectures, following in the footsteps of his great-grandfather who had been one of the RSE's founder members. He himself became a Fellow (FRSE) in February 1821. His proposer had been Sir George S. Mackenzie (1780–1848) of Coul in Ross-shire, the agriculturalist, geologist, and chemist, who proved that diamond was simply a form of carbon.

The first RSE meeting which James attended took place a week before Christmas 1843, when two papers were presented. Professor Allen Thomson (1809–1884) spoke on issues of sound and hearing. Dr Thomas Traill (1781–1862) lectured on the luminescence of the sea.[22] The meetings were held on Mondays, in what is now the Royal Scottish Academy Art Gallery, at the foot of the Mound. Until 1911 the building was known as the Royal Institution and was home to various societies, with one wing reserved for the RSE.[23] Getting there from Heriot Row for these late-afternoon occasions required less than

20 Garnett, 'James Clerk Maxwell', *Nature,* Vol. 21/524 (13 Nov 1879), 44.

21 C&G1, 84, 122, 486. Anon, 'Scientific Worthies: James Clerk Maxwell', *Nature,* Vol. 24/626 (27 Oct 1881), 601.

22 *PRSE,* Monday, 18 Dec 1843.

23 Knott, *Life and Scientific Work of Peter Guthrie Tait,* 29, gives a detailed description of the RSE's premises.

a ten-minute walk on a fair day, or a quick carriage ride in inclement weather—not unknown in Edinburgh.

The RSE was founded in 1783 as a counterpart to the Royal Society in London. It focussed mainly on developments in natural philosophy which, in the mid-nineteenth century, was a broad term for what today might include science, mathematics, the philosophy of science, and even natural theology.[24]

Though the RSE also hosted lectures on literature and the arts, science was its mainspring. It was this which fascinated John Clerk Maxwell and what he wanted James to be interested in. Attendees and Fellows did not have to be professional scientists but they did have to be attracted by science, especially the physical sciences of geology, physics, chemistry, and related disciplines. At the time, the exclusively male RSE was one of the most prestigious gentlemen's clubs in the social life of the capital. Years later, James reflected on its atmosphere, as he had known it as a young man:

> The Royal Society of Edinburgh used to be a very sociable body, but it had several advantages. Most of the fellows lived within a mile of the Society's rooms. They did not need to disturb their dinner arrangements in order to attend. Many of them were good speakers as well as sensible men, whose mode of considering a subject was worth hearing, even if not correct. The subjects were not limited to mathematics and physics, but included geology, physiology, and occasionally antiquities and even literary subjects. Biography of deceased fellows is still a subject of papers.[25]

He lamented the increased specialisation of disciplines,

> [Nowadays] those who cultivate the mathematical and physical sciences are sometimes unable to discuss a paper, because they would require to keep it some days by them to form an opinion on it, and physical men can get up a much better discussion about armour plates or the theory of glaciers than about the conduction of heat or capillary attraction.

---

24 Cf. Campbell & Smellie, *The Royal Society of Edinburgh (1783–1983): The First Two Hundred Years.* Maxwell's friends, Peter Tait and William Thomson, wrote that the term 'Natural Philosophy was used by Newton, and is still used in British Universities to denote the investigation of laws in the material world.' (T&T, *Treatise on Natural Philosophy,* Vol. 1, Preface, v).

25 JCM, *Letter to H.R. Droop,* 19 July 1865 (C&G1, 343; Harman, II/1,226).

A notable proportion of RSE attenders were clergy, or sons of clergy, including James' own ministers of St Andrew's Church. The list of FRSEs included ecclesiastical luminaries such as Dean Ramsay of St John's Episcopal Church, Archdeacon John Williams (1792–1858) the Rector of Edinburgh Academy, and Dr Thomas Chalmers (1780–1847) the leader of the evangelical cause in Scotland, prominent Free churchman, and a fine mathematician in his own right. Throughout his youth James met men at the RSE who were passionately religious and equally passionately interested in science.[26] Though we can draw only very broad conclusions from the RSE membership lists, it is striking that, during the period 1842 to 1850—from James aged eleven to nineteen when his mind was highly impressionable—several of the worshippers whom he met at church on Sunday he saw at RSE meetings on Monday. It was natural for him to detect little tension between faith and science. This reinforced his lifelong assumption that faith and science complemented one another, with science unveiling the handiwork of the Creator.

In 1846, when he was only fourteen, the RSE was the forum for the unveiling of James' own mathematical brilliance. It sprang from his interest in the attempts of the artist David Ramsay Hay (1798–1866) to use mathematical principles to explain beauty in form and colour. Hay was a celebrated figure in Edinburgh society, who had been engaged by the novelist Sir Walter Scott (1771–1832) to advise on and execute the interior design and decoration of Scott's home at Abbotsford, near Melrose. Hay was also involved in the redecoration of the Palace of Holyrood House in Edinburgh for Queen Victoria.

Hay regarded the ellipse as the primary geometrical form which the eye interprets as pleasing and beautiful.[27] Stimulated by Hay's ideas, on Monday the sixth of April, James' first academic paper *'On the description of Oval Curves, and those having a Plurality of Foci'* was read to the Society by James Forbes (1809–1868), who was professor

26 Cf. RSE, *Index of Former Fellows of the RSE, 1783–2002*. From 1842–1850 the average number of FRSEs at any one time was 280. See also: Campbell & Smellie, *The Royal Society of Edinburgh*, 16.

27 C&G1, 73. Hay, *Proportion: Or the Geometric Principle of Beauty Analysed*, (1843), 67. Cf. Hay, *First Principles of Symmetrical Beauty*, 21, published a year after James' RSE paper, where Hay discusses the oval shape of the human face as integral to its beauty.

of natural philosophy at the university.[28] James was deemed too young to present the paper himself.[29] The method of physically drawing a regular ellipse by using pencil and string looped around two nails (and representing it mathematically) was already known, but James' ingenuity took this further. Forbes' admiring verdict was that the paper contained some genuinely original insights, plus replication of some of Rene Descartes' work on ellipses. Significantly, both the original insights and the sections paralleling Descartes were products of James' unaided thinking. At the time he had had no instruction in mathematics beyond Euclid and a little algebra.[30] From that moment Forbes became James' enthusiastic supporter. In Raymond Flood's opinion, this first paper displayed not only James' ability to generalise, but his predilection for geometric thinking.[31]

Alongside the RSE, John Clerk Maxwell was a member of the Royal Scottish Society of Arts, to which also he took his son. The RSSA had been founded in 1821 as *The Society for the Encouragement of the Useful Arts in Scotland*, and was incorporated by Royal Charter in 1841. It focussed specifically on technology, engineering, and manufacture, which were known as the *useful* arts as opposed to the *fine* arts. John Clerk Maxwell chaired some of its meetings. For example, he presided on Monday the twenty-third of March 1846, when Gordon Stuart from Fife described developments at his flax-spinning works at Balgonie.[32] James, with his sponge-like ability to absorb new ideas, was enthused by the presentations which he heard at the RSE and RSSA, as well as at other meetings thought by his father to be of interest.

---

28 *PRSE*, Vol. II. 89-93. See Harman, I, 35–42.

29 Forbes was well-known to the Wedderburns. He visited Jemima and Hugh Blackburn on their return from honeymoon in Switzerland to look at Jemima's glacier sketches to see how the glacier had changed since his own visit. Forbes had written an important work on the viscosity of glaciers. Fairley, *Jemima*, 65, wrote: 'Science was changing, a fact of which Hugh was aware. However, he, like Clerk Maxwell, had no difficulty in assimilating these contemporary discoveries to his faith.'

30 J.J. Thomson, 'James Clerk Maxwell', *A Commemoration Volume*, 4.

31 Flood, 'Introduction', in Flood (ed.), *JCM: Perspectives on his Life and Work*, 4.

32 *Fife Herald*, 21 Apr 1846.

The Edinburgh Academy

Edinburgh Academy was founded in 1824, only seventeen years before James entered its doors. Though the city already had well-established schools, including the Old High School and George Heriot's School, the new Academy boasted influential supporters such as the redoubtable High Court Judge Lord Henry Cockburn (1779–1854) and Sir Walter Scott. From its inception, the Academy placed high value on teaching Greek and Latin. Cockburn, in *Memorials of His Time*, recalled that he and his friend Leonard Horner (1785–1864) had often discussed the cause and remedies of the 'decline of classical education in Scotland' and that one day, atop the Pentland Hills, they had resolved to set about the establishment of a new school in Edinburgh. Building began in 1823.[33]

Though Sir Walter Scott was a former pupil of the Old High School he was enthusiastic about the Academy, becoming the first Chairman of its Board of Governors and giving the keynote speech at the school's opening. Sir Walter maintained a keen interest in the Academy until his death in 1832, referring to it several times in his *Journal*. In one entry he reflected on the use of physical chastisement at the school: 'I am an enemy to corporal punishment, but there are many boys who will not attend without it. It is an instant and irresistible motive, and I love boy's heads too much to spoil them at the expense of their opposite extremity!'[34] Whether or not James experienced this branch of the Rector's despotism is unknown! But one of James' future scientific colleagues, Fleeming Jenkin (1833–1885), most certainly did, receiving a severe beating from the headmaster for something of which he forever claimed innocence![35]

On more elevated issues, Sir Walter took an active part in drawing up a report by the school's Directors on the pronunciation of Latin.[36] Himself a lawyer, Sir Walter was anxious that any boy from

33 Cockburn, *Memorials of His Time*, 360.

34 Scott, *Journal*, 13 Dec 1826.

35 R.L. Stevenson, *Memoir of Fleeming Jenkin*, 37.

36 Scott, *Journal*, 25 Jan 1827. Cf. Fergusson, *Chronicles of the Cumming Club*, 16: 'Our pronunciation of Latin was in accordance with what was known as the Oxford method. In fact, it was the 'nippit fit' and 'clippit fit' speech condemned by King James VI . . . The broader vowel sound now generally in vogue, and then practised in most Scots schools, we considered somewhat vulgar.'

the Academy who chose a career in Law would be understood in court when he used the Latin tongue. Over the years the Academy established a high reputation in classics, with James becoming proficient in Greek and Latin, though he was modestly dismissive of his abilities. He was a year younger than most of his schoolmates, but finished his school course not only first in mathematics and English but very nearly first in Latin, despite his attendance at school during his senior year (1846/47) being interrupted by health problems.[37]

Situated in the capital's New Town district in Henderson Row, the Academy was a new school for a new era. The Wedderburn house was close by and getting to the school was a short downhill walk which was nothing for a lad used to roaming the countryside around Glenlair. Disastrously, on his first day James was clad in the same father-designed, home-made, rustic, well-wearing clothes he had worn in Galloway. This, plus his broad accent, earned him the name 'dafty.' He was remarkably unaffected by this reception and when teased too far stood up for himself. Schoolmates remembered him as more of a loner than a boy at the centre of escapades, though in his adult life he could be gregarious, talkative, and not averse to playing jokes on friends. One Academy contemporary recalled him as a boy with a fine high forehead, who preferred to wander about alone, watching with deep earnestness the butterflies in the sunshine.[38]

He settled into school life and quickly excelled in Scripture, Biography, and English. It was later that his mathematical ability shone through, winning the Academical Club Medal for Geometry and the Silver Medal for Mathematics. His responses in class were inhibited by a speech-hesitation, which prompted an ingenious solution. The classroom seating was in horseshoe fashion, with the teacher's desk at one end and a large fireplace at the other.[39] One side of the room had a large window made up of a series of glass frames, and James mentally assigned some words of what he wanted to say to each of the frames; then, when asked to respond, he looked at the window and 'read-off' his reply from it. His only worry was that he might be asked to sit where he could not see the window.[40]

---

37 Robert Walker, 'James Clerk Maxwell', 197.

38 Fergusson, *Chronicles of the Cumming Club*, 24.

39 Ibid, 21.

40 C&G1, 67.

Rector of the Academy was a Welshman, Archdeacon John Williams. The rotund Williams was churchman, scholar, one-time tutor to Sir Walter Scott's second son Charles (1805–1841), and close friend of John Gibson Lockhart (1794–1854) who was Sir Walter's son-in-law and biographer. Williams was invited to be the Academy's first Rector and remained in post until 1847, apart from a hiatus in 1828 when he was tempted briefly to a professorship in Latin at the University of London. His passion was Celtic culture which Scott encouraged. When Sir Walter died in 1832 it was Williams who read the burial service at Dryburgh Abbey in the Scottish Borders.

Williams was obsessive about the early history of the Celts, and about Welsh language and literature. He published voluminously and read papers to the RSE. Though James had limited direct contact with Williams as a teacher—Williams only dealt with the advanced classics classes—it was Williams who moulded the ethos of the school.

At the start of each day he stood in the centre of his classroom, wrapped in his college gown, and holding his felt hat in one hand and staff in the other. With his head bowed in prayer he repeated the Anglican collect of the day from the Prayer Book, 'from which he never allowed himself to wander by one syllable, generation after generation of pupils.'[41] The Rector was a fine scholar, but smugly aware of his abilities. Former pupils told the tale—true or not—that he never walked on the path under the basaltic formation in Edinburgh's Holyrood Park, known as 'Samson's ribs', because there was an old saying that these great columns would fall on the wisest man in Edinburgh![42]

Attending Edinburgh Academy around the same time were boys of the Mackenzie family from Peebles, who were James' distant relatives through marriage.[43] Charles Mackenzie—six years older than James—was destined to be the most famous of the Mackenzie siblings, becoming a pioneering Anglican missionary bishop in southern Africa. Most of the Mackenzie boys went to Cambridge, and it was this family which was consulted by John Clerk Maxwell when the possibility of James also going there was raised. Along with the Mackenzies, some of James' school friendships lasted his lifetime. The most notable of these were with Lewis Campbell and Peter Guthrie Tait.

41 Fergusson, *Chronicles of the Cumming Club*, 66f.

42 Ibid., 69.

43 Janet Isabella Wedderburn (1815–1852), married James Hay Mackenzie (1810–1865) in 1838.

Lewis Campbell (1830–1908)

Lewis Campbell lived with his mother, the authoress Eliza Morrison (1796–1864), and her second husband at 27 Heriot Row, only a few steps from Aunt Isabella's house at No. 31. After attending Edinburgh Academy, Campbell had a brilliant student career at Oxford where he came under the influence of the liberal theologian, Benjamin Jowett (1817–1893). Lewis was later co-editor of Jowett's *Life and Letters,* which reveal Jowett's preference for faith to be based on a universal spirituality rather than in revelation.[44]

After graduating from Oxford, Campbell took Holy Orders in the Church of England. He was ordained a Deacon in September 1856, became a Priest in December 1857, and was appointed Vicar of Milford in Hampshire in 1858. In May of the same year, he married Frances Pitt at the Parish Church, Brighton, with James Clerk Maxwell his Best Man. On the Sunday before the wedding, James helped his busy friend by taking a class for youngsters at the church.[45] A few weeks later, the freshly married Campbells travelled north to Aberdeen to attend James' own wedding to Katherine Dewar.

Campbell became professor of Greek at the University of St Andrews in 1863, holding the chair until retirement in 1894. Post-retirement, he was the Gifford Lecturer from 1894 to 1896, delivering a series of lectures on 'Religion in Greek Literature.' Founded through the Will of Lord Adam Gifford (1820–1887), the prestigious Gifford lectures were aimed at advancing the cause of Natural Theology: that is, a knowledge of God which is not entirely based on claims of special revelation from God, but one established through reason, ethics, and observation of the natural order. Not every Gifford Lecturer (most notably Karl Barth in 1937 and 1938) has been as convinced of the value of Natural Theology as was Lord Gifford, but the broad terms of Gifford's Will have allowed his trustees to operate with generous criteria.

Campbell's Gifford Lectures developed the then popular argument that there exists across all humanity a common religious consciousness which manifests itself in the religion of each culture, and

---

44 Jowett, *Letters to [Unknown Recipients]*, in Abbott & Campbell (eds.), *The Life and Letters of Benjamin Jowett*, 363, 369.

45 JCM, *Letter to K. Dewar,* 9 May 1858, (C&G1, 310).

that the yearning of the human spirit for the divine is the decisive step to knowing the God who has created all men and women in his image. These were Campbell's thoughts after a lifetime of scholarship in the Greek classics and may well have reflected ideas held since he was a youth. This Romanticist view—that the divine has made itself known to all cultures through a received common religious consciousness—was also held by Rector Williams of the Academy. Williams' notions concerning a common human religious consciousness no doubt influenced some of the young minds under him.[46]

Years later, it was Campbell, along with James' scientific assistant William Garnett, who wrote the first *Life* of his late friend. In its pages Campbell reminisced on his boyhood visits to Glenlair, recalling the devoutness of the Maxwell household. He remembered the three-mile Sunday drive to Corsock church, where the 'absolute gravity' of James countenance was a study in itself.[47] That was followed by a walk home beside the river, past the Kirk pool—renowned for bathing—accompanied by 'conversations of a more earnest kind.' Campbell wrote of how, at church, James sat 'preternaturally still', with one hand lightly resting on the other, not moving a muscle however long the sermon might be.

Unlike cousin Jemima, who bristled at the Presbyterian form of worship, James enjoyed listening. Campbell also recalled that later in the week James might show by some remark, that the whole service, whether good or bad, had been photographed on his mind. In the biography Campbell's aim was to honour his lost friend, and there are obvious hagiographical elements in what he wrote. But Campbell was not as sympathetic to evangelical expressions of faith as James was—perhaps a legacy of Jowett's influence—and some of his comments in the *Life*, plus his selection of material, sprang from that perspective.

Peter Guthrie Tait (1831–1901)

James and Peter Tait were a formidable challenge for any teacher. On one occasion, the class was to be taught some new mathematics, but the unfortunate classics master assigned to the course knew less

46 Williams, *Essays on Various Subjects,* vii, 217, 219f, 282f.

47 C&G1, 83.

than the two boys.[48] Happily, other tutors were able to stimulate their exceptional talents. Worthy of mention is the official mathematics teacher, Mr Gloag, who combined a gift for teaching with humorous peculiarities of tone and manner. Though sometimes impatient 'he had a kind heart', and the boys liked him all the more because they were able to mimic him! Importantly, Gloag was rigorous. He never allowed his pupils to miss a step in a proof, and he made them engage in mental deductions which had to be puzzled out in the abstract, 'entirely without help.'[49]

Under Gloag's skilled guidance, James and Peter Tait excelled in mathematics. In due course, Peter became professor of natural philosophy at Edinburgh in 1860, beating James to the post. He specialised in thermo-electrics, materials science, early investigations into knot theory, and practical applications of physics. Along with William Thomson (Lord Kelvin) (1824–1907), Tait published in 1867 what became a standard textbook, *Treatise on Natural Philosophy.* Its 1879 edition was reviewed by James for *Nature* a few months before his death.[50]

Tait was a devout member of the Scottish Episcopal Church. In 1902 his daughter Edith (1865–1948) married the Rev Harry Reid (1866–1943) who went on to become the Bishop of Edinburgh. Tait also contributed articles on popular scientific topics to the *Good Words* religious magazine, which, at its peak, was published twice a month with a circulation of 100,000 in Scotland alone. In 1874 he contributed a series on Astronomy to *Good Words*, which sprang from lectures delivered to the 'Industrial Classes' in the Museum of Science and Art, which is now the Royal Scottish Museum.

Tait's passion was golf. He was introduced to the game on Edinburgh's Bruntsfield Links, and the entire family became enthusiasts, with extended holidays at St. Andrews each summer, playing rounds with both Old Tom Morris and Young Tom Morris. According to family tradition, Tait could on occasion play five times a day, starting at 6 am.[51] His son, Freddie Tait (1870–1900), was

---

48 Ibid., 86.

49 Ibid., 68.

50 JCM, 'Review of Thomson & Tait, *Treatise on Natural Philosophy*', *Nature*, Vol. 20/505, (3 July 1879), 213–216.

51 J.K. Robertson, *St. Andrews: Home of Golf*, (1974), 20-22.

British Amateur Champion twice (1896 & 1898), third in the British Open twice (1896 & 1897), and leading amateur in the Open thrice (1894, 1896 & 1899). Freddie also assisted his father's mathematical investigations into ball flight, the results of which were published in 1896 as a paper on the trajectory of a golf ball.[52] Tait showed that a non-spinning golf ball travelled less far and hung in the air for a shorter time than a spinning one. Another time, Freddie drove golf balls for an hour as his father calculated the launch speed, settling on 215 feet per second, or 147 miles per hour. Tait reckoned that the constitution of a golf ball gave it a limit of 190 yards for carry through the air, though that seemed disproved in January 1893 when Freddie carried a gutta-percha ball 250 yards at the thirteenth hole of the Old Course, before it rolled to a total distance of 341 yards! However, in his original article Tait had stated that the physical limit was for a golf ball with *no spin*.[53] Freddie was killed at Koodoosberg, South Africa, during the Second Boer War.

In 1871, following the British Association for the Advancement of Science (BAAS) gathering in Edinburgh, the Tait family went to St Andrews along with 'Darwin's Bulldog' T.H. Huxley (1825–1895), who was a fellow-devotee of the game.[54] The theist Tait and the materialist Huxley were friends, dinner-companions, robust conversationalists, holders of diametrically opposing views on religion and science, and keen golfing adversaries.

Huxley played on the links every afternoon during his two-month stay, living in the house known as Castlemount, near the Castle Gate. Ten-year-old Jack Tait, Peter Tait's eldest son, frequently teamed up with the young biblical scholar William Robertson Smith (1846–1894) against Huxley and various partners. St Andrews used to have the custom that the weaker player made the easy drive at the first hole, and on one occasion Jack was asked to do so. He refused, protesting he was 'not the biggest duffer of the party'! This amused

52 P.G. Tait, 'On the Physics of Golf', *Nature*, Vol. 42/1087, 420–423 (28 Aug 1890); & P.G. Tait, 'Some Points in the Physics of Golf', *Nature*, Vol. 48/1235 (29 Jun 1893), 202–204. Cf. Denley & Pritchard, 'The Golf Ball Aerodynamics of Peter Guthrie Tait': *The Mathematical Gazette*, Vol. 77, No. 480 (Nov 1993), 298-313.

53 Knott, *Life and Scientific Work of Peter Guthrie Tait*, 26f.

54 Ibid., 56.

Huxley, who consented to go first and 'topped his ball gently towards the road.' On the same holiday, Peter Tait organised a night match using phosphorous coated golf balls. The golfing foursome comprised himself, his wife Margaret (1839–1926), his brother-in-law Professor Alexander Crum Brown (1838–1922), and Huxley.[55] The experiment went well until Crum Brown's glove went on fire!

Peter Tait's friendship with Huxley was one of several links between James' theistic friends on the one hand and Charles Darwin and his evolutionist supporters on the other. Another confrère was Richard Litchfield (1832–1903) who married Darwin's daughter Henrietta. Yet another was Dean Farrar (1831–1903) who conversed frequently with Darwin about science and faith, and who arranged for Darwin's funeral to be in Westminster Abbey, acting as a pallbearer. A further link was through Harvey Goodwin (1818–1891), James' minister in Cambridge, who preached a tribute sermon in Westminster Abbey the Sunday after Darwin's funeral.

There was also William Whewell (1794–1866) Master of Trinity, who encouraged Darwin to become secretary to the Geological Society in London after Darwin returned from his voyage on the *Beagle*. When Darwin's *On the Origin of Species* was published in 1859, its title page bore a quotation from Whewell to the effect that the material world was founded on God-given comprehensible laws which are amenable to human investigation and understanding, and that the natural world is not the product of disconnected, discontinuous, unpredictable divine fiats.[56] Whewell later queried Darwin's natural selection mechanism as the driver of biological evolution: doubts which James shared. Closer to home, James' cousin Jemima corresponded with Darwin's cousin, Francis

55 The organic chemist Crum Brown married Jane, sister of Margaret.

56 Darwin cited Whewell: 'But with regard to the material world, we can at least go so far as this—we can perceive that events are brought about not by insulated interpositions of Divine power, exerted in each particular case, but by the establishment of general laws.' This was quoted from Whewell, *On Astronomy and General Physics*, (1834), 356/357. Whewell's book was Treatise III, of the: *Bridgewater Treatises on the Power, Wisdom, and Goodness of God as manifested in the Creation*. The Earl of Bridgewater's project aimed at demonstrating not only the existence, but the moral nature of God as manifested in the creation.

Galton (1822–1911), on ornithological issues.[57] In 1873, Maxwell and Huxley were appointed joint scientific editors of the ninth edition of the *Encyclopaedia Britannica*, with each contributing several articles. Maxwell was the physical sciences editor, while Huxley covered biology and zoology.[58]

We consider some of these links further in later chapters, but even at this stage we note that although the nineteenth-century debate between theists and evolutionists has generally been represented as a spiteful and hostile conflict with little meeting of minds (e.g., Wilberforce *versus* Huxley), there was a considerable degree of social and professional interaction between the parties, notwithstanding heated arguments. Huxley was a great conversationalist, and a much sought-after dinner party guest by hosts and hostesses across a spectrum of opinions. He was also tremendously kind. On one occasion he arranged for the young mathematician, W. K. Clifford (1845–1879), to go to a warmer climate for his health, and, when Huxley invited fellow scientists to contribute, James Clerk Maxwell gave a generous five pounds.[59]

Peter Tait was grief-stricken by Freddie's death in South Africa. The broken-hearted father died a year later on the second of July 1901, with his funeral service in St John's Episcopal Church conducted by Canon Cowley-Brown (1832–1924) and his daughter's future husband, Harry Reid. When university classes resumed after the summer, Robert Flint (1838–1910), the divinity professor, opened the session with an extended tribute to his late friend and fellow believer.[60] The connection at Edinburgh University between Divinity and Natural Science was more than a personal bond between two professors. Until

---

57 Fairley, *Jemima* ,54. Jemima's drawing from life, of a cuckoo deliberately ejecting a meadow pipit from its nest, helped settle the question as to whether such ejections were accidental or deliberate.

58 Cf. *Encyclopaedia Britannica,* 9th ed., Vol. I, vii. The *Peeblesshire Advertiser,* 15 Dec 1888 reported fulsome tributes to the late James Clerk Maxwell at a dinner to celebrate the completion of the Encyclopaedia's ninth edition, given by its chief editor, W. Robertson Smith, the university librarian, in the hall of Christ's College, Cambridge.

59 Marston, 'Maxwell, Faith and Physics', in Flood (ed.), *JCM: Perspectives on his Life and Work,* 281.

60 Flint, 'The late Professor Tait: An Appreciation'. (Knott, *Life and Scientific Work of Peter Guthrie Tait,* 44–46).

1892 every Arts student, including Divinity students, had to study some natural philosophy, ensuring an all-round education for each graduate. Tait loved giving these lectures.

Tait admired his friend James Clerk Maxwell as a Christian. He wrote in the *Encyclopaedia Britannica* that, in private life, James was one of the most lovable of men, 'a sincere and unostentatious Christian', though one who despite being 'perfectly free from any trace of envy or ill-will' yet showed on fit occasions 'his contempt for that pseudo-science which seeks for the applause of the ignorant by professing to reduce the whole system of the universe to a fortuitous sequence of uncaused events.'[61]

This confirmed that James, like Tait himself, rejected any quick *Deus ex Machina*, or *God of the Gaps* solution to a problem in science. Instead, natural causes could and should always be sought. For James, the ultimate cause behind reality was God: but he abhorred a too hasty appeal to God as a *Deus ex Machina* to fill a current gap of knowledge. He viewed that as injurious to both science and faith, and as inhibiting enquirers from seeking the God-given natural causes which were there to be found.

Maxwell held that the ultimate cause behind all things was God, but he worked with the assumption that there were always more natural causes in the chain of causation to be discovered. Hence there should be no bypassing the aim of finding and expounding natural explanations which, for James, honoured God, since they revealed how God had made the world. Yet, within that context, he also believed that the notion of a final cause (in God) had 'been a great help to enquirers into nature.' In other words, belief in God as Creator and as Ultimate Cause gave confidence that nature had a logic and rationality, making the whole scientific enterprise possible. God as Ultimate Cause gave sense to the *whole*. Moreover, though each component *part* of nature might be explicable naturally, the *wholeness* of nature, including its origin, order, and rationality, was incomprehensible without Deity.[62]

---

61 Tait, 'Maxwell, *Encyclopaedia Britannica* (9th ed.), Vol. 15, 647. Tait may have had in mind sentiments expressed by Maxwell on various occasions (e.g., his 1853 talk to the Apostles' Club) in which he lambasted the method used by 'spiritual theorists' to make their ideas seem scientific by imitating the phraseology of science and combining its facts with their own speculations!

62 Gaal, *The Faith of a Scientist: James Clerk Maxwell.*

For Tait and Maxwell, the conviction that natural causes could always be found for natural phenomena did not, in any way, cut across their belief that the universe and its order was contingent upon, and sustained by, the creative action of God. Yet they were not deists who believed that God made the universe and then left it to run on its inexorable laws. They held that when God interacted directly with the created order—such as in the incarnation—then natural causes on their own provided insufficient explanation. As James later discussed in his letters to Bishop Ellicott, he viewed the created order as being upheld by God moment to moment in Christ, with a continual dynamic interaction between the Creator and his creation, even though the Creator had granted the created order its own existence and order.[63] On the one hand, precisely because God had granted the created order a logic and intelligibility, it was explicable from within itself. On the other hand, such explanations halted at the unbreachable boundaries of science, especially questions concerning the origin of the 'stuff' of the universe, and the origin of its logic and structure.

Did James share all of Tait's religious ideas? Not quite. In 1874, Tait and his friend Balfour Stewart (1828–1887) co-authored *The Unseen Universe*. The book was an instant best-seller. The authors argued—as scientists—for a rational faith and for the immortality of the soul. The volume's popularity spawned a follow-up publication, *Paradoxical Philosophy: A Sequel to the Unseen Universe,* which James reviewed for *Nature* magazine. A draft copy of *Paradoxical Philosophy,* with some of James' thoughts preserved in comments jotted in the margins, is held by the National Library of Scotland.[64] We reflect on James' views concerning Tait and Stewart's hypotheses in our later chapter 'The Unseen Universe'.

63 JCM, *Letter to Ellicott,* 22 Nov 1876, (C&G1, 393: Harman, III/1, 417).

64 NLS: Acc.10000/399, 'Balfour Stewart & P.G. Tait, *Paradoxical Philosophy* with annotations by JCM.'

Chapter 4
One Faith, Two Churches

James' Presbyterian father and his Episcopalian Aunt Jane were anxious for him to appreciate both traditions. During his years in Edinburgh, he was given full exposure to each form of church life, with St Andrew's Church of Scotland on George Street and St John's Episcopal Church on Princes Street enriching his Christian experience. Yet, these two congregations not only nourished his growing faith and piety but introduced him to ecclesiastical controversy. The Presbyterian earthquake took place when he was a pupil at the Academy, and the Episcopalian turmoil erupted when he was a student at the university.

THE CHURCH OF SCOTLAND

One of the most dramatic events of nineteenth-century Scottish life, secular or sacred, took place during James' schooldays. This was the Disruption of 1843, when the Church of Scotland was torn apart, and when one third of its ministers and elders left to form the Free Church. In our more secular times it can be hard to appreciate how seismic the issue was or how it affected every sector of society. But it did. And the Maxwells' Edinburgh church of St Andrew's was at the epicentre of events. By the 1840s its senior minister, the Revd Dr David Ritchie, who had baptised James, was an old man in indifferent health. He had retired from his university chair in 1836, but continued as one of the St Andrew's ministers alongside his younger colleague, John Bruce. In the crisis of 1843, the congregation's two ministers took different paths, making St Andrew's a symbol of the rupture affecting the Kirk the length and breadth of the land.

The Patronage Issue

In May 1843 the annual General Assembly of the Church of Scotland took place in St Andrew's Church, and the issue of patronage came to a head. Patronage was the system whereby a new minister of a parish was nominated by a local patron. The patron might be a member of the aristocracy, a landowner, or in some cases a corporate body such as the burgh or town council. There was a strong feeling, especially amongst evangelically inclined church people, that ministers should be chosen by the congregation or, at the very least, be approved by the congregation before being appointed. The fact that some patrons were Episcopalian, or even Roman Catholic, and had patronage powers only because of historical or legal reasons relating to land ownership in a parish, added fuel to the discontent. There had been objections to patronage ever since 1712 when the Government of Queen Anne re-imposed the system on the Church of Scotland without any consultation with, or agreement from, the Kirk. The Kirk repeatedly asked Parliament at Westminster to repeal the Patronage Act, but without success.

In the 1830s the evangelical party within the Kirk gained ascendency, and it initiated attempts to dilute the patronage system and give congregations a meaningful say in appointing a minister. In 1834 the General Assembly passed a Chapels Act and a Veto Act. The first of these helped to consolidate an evangelical majority in the General Assembly. The second ruled that a patron's nominee could only become minister of a parish if a majority of the male heads of households of the congregation concurred. This prompted every congregation of the Church of Scotland, including that of St Andrew's, to draw up a list of all the male communicants.[1] Previously, very few membership lists had been kept.

In due course the civil courts ruled that both Church Acts, which had been designed to mitigate the patronage system, were illegal. This brought fundamental principles into play. Was the Kirk, in its national role as the State-recognised established Church within Scotland, controlled by the machinery of the State? Or was the Kirk free to make its own decisions in spiritual matters without needing the approval of the State? In the evangelical mind this transmuted into critical theological questions. Was Christ head of the Church? Was the State head of the Church? Had the 'Crown Rights of the Redeemer' been abrogated? It was not disputed that the Kirk had to obey the ordinary laws of the land, but when it came to spiritual matters—such as determining how its ministers were appointed—why should the State set the rules for that process?

1 Christie, *The Story of St. Andrew's, Edinburgh*, 34.

Impasse between Kirk and Parliament resulted in the General Assembly of 1842 submitting a *Claim of Right* to the Government in London, asking Her Majesty's Government to acknowledge that the Kirk had an inherent right to solely determine any spiritual matter, such as the calling of a minister to a parish. However, as the 1843 General Assembly drew close, no such acknowledgement had been given. Westminster was silent. Hence, when ministers and elders gathered for the 1843 General Assembly there was feverish excitement. How would the Kirk respond?

This was the burning question throughout Scotland. It dominated church meetings. It was the hot topic in household conversations across society and across denominations. For members of St Andrew's, it would be the talk of the congregation, not only in the days immediately preceding the General Assembly but for weeks beforehand. Even the Episcopalian Wedderburn and Cay households would be abuzz with discussions, with eleven-year-old James fully aware of issues and events.

The 1843 General Assembly

The General Assembly met in St Andrew's Church on the eighteenth of May 1843, with the Revd John Bruce, the congregation's younger minister, one of its members that year. How would he respond? He was known to be sympathetic to the reformist party, and all knew what was at stake. There was the possibility of a major schism, even bigger than two secessions from the Kirk in the eighteenth century over similar issues. Preliminary meetings had already taken place. Options had been prepared. Big announcements were expected. All eyes were on Thomas Chalmers who was the leader of the evangelical grouping, and on David Welsh who was the Moderator, or chairman, of the General Assembly. Welsh had been Assembly Moderator in 1842 and was due to preside at the start of the 1843 General Assembly until the next Moderator was elected. That was the custom. The election of a new Moderator was normally the first piece of business. But things turned out differently.

Before the Assembly was formally constituted, Welsh stood up and read a *Protest*, signed by two-hundred ministers and elders. The *Protest* stated that, since the State had rejected the *Claim of Right*, a free Assembly of the Kirk was impossible, and the signatories could not in conscience continue their connection with the State because the conditions laid down by the State were derogatory to the 'Crown Rights of the Redeemer.' Welsh and Chalmers then walked out of St Andrew's, along with one hundred and twenty-one ministers and seventy-three elders. They marched along George Street and downhill to the Tanfield Hall at Canonmills, to the cheers of the immense crowd which had gathered in anticipation. In the Tanfield Hall the first General Assembly of what was now called the Free Church of Scotland took place, with Chalmers elected as its first Moderator. At Tanfield, a further meeting saw the formal signing of the *Act of Separation,* and ultimately four hundred and seventy-four Church of Scotland ministers, out of a total of around twelve hundred, left the Kirk to join the Free Church. In some areas, especially the Highlands and Islands, almost the entire church membership went over to the Free Church. No part of the nation was unaffected.

On the day of the Disruption, Jemima Wedderburn was sitting on Edinburgh's Calton Hill with an unobstructed view along George Street. She witnessed the walkout and sketched the procession on her artist's pad: 'I saw a black train of ministers march out of St Andrew's Church, George Street, in Edinburgh. I was on the Calton Hill at the time. They gave up their livings which was noble of them, though the cause was not worth it. I could not see the harm of patronage.'[2]

Directly across the Street from St Andrew's was Mr Black's Junior School, whose pupils also witnessed the dramatic walkout: among them was the young Archibald Geike (1835–1924), the future geologist, who reappears later in our narrative.[3] Even closer to the action was Professor James Forbes, family friend and James'

---

2 Fairley, *Jemima,* 15, 125.

3 Geike, *A Long Life's Work,* 4: 'On the 18th May, 1843, the town was crowded with black coats—ministers and elders from every parish in Scotland, all in a ferment of excitement, outside and inside the church . . . I was too young to understand the cause of this trepidation. But no schoolboy could fail to be impressed by the sight of the long procession of black coats, filing out of the building and marching off to turn down the hill to the large hall which had been engaged at Canonmills. We were witnessing the actual Disruption of the Kirk of Scotland. The processionists whom we watched, in their sombre attire, looked like a gigantic funeral, but they were on their way to inaugurate a new ecclesiastical assembly, and to found the Free Kirk of Scotland.'

future champion at the RSE. Forbes was a devout Episcopalian who originally wanted to be a Church of England clergyman but had been dissuaded by his father.[4] Forbes had managed to secure a seat within St Andrew's church as a member of the public. Having witnessed the excitement of the Disruption drama at first hand, Forbes made the day even more memorable by asking Alicia Wauchope (1821–1885) to be his wife. She accepted![5]

The same afternoon, a group of Edinburgh Academy schoolboys were wending their way homeward when they encountered 'a long line of clergymen, three abreast, marching down the centre of the road, some earnestly conversing, others walking silently with downcast heads, all of them solemn and anxious.'[6] The young Academicians had met the procession of Free Church Fathers on their way to the Tanfield Hall.

Not only was St Andrew's the scene of the split but its younger minister, John Bruce, left to join the new Free Church whilst their older minister, David Ritchie, stayed. For John Bruce the event was traumatic.[7] The whole incident took place in his church where, week by week, he led the congregation in worship. Though Bruce did not participate in the immediate exit led by Welsh and Chalmers, by the weekend he had joined the breakaway movement, prominently sharing a platform with its leaders on the Sunday.[8] On the actual day, he was undecided. Along with several others he waited until the Queen's traditional 'Gracious Letter to the General Assembly' was read. He hoped there might be some indication in it, that, although the Government had not responded formally to the *Claim of Right*,

---

4 Shairp (etc.), *Life and Letters of James David Forbes,* 16, 54.

5 *Ibid.,* 156.

6 Fergusson, *Chronicles of the Cumming Club,* 30.

7 Like Chalmers and the great majority of the Free Kirk ministers who left the Church of Scotland in 1843, John Bruce did not want a complete divorce of Church from State, merely that the Church should be free of State interference in spiritual matters. Chalmers and Bruce believed that in a Christian country the national Government had a duty to aid the established Church and ensure adequate provision for religious ordinances, through building places of worship etc. In 1835 Bruce had published a monograph, *The Lawfulness of the Church accepting an Endowment from the State* (Edinburgh: Fraser, 1835). This remained his position after the Disruption

8 *Scotsman,* 24 May 1843.

it might be prepared to meet with the Kirk and discuss the issue. When no indication of this was detected in Her Majesty's Letter, his mind was made up. He added his name to the *Protest*, left his manse, his church, his parish, and his livelihood. In leaving, he took a substantial number of wealthy and influential families with him, and they built the new St Andrew's Free Church at 80 George Street. When he died in 1880, he was in his 86th year and the oldest of the Free Kirk ministers.[9]

Within a week the St Andrew's congregation had lost a minister. But the Maxwell family did not follow John Bruce. Nor did David Ritchie. It was not that Ritchie was insensitive to the anti-patronage cause: he agreed with its fundamental principle that no patron should have the power to impose a minister onto an unwilling congregation. At a meeting of the Presbytery of Edinburgh in 1834, he had argued strongly that no minister should be intruded into a parish contrary to the will of the congregation, and that the dissent of a majority of the male heads of families in a congregation should suffice to set aside a patron's presentee.[10] Despite this, Ritchie was one of a body of evangelically inclined ministers who did not follow Chalmers and Welsh. By 1843 he was eighty years old, and far from well. He died the following January.

What was James' reaction to these events? At school, some of the Academy teachers joined the Free Church. At home, he would listen to adult conversation, ask questions, and form opinions. Though the Maxwells stayed firmly with the Church of Scotland, it was a sore time for James' father, John Clerk Maxwell. The Disruption affected even distant rural Galloway, with a split in the membership of the new church at Corsock which he had worked so hard to establish. Some remained with the Kirk, whereas others seceded and formed a Free Church congregation in nearby Kirkpatrick Durham and in time erected their own church in Corsock itself. Within two years, a Free Church school was established in adjacent Castle Douglas under the auspices of

9 Bruce was close to Chalmers. In 1847 it was he who was asked to preach a tribute sermon in Thomas Chalmers' Morningside Free Church on the Sunday immediately following Chalmers' death. Cf. John Bruce, *Sermon preached in Morningside Free Church, 6th June 1847*, (Edinburgh: Johnstone, 1847).

10 *Caledonian Mercury*, 3 May 1834.

the Free Church Presbytery of Kirkcudbright.[11] A few years later, the new proprietor of Corsock House—previously owned by the Fletcher family—was the powerful Free Churchman Alexander Murray Dunlop (1798–1870), who was the dominant legal mind behind several of the Disruption documents, including the *Claim of Right,* the *Protest,* and the crucial *Act of Separation.*

John Clerk Maxwell was a traditionalist. He thought that the old system worked well. He was committed to it. In his heritor's role he had contributed personally far beyond his obligations and did so gladly. He felt that the enthusiasm surrounding the Disruption was misplaced and he was wounded by seeing the new congregation at Corsock rent asunder. It all seemed so wrong. Unlike some, he did not regard the Disruption as a catalyst towards deeper faith and piety. Instead, he saw it as a cause of unnecessary division.

His feelings were expressed in remarks recalled by James some years later, and recorded in a letter written by F.J.A. Hort (1828–1892) to Lewis Campbell.[12] Hort claimed that James was of the view that the ferment about the Free Church movement had one very bad effect which was that, 'quite young people were carried away by it; and, when the natural reaction came, they ceased to think about religious matters at all, and became unable to receive fresh impressions.' According to Hort, James elaborated: 'My father was so much afraid of this, that he placed me where I should be under the influence of Dean Ramsay, knowing him to be a good and sensible man.' Hort added that James' comment on the Disruption came in the context of discussing the perilousness of strong religious excitement in early youth, which could lead to spiritual exhaustion and a permanent religious insensibility which were 'apt to follow the dying-out of the original fervour.'

If Hort's recollections are correct, then they reflect James' later opinion, which was also influenced by some aspects of the Revivalist Movement of the late 1850s which he witnessed first-hand in Aberdeen. Hort claimed that James spoke with thankfulness of his own escape from a similar danger. The hazard for James, was not evangelicalism as such, but over-heated emotions attached to movements.

11 *Dumfries and Galloway Standard,* 10 Sep 1845.

12 F.J.A. Hort, *Letter to Campbell,* 4 Feb 1884, (C&G1, 420).

Thomas Jackson Crawford

St Andrew's congregation lost John Bruce to the Free Kirk in 1843, and David Ritchie to death in 1844. Its new ministers were the theologically evangelical Thomas Jackson Crawford (1812–1875) who is mentioned in Campbell and Garnett's *Life*,[13] and the Revd Thomas Clark (1790–1857). Thomas Crawford was appointed professor of divinity at Edinburgh University in 1859 after which he resigned his ministerial position at St Andrew's, though he remained an active member of the congregation. He became an FRSE, and was yet another of James' spiritual mentors who took an interest in the scientific ideas of the day.

Crawford was James' Edinburgh pastor for the remaining six years of James' teenage years, a critical period for a sharp young mind. Throughout that formative period—as schoolboy until 1847, and as university student until 1850—this thoughtful young man listened attentively each Sunday, considered what he heard, and absorbed it into his developing Christian faith. Crawford was also convenor of the General Assembly's committee with responsibility for encouraging family worship, as well as being an enthusiastic supporter of the Moody and Sankey evangelistic campaigns in the 1870s.[14]

What was the theology which underpinned Crawford's preaching? He authored several books, including: *The Fatherhood of God* (1867), *The Doctrine of Holy Scripture respecting the Atonement* (1871), and his most famous work *The Mysteries of Christianity* (1874) which was still in print in the early twenty-first century through Banner of Truth publications.

In *The Mysteries of Christianity*, originally delivered as the 1874 Baird Lectures, Crawford set out his Reformed credentials, defending Christianity as a supernatural, trinitarian faith, with the incarnation, atonement, and sovereign work of the Holy Spirit at its heart. He also carefully expounded the critical Reformation doctrine of justification solely by God's grace through faith: this was the issue which his predecessor, David Ritchie, had written on with unfortunate ambiguity in the 1830s.[15] Presumably that crucial

13 C&G1, 55.

14 Crawford signed the 'United Prayer for Scotland Statement'. (*Huntly Express*, 3 Jan 1874.)

15 Crawford, *The Mysteries of Christianity*, 173–195.

doctrine, explained so clearly in *The Mysteries*, would be expressed with equal clarity from the St Andrew's pulpit. The doctrine of justification is kernel to evangelical Protestantism. In later years, some who had the greatest spiritual impact on James, such as thc Anglican Charles B. Tayler and the Baptist John Offord, were men for whom that doctrine was at the forefront of their preaching. For them, it was not simply something to be assented to intellectually, but was the practical driver of their ministry. It was the foremost tool in their communication of the faith, and integral to their pastoral and practical application of the Gospel.

Crawford had another important emphasis. In contrast to some Scottish preachers of the time, he taught that the primary relationship between God and humanity was the *filial* bond of sonship rather than the *legal* obligation of the law. This lay behind his book *The Fatherhood of God* in which he took issue with Robert Candlish (1806–1873), one of the key thinkers of the Free Church. Some interpreters of the *Westminster Confession of Faith*—the primary doctrinal standard for both the Church of Scotland and the Free Church—held that the core principle in the relationship between God and humanity was humanity's need to keep the Law of God, and that any *filial* bond was consequent on that *legal* obligation being fulfilled through the atonement. For Crawford, the primary relationship between God and humanity, initiated at the creation, was *filial*; for Candlish, the primary relationship was *legal.* This affected how each understood the atonement. Emphasis on the *filial* meant that men and women were already sons and daughters of God in virtue of creation, and therefore God's intention of salvation was directed towards *all* his children. A *father* would seek to try to save *all* his errant children, whereas a *lawmaker* might not.

The tension between the two approaches spilled into an extended public debate in print between the two men.[16] In 1856 the Revd John MacLeod Campbell (1800–1872) had published his historic book *The Nature of the Atonement* which was predicated on the same distinction between the *filial* and the *legal.* MacLeod Campbell took the *filial* relationship as the starting point of all Christian theological

16 Cf. T. J. Crawford, *The Fatherhood of God: Considered in its General and Special Aspects and Particularly in Relation to the Atonement, with a Review of Recent Speculations on the Subject, and a Reply to the Strictures of Dr. Candlish,* (1868, 3rd ed.).

thinking, especially in the doctrine of salvation. James became familiar with MacLeod Campbell's book but had reservations about some of MacLeod Campbell's argument, though his detailed comments have not been preserved.[17] Crawford's public dispute with Candlish occurred after James' time in Edinburgh, but the principles expressed were part of Crawford's life-long understanding of the faith.

One of James' later letters to his wife Katherine may reflect a Crawford impact. The missive includes one of James' insightful theological comments, revealing his Christocentric piety, and conveying how he understood Jesus as Son of Man and Son of God. He viewed the humanity of Christ as enabling men and women to know God (revelation), and the divinity of Christ as enabling God to change and transform them (salvation):

> I can always have you with me in my mind, why should we not have our Lord always before us in our minds for we have His life and character and mind far more clearly described than we can know anyone here? If we had seen Him in the flesh we should not have known Him any better, perhaps not so well. Pray to Him for a constant sight of Him, *for He is man that we may be able to look to Him; and God, so that He can create us anew in His own image.*[18] (Italics ours)

In James' mature faith, Jesus Christ *as man* enabled humanity to know God. Jesus Christ *as God* enabled humanity to be transformed and recreated. But from where did he get this interpretation of the dual nature of Christ? It was not from the *Shorter Catechism*, nor from Dean Ramsay's *Catechism*, both of which linked the humanity of Christ to salvation rather than to revelation. He may have taken it from Crawford, who, in *The Mysteries of Christianity* wrote that the humanity of Christ 'gives us a clearer and fuller manifestation of God than we can otherwise attain' and that it 'shows us divinity through the medium of humanity.'[19]

Interpreting the nature and character of God, through what is seen and experienced in the humanity and ministry of Christ, was an important emphasis for Presbyterians such as Crawford, over

17 C&G1, 431

18 JCM, *Letter to Katherine*, 28 June 1864, (C&G1, 339).

19 Crawford, *The Mysteries of Christianity*, 164–166.

against some theologians who pointed to hidden attributes of God. For Crawford, God is what we see and encounter in Jesus Christ. As James expressed it: God became man 'that we may be able to look to him': we may love and trust God through knowing what he is truly like in Christ.

THE SCOTTISH EPISCOPAL CHURCH

As F.J.A. Hort recalled in his letter to Lewis Campbell in which he discussed John Clerk Maxwell's reaction to the Disruption, one of the immediate consequences of the events of 1843 was James being given more exposure to Episcopalianism. Each Sunday, after attending St Andrew's Church of Scotland in the morning, he went to St John's Episcopal Church in the afternoon, with Aunt Jane's encouragement and his father's full support. This resulted in James becoming acquainted not only with the *Westminster Shorter Catechism* of Presbyterianism, but with Dean Ramsay's *Catechism* for young people of the Episcopalian tradition.

The affable and much-beloved Edward Bannerman Ramsay came to St John's in 1830 where he remained until his death in 1872. He was an FRSE, serving as the Society's vice-president, and yet another churchman whom James saw at its meetings. In 1841, Ramsay became Dean of Edinburgh Diocese. As a man of humour as well as piety, he is best remembered for his *Reminiscences of Scottish Life and Character*, first published in 1857.

Ramsay's 'Catechism'

In 1835 Dean Ramsay composed a catechism for young people, which set out Christian faith from the Episcopalian perspective. It had the impressive title: *A Catechism Compiled and arranged for the Use of Young Persons belonging to the Congregation of the Episcopal Chapel of St. John the Evangelist, Edinburgh.* It was this which he taught to James and other members of his young people's class.

In its pages Ramsay set out not only traditional Christian doctrine, but a strong defence of Episcopalian church order. At the same time, he took pains to stress that other denominations were also true Christian churches: an emphasis not always present in publications of the time. For example, in considering how Episcopalians should look on those who differed from them on matters such as the Order of

Bishops, Apostolic Succession, and the need for an Episcopal Polity, the simple answer from Ramsay's *Catechism* was: 'With candour and Christian kindness.' Similarly, when considering whether Episcopalians should deny salvation to those who were outside the 'true' Church (i.e. a church with an episcopal structure), the *Catechism's* response was a decisive: 'Certainly not!', with references to Matthew 7:1 and Romans 14:4.[20] What James absorbed from Ramsay's *Catechism* was a broad-minded approach to matters of denominational difference and identity, yet without sacrificing any of Ramsay's deeply held beliefs. It was an attitude practised by James himself for the rest of his life.

Not everyone in the hierarchy of Scottish Episcopacy had Ramsay's openness. In one of James' letters to Aunt Jane he noted: 'I see that the Revd Berkeley Addison is in trouble about the Scottish Reformation Society, for associating with non-Episcopal clergymen.'[21] Berkeley Addison (1816–1882) was a young and intensely evangelical Episcopalian who was part of the St John's team of clergy, serving as Dean Ramsay's curate for over twelve years and well-known to the young people of the congregation.[22] After Addison spoke at meetings of the Protestant Scottish Reformation Society, he was severely rebuked by leading Episcopal clergy for doing so.[23] Bishop Charles

20 Ramsay, *Catechism for the Use of Young Persons,* Part V, Q&A67; Q&A 69. Matthew 7:1 (AV): 'Judge not, that ye be not judged.' Romans 14:4 (AV): 'Who art thou that judgest another man's servant?'

21 JCM, *Letter to Jane Cay,* 12 Nov 1853, (C&G1, 194).

22 Obituary: 'Canon Berkeley Addison', *Newcastle Journal,* 14 Jan 1882. Cf. *The Morning Post,* 18 Jan 1882: 'A leading member from the ranks of the Evangelical party in the diocese of Durham, Canon Addison graduated at Cambridge and was ordained deacon in 1839. After serving curacies at Brighton and Kensington, he accepted the post of assistant to the well-known Dean Ramsay in Edinburgh, where he was exceedingly popular both as a preacher and a controversialist.'

23 *The Scottish Magazine & Clergyman's Review,* (New Series, Vol. III, Edinburgh: Lendrum, 1854), 568–589. Cf. *Scottish Press,* 28 Oct 1853: 'Bishop Wordsworth of St Andrews has recently written a letter in the shape of a Pastoral to his Clergy, reprobating the Rev Berkeley Addison's "conduct" for taking part in "certain public meetings" in connection with the Scottish Reformation Society, as inconsistent with his position and duties in the church of which he is a minister; and he is particularly indignant that these "certain public meetings" were held in his diocese without his leave.'

Wordsworth (1806–1892) was insistent on the superiority of the Episcopal system, and condemned Addison publicly. In Wordsworth's mind, not only was the Scottish Reformation Society a decidedly non-Episcopalian body but it adhered to Calvinist doctrines which were at odds with Episcopalian teaching on, for example, baptismal regeneration. The Berkeley Addison incident was a salutary reminder that Dean Ramsay's personal commitment to an ecumenicity of spirit was not universal across his denomination.

Concerning specific doctrines, Ramsay's *Catechism* had a section on 'Justification', in which Ramsay explained that salvation was by grace alone and received through faith alone. This would be music to the ears of a Presbyterian such as John Clerk Maxwell.[24]

Ramsay was certainly viewed by his contemporaries as an Episcopalian with evangelical doctrine in his heart. The strongly evangelical D.T.K. Drummond (1805–1877), who left the Scottish Episcopal Church because of the stance taken by its bishops on several issues, wrote that Ramsay's ministry in Edinburgh was coincident with the revival of evangelical religion in the city. Drummond had no hesitation in affirming that Ramsay preached, 'the great doctrines of the fall of man; his utter depravity; justification by faith in Christ; regeneration and sanctification by the Holy Ghost.'[25] Drummond stated that when Ramsay first expounded these tenets, they were strange and new to his Episcopalian audiences. He recalled Ramsay's 'enlarged and liberal catholicity: while at the same time repudiating anything approaching to latitudinarian comprehensiveness.' Ramsay himself, in his 1866 lectures on preaching, emphasised that the Church was commissioned to proclaim the Gospel of Jesus Christ to a lost and guilty world, and not just to expound morality.[26] But his own

24 Ramsay, *Catechism for the Use of Young Persons,* Part II, ix, deals with 'Justification'. Q&A 2: 'In justification, is the sinner actually made righteous, and therefore *titled* to his pardon and the favour of God? No: but for Christ's sake he is acquitted, and God receives him *as if he were* righteous.' Q&A 3,4: 'What is the real or meritorious *cause* of the sinner being justified? The merit of our Lord and Saviour Jesus Christ. How is the blood of Christ made *effectual* for our justification? By faith.' Ramsay drew heavily on the Thirty-Nine Articles of the Anglican tradition.

25 Drummond, *Discourse in Memory of Dean Ramsay,* 1872 (Cited by Rogers, *Memorials and Recollections,* 58–60).

26 Ramsay, *Pulpit Table-Talk,* 9.

method was to win souls by focussing on the love of Christ rather than on the 'terrors' of the Law.[27]

After James left Edinburgh, Dean Ramsay kept his eye on the Galloway youngster. During a visit to Cambridge in the early 1850s he wrote to Jane Cay: 'I had great pleasure in seeing your nephew, young Clerk Maxwell. He is shrewd and cautious. He seems to like Cambridge and, I doubt not, will distinguish himself. He is sparing in his words, but what he says is to the point.'[28]

The Gorham Business

If, in 1843, Scottish Presbyterians were rocked by the earthquake of the Disruption, then in 1847 it was the turn of Episcopalians to feel the tremors of ecclesiastical upheaval. The issue was a controversy in the Church of England, known as the Gorham Business.[29]

George Cornelius Gorham (1787–1857) was a Church of England clergyman who appealed to the secular courts after being blocked from a parish. In 1847 he was presented by the Lord Chancellor to the vicarage of Brampford Speke near Exeter, but Bishop Henry Phillpotts (1778–1869) refused to appoint him. Phillpotts had discovered that Gorham rejected the doctrine of baptismal regeneration, insisting instead that it was an individual's later personal adoption of the promises made on his or her behalf at baptism which were necessary for true spiritual rebirth. Phillpotts, as a high churchman, declared Gorham a Calvinist who was out of step with traditional Anglican churchmanship and hence unsuitable for the post. The subsequent controversy became a *cause celebre*, with Phillpotts mentioned by Anthony Trollope in his 1855 novel *The Warden*, in which the bellicose and combative Archdeacon Grantly has a bust of Dr Phillpotts in his study. Trollope's fictional Archdeacon clearly admired Phillpotts' insistence on authority over underlings!

In the real world, Gorham appealed against Phillpotts' barring of his appointment. He went first to the ecclesiastical Court of Arches, but that court sided with the bishop and awarded costs

27 Ramsay, *Pulpit Table-Talk*, 28f.

28 Ramsay, *Letter to Miss Cay*, Undated, (C&G1, 163).

29 JCM, *Letter to Campbell*, 22 March 1850, (C&G1, 130).

against Gorham. He then took his case to the Judicial Court of the Privy Council which was a secular court. This controversial move raised the question: could or should a secular court decide on a doctrinal issue? Nevertheless the case was heard, and in March 1850 the Court of Arches' decision was reversed. Bishop Phillpotts instantly rejected the Judicial Court's verdict (commonly termed the *Judgment*) and threatened to excommunicate anyone who dared admit Gorham to the parish, even if it were the Archbishop of Canterbury himself!

Fourteen other leading Anglicans supported the bishop, calling on the Church of England to disclaim the interpretation of baptism implied by the *Judgement*. When they received little support, some joined the Roman Catholic Church.

Gorham's appeal to a secular court raised fundamental questions, both legal and theological. Legally: what role, if any, should secular law courts take in questions of doctrine? In Gorham's case the secular court sided with what might be labelled the 'evangelical' position. But was appeal to a secular court appropriate? Theologically: was belief in baptismal regeneration a required doctrine for Anglicans?

The case was keenly followed north of the border, where the question of the propriety of secular courts interfering with spiritual judgements had already been raised in connection with the 1843 Disruption. Apart from constitutional issues, the Gorham Business centred on the important doctrinal issue of spiritual regeneration/rebirth, and it was this which appears to have sparked James' interest. Did spiritual regeneration/rebirth come through the act of baptism as Phillpotts insisted? Or did it follow personal faith as Gorham believed? In a letter to Campbell, penned soon after the Privy Council's *Judgement* in favour of Gorham, James wrote:

> I postpone my answer to you about the Gorham business till another time, when also I shall have read Waterland on *Regeneration*—which is with Mrs. Morrieson—and some Pusey books I know. In the meantime, I admire the *Judgement* as a composition of great art and ingenuity.[30]

30 Ibid. '*Judgement*' was a reference to the Privy Council legal judgement.

James may well have read the Privy Council *Judgement* in the *Scotsman*.[31] The mention of 'Pusey books' in his letter was a reference to the high church Edward Pusey (1800–1882) who had an exalted view of the sacraments, was uncomfortable with many aspects of the Reformation, and who wanted the Church of England to readopt the beliefs and practices of the Church Fathers as these were interpreted by himself.[32]

In his letter to Campbell, James also cited the eighteenth-century divine Daniel Waterland (1683–1740), the full title of whose treatise was: *Regeneration Stated and Explained according to Scripture and Antiquity with a summary view of the Doctrine of Justification*. Lewis Campbell's mother, Mrs Morrieson, was also reading the book. In studying Waterland, James was digging into Protestant doctrine on spiritual rebirth, doctrine which starkly contrasted Puseyism. In Reformed theology any notion of baptismal regeneration was rejected, with spiritual rebirth entirely predicated on God's grace received through faith.

James would also be aware of what the *Shorter Catechism* and Dean Ramsay's *Catechism for Young People* stated concerning baptism. The *Shorter Catechism* spoke of baptism as the sacrament which signifies and seals a person's ingrafting into Christ, but did not regard spiritual regeneration as occurring through the act of baptism itself.[33] Dean Ramsay's *Catechism* employed words with which any Presbyterian would be comfortable, such as Augustine's statement that a sacrament is an outward and visible sign of an inward and invisible reality.[34] Critically, nowhere did Ramsay state that the act of baptism

---

31 Cf. *Scotsman*, 13 March 1850: (Syndicated Report) 'GORHAM v THE BISHOP OF EXETER: The judgment in this important appeal . . . was pronounced on Friday by the Judicial Committee of the Privy Council. . . . Lord Langdale read the printed judgment of the committee. After setting forth the nature of the appeal made by Mr Gorham against the decision of the Court of Arches, and stating that in the judgment about to be delivered the Archbishops of Canterbury and York had expressed their approbation, while the Bishop of London dissented.'

32 Pusey, *Tracts for the Times: Scriptural Views of Holy Baptism*, (1839).

33 *Westminster Shorter Catechism*, Q&A 91, 94.

34 Ramsay, *Catechism for Young People*, Sect. XXV, Q&A 7.

automatically imparted spiritual regeneration.[35] What the Gorham Business showed was that Anglican churchmen could sign up to the Thirty Nine Articles and use the same liturgical vocabulary, but have diverse interpretations.

The Scottish Episcopal hierarchy were of one mind. Though they knew that the Archbishops of Canterbury and York both accepted the Privy Council *Judgement*, all the Scottish bishops sided with Phillpotts in abhorring the decision of the Judicial Committee. In a letter to *The Times*, reprinted in Scottish newspapers, they thanked the Bishop of London—a leading supporter of Phillpotts—for not concurring in a decision, 'whereby it is propounded that the regeneration of infants in holy baptism is not the clear and unquestionable doctrine of the Church of England' and for not accepting 'that a minister of that Church may reasonably deny that infants are regenerated by holy baptism.'[36]

Dean Ramsay did not contradict his bishops in public, but it is unlikely he ever taught a theology of baptismal regeneration in the classes which James attended. Given James' awareness of Episcopalian doctrine, given the wide coverage of the Gorham Business in the Scottish press, and given the certain discussions on the matter in the Wedderburn household, it is no surprise he researched further.

There was another important theological issue of which James became aware whilst a teenager. It was David F. Strauss's attempt to interpret the Gospels as simply myth. Strauss (1808–1874) promoted this view in his 1834 book *Das Leben Jesu*, which had been translated into English in 1846 by the novelist George Eliot. James' reference to Strauss comes in an 1849 letter to Campbell, in which he refers to the American, Theodore Parker (1810–1860) who parodied Strauss' approach.[37] Parker showed that Strauss's principles for interpreting documents could produce any result which a person

35 Ibid., Sect. XXV, Q&A 17, 23, 24, 25, 26, 28, 29, 30. Ramsay's *Catechism* used wording similar to that found in Presbyterian documents: in baptism we are made 'Children of grace'; we become 'heirs or inheritors of the grace of God through Christ'; we are engrafted 'into the body of Christ's Church.'

36 *Caledonian Mercury*, 28 March 1850.

37 JCM, *Letter to Campbell*, 19 Oct 1849, (C&G1, 125).

wanted, illustrating his argument by 'proving' that the Declaration of American Independence was a fictional and mythological event! James approved of Parker's argument. He told Campbell that Parker showed 'the way to dissolve any given historical event in a mythical solution.' At age eighteen, James was alert to continental liberalism, as well as to home-bred theological controversies.

Theological considerations, especially the influence of high church Puseyism at English Universities—brought into focus by the Gorham Business—were among the factors which James and his father considered, long and hard, before he committed to Cambridge.[38] Though John Clerk Maxwell was content for his son to be influenced by Dean Ramsay, Puseyism was a different matter. Despite these hesitations, and after discussions with academic friends and members of the Mackenzie family, it was decided that James would go south. By this point his father had abandoned his original hope that James would study Law and be called to the Scottish Bar.[39] But before Cambridge there was Edinburgh University.

38 C&G1, 131.

39 Ibid., 130, 146.

Chapter 5
Edinburgh Student

James Clerk Maxwell entered the University of Edinburgh in 1847 as a private student, taking classes but without the intention of graduating. Free from the burden of degree examinations he dipped into numerous fields of intellectual activity.[1] The brilliant Peter Tait started at Edinburgh along with his friend, but after one session Tait transferred to Peterhouse College, Cambridge. James also had an opportunity to go to Cambridge but waited until 1850.

He attended James Forbes' lectures on natural philosophy for two years and occasionally during his third year. Forbes gave him use of the experimental apparatus in the department to pursue his own investigations, and Forbes' only criticism of his favourite student related to James being a 'little uncouth in manners.'[2] Having been brought up in rural Glenlair where his wont was to run, jump, and climb, he appears to have done the same when getting to his seat in class, clambering over the backs of the benches. At Cambridge he had similar gymnastic impulses, though more often confining these to open air activities. This 'uncouthness' was a part of his general demeanour: never deliberately impolite, but awkward and rustic in behaviour and speech.

Philip Kelland's mathematics lectures were attended for two years. Like Forbes, Kelland (1808–1879) was an Episcopalian, and, when Dean Ramsay died, it was Kelland who preached a commemorative sermon in St John's.[3] Pure chemistry was studied under William Gregory (1803–1858), and practical chemistry under Alexander Kemp (1822–1854). Moral philosophy came from John Wilson (1775–1854) who, under his pseudonym of Christopher North, was a regular contributor to journals such as *Blackwood's Edinburgh Magazine,* and a prominent figure in the capital's intellectual life.

---

1 C.G. Knott, 'James Clerk Maxwell', [Stevenson, *Famous Edinburgh Students*], 173–178.

2 Tait, *Letter to Campbell,* 14 Feb 1849; Forbes, *Letter to Whewell,* 30 Sep 1850; & Whewell, *Letter to Forbes,* 20 Apr 1852. (All in Harman, I, 8f.).

3 Rogers, *Memorials and Recollections of the Very Reverend Edward Bannerman Ramsay, Dean of Edinburgh,* 55.

SCIENCE

When James was at Trinity College, Cambridge, he hung portraits of Professor James Forbes and Professor George Wilson (1818–1859) in his room.[4] Forbes we have already met, but who was Wilson? George Wilson held the Chair of Technology at Edinburgh and was another evangelical who impinged on James' life. During a period in London during the late 1830s, Wilson befriended David Livingstone (1813–1873) who was completing preparations for service overseas with the London Missionary Society. Their bond endured, with letters addressed to Wilson from Livingstone in Africa still arriving after Wilson's death. Livingstone had intended to bring back specimens for Wilson's Industrial Museum of Scotland.[5]

After completing his doctoral studies in London, Wilson returned to Edinburgh in 1839 where he was appointed Regius Professor. In Edinburgh he became a member of the Congregational Church whose minister was William Lindsay Alexander (1808–1884).[6] The Congregationalist Alexander and the Episcopalian Dean Ramsay were long-standing friends, and when Dean Ramsay died in 1872 he left instructions for Alexander to lead a service in the Dean's own home prior to the formal funeral in church.[7]

Marston speculates that a series of Wilson's letters (1847/48) to a young man in poor health, in which Wilson combines discussion of scientific interests with expositions of the Bible, may have been written to James Clerk Maxwell. If so, these would shed considerable light on James' faith. However, the letters were more likely addressed to Daniel MacMillan (1813–1857) from the Isle of Arran, who was then living in London and Cambridge, and who co-founded the MacMillan publishing dynasty.[8] This emerges from a comparative study of: Jessie Wilson, *Memoir of George Wilson* (1862); George Wilson, *Counsels of an Invalid: Letters on Religious Subjects* (1862); and Thomas Hughes, *Memoir of Daniel MacMillan* (1883). Wilson befriended several promising young men, helping them spiritually as well as encouraging their scientific careers. The geologist Archibald Geike was also a

4 C&G1, 203.

5 Wilson, *Memoir of George Wilson,* 165.

6 Ibid., 321.

7 Ramsay, *Memorandum,* cited in: *Scotsman,* 3 Jan 1873.

8 Marston, 'Maxwell, Faith and Physics', 261.

recipient of his kindness and spiritual guidance, referring to Wilson in the warmest of terms in his autobiography.[9] Another whose life was impacted by Wilson, both academically and religiously, was John Cairns (1818–1892) who became the leading theologian of the United Presbyterian Church in Scotland and a respected interpreter of Sir William Hamilton's philosophy, of which more below.[10]

For James to have an image of Wilson on his wall in Cambridge indicates how important Wilson was to him. Apart from spiritual issues, they had a mutual interest in the study of colour-blindness, with Wilson and Maxwell corresponding on the topic and referring to one another's work in their publications.[11] There was also a shared interest in the eighteenth-century scientist Henry Cavendish (1731–1810), of whom Wilson wrote a biography.[12] In later years, when James became Director of the Cavendish Laboratory in Cambridge, he edited and published Cavendish's scientific papers, and his initial interest may have been sparked by Wilson.

Alongside the university work, which he found easy, James immersed himself in private study and research. He dabbled with electrical and magnetic devices of his own construction, and one of his chief interests was the properties of polarized light. He experimented by shining polarized light through shaped blocks of gelatine, observing the coloured fringes within the jelly. This led him to photoelasticity, a means of determining how stress is distributed within a physical body.[13] Still in his teens, he submitted two more papers to the RSE.

---

9 Geike, *A Long Life's Work,* 26f, 30, 70. Geike reveals that Alexander Macmillan (Daniel's brother) was also a Wilson contact.

10 Cf. John Cairns, *Letters to George Wilson,* 17 Dec 1856, 25 Nov 1857, 5 Aug 1858, etc. [MacEwen, *Life and Letters of John Cairns*].

11 Wilson, *Researches in Colour Blindness,* 154–159 (cf. 173ff), cites a lengthy letter from Maxwell: JCM, *Letter to George Wilson,* 4 Jan 1855, 'Theory of Colours in Relation to Colour Blindness' (Harman, I, 267–274). In the spring of the same year, Maxwell referred to Wilson's work in a paper for the RSE (19 Mar 1855): 'Experiments on Colour, as perceived by the Eye, with Remarks on Colour-Blindness', (RSE, *Transactions,* Vol. XXI, part II, 275–298; Harman, I, 287–289). In the autumn, Wilson sent Maxwell a copy of his work: cf. JCM, *Letter to John Clerk Maxwell,* 5 Oct 1855 (C&G1, 217; Harman, I, 328).

12 Wilson, *The Life of the Honourable Henry Cavendish,* 1851.

13 C&G1, 106-107.

His *Theory of Rolling Curves* was presented by Professor Kelland in February 1849; and his *The Equilibrium of Elastic Solids*—after edits suggested by Forbes—was delivered by James himself in the spring of 1850.[14] Following the latter lecture Forbes wrote again to him, stressing that the paper required further revision and that its reasoning had to be made simpler and more obvious if it were to be published.[15] Even Kelland had struggled to follow its logic!

In the summer of 1849 there was a family holiday on Scotland's west coast, during which he calculated that a body, deprived of friction, would take eighty-one seconds to slide on an inclined plane from the summit of Ben Nevis to Fort William.[16] It was after his return to Edinburgh that he set about writing his paper on *The Equilibrium of Elastic Solids.*

The following year, the British Association for the Advancement of Science met in Edinburgh. The British Association had been formed in 1831 by the clergyman William Vernon Harcourt (1789–1871), following a suggestion by the Scottish scientist Sir David Brewster (1781–1868). Harcourt and Brewster regarded the London based Royal Society as too stuffy and so they created what they hoped would be a more vigorous and democratic organisation which would take scientific conferences to every corner of Britain and Ireland.

James attended several of the 1850 British Association lectures, including one on optics given by Brewster. It was a short paper dealing with the polarizing structure of the eye and referred to the phenomenon called Haidinger's Brushes, first described in 1844 by the Austrian physicist Wilhelm Karl von Haidinger (1795–1871), who had lived in Edinburgh between 1823 and 1826. The phenomenon involves the eye 'seeing' a yellow horizontal block—sometimes in a bow-tie shape—against a blue sky when a person is facing away from the sun. Brewster admitted that neither he nor Haidinger could explain its cause. James had been interested in the phenomenon even

14 JCM, 'On the Theory of Rolling Curves', *TRSE*, Vol. xvi (1849), 519–540; (Cf. Harman, I, 116). JCM, 'On the Equilibrium of Elastic Solids', *TRSE*, Vol. xx (1850), 87–120; (Cf. Harman, I, 133–183). In the RSE *Transactions,* Maxwell's, 'Theory of Rolling Curves' is recorded as having been 'communicated by the Rev. Professor Kelland'; whereas Maxwell's, 'On the Equilibrium of Elastic Solids', is simply attributed to 'James Clerk Maxwell, Esquire.'

15 Forbes, *Letter to JCM,* 4 May 1850, (C&G1, 137f; Harman, I, 186 fn.).

16 JCM, *Letter to Campbell,* 6 July 1849, (C&G1, 124; Harman, I, 117).

before he visited Nicol's laboratory in 1847, and at Brewster's lecture he spoke from the audience describing his own experiments and querying some of Brewster's ideas.[17]

The newspapers noted that this initiated a conversation between the teenage James, Sir David Brewster, the Astronomer Royal Sir George Airy (1801–1892), and Professor George Stokes (1819–1903) who was the Lucasian Professor of Mathematics at Cambridge.[18] Though he was confident in his science even in such august company, James was full of nerves and spoke in sudden gushes with notable pauses between his words. Despite his broken delivery the audience was impressed. This was his first encounter with the legendary Brewster. They would cross paths several times in the future.

PHILOSOPHY

A major influence at Edinburgh University was Sir William Hamilton (1790–1865), whose enthralling lectures on logic and metaphysics took place in his classroom at the top of a steep flight of stairs.[19] Sir William's wife had tried to persuade the university to give him another room, but without success. The near-perpendicular staircase not only impeded the ageing professor's progress—especially following a stroke in 1846—but once caused even the youthful James to sprain his ankle badly, though he hobbled on upwards to sit through the class.[20] James loved the subject and esteemed his teacher, even though by the time he attended the classes, Hamilton only delivered part of the lecture himself and then sat in his chair whilst his assistant, Alexander Nicolson, read the rest.[21]

Archibald Geike heard the same lectures a few years later. He recalled Sir William being wheeled into his classroom: 'the impression made by his grave pale face, massive brow, and look of weariness remains ineffably in my memory.'[22] Before the stroke Hamilton used to stride into class clad in his flowing academic gown and, without sitting down but standing behind his high desk, began to read his hour-long lecture, 'often waving his hand to stay the applause' with

17 C&G1, 84.

18 *London Express,* 5 Aug 1850. Cf.C&G1, 489f.

19 *Scottish Guardian (Glasgow),* 17 May 1856.

20 C&G1, 108.

21 Veitch, *Memoir of Sir William Hamilton,* 281.

22 Geike, *A Long Life's Work,* 37.

which his admiring students greeted him.[23] Hamilton had his own way of examining students. Instead of asking specific questions, he invited them to give an account of his lectures on a special subject. Volunteers were allowed to proceed for ten or fifteen minutes without interruption or correction until their memory was exhausted. Then he requested additional information, at which point other students might give the essence of a relevant book which they had been reading.[24]

Before Hamilton gained the chair in 1836, his suspected atheism or agnosticism was highlighted by opponents. This was brushed aside by supporters who pointed to numerous references to the Deity in his writings.

Notwithstanding these claims of religious orthodoxy, the Revd John Cairns, one of Hamilton's devoted admirers, coupled admiration with deep concern for his mentor's lack of an assured faith. Cairns wrote to Sir William later in life, expressing a wish that his honoured master 'were one with him in the exercise of the convictions and the enjoyment of the comforts of living Christianity.'[25] Sir William replied in warm and kindly terms but without expressing any religious principles.[26] Whatever his personal position, his ancestors had impeccable credentials as prominent Covenanters, leading the Cameronian faction at the 1679 battles of Drumclog and Bothwell Bridge.

On the eve of the Disruption in 1843, Hamilton published a tract, *Be not schismatics: Be not Martyrs by Mistake*, in which he argued that secessionists from the established Church were contradicting the policy of their revered forefathers in the Reformed faith. He displayed detailed knowledge of theology and church history, though his interpretation was challenged by Free Kirk supporters. Ironically, given his covenanting heritage, Hamilton was laid to rest in the vaults of St John's Episcopal Church in Edinburgh, as the family was now Episcopalian.[27]

---

23 Note from John Cairns (1818–1892), cited by Veitch, 'Memorials of Sir William Hamilton', 227.

24 MacEwen, *Life and Letters of John Cairns,* 69.

25 Cairns, *Letter to Sir William Hamilton,* 16 Nov 1848, (MacEwen, 291f.).

26 Hamilton, *Letter to John Cairns,* 4 Dec 1848, (MacEwen, 293).

27 *Inverness Courier,* 15 May 1856.

James Clerk Maxwell viewed Hamilton's philosophy lectures as the most substantial which he attended at Edinburgh, and fed on their insights for the rest of his life.[28] (He was also influenced by William Whewell of Cambridge, especially on the use of inductive thinking.[29]) Hamilton stood in the line of Scottish Common Sense philosophy, albeit with modifications. Three years after his death, an edition of his lectures was published in Edinburgh and reviewed for the *Scotsman* newspaper by John Cairns, who summarised Hamilton's philosophy under three heads.[30] First: his vindication of the doctrine of Common Sense. Second: his discussion of the theory of perception in relation to our belief in an external world. Third: his enunciation of the Law of the Conditioned as bearing on our knowledge of the absolute and infinite (i.e. nothing infinite could be known by the finite).[31]

Cairns remarked that the first two heads reflected Scottish philosophical thinking, whilst the third indicated German (Kantian) influences. Cairns stressed that the impulse which Hamilton 'gave to the third was the most powerful'. It was impressed on each of Hamilton's students that there could be no knowledge of the infinite or of the

---

28 JCM, *Letter to Campbell,* 16 Nov 1847, (C&G1, 115; Harman, I, 69). Harman, I, 4 points out that in the paper on 'Rolling curves', Maxwell used classifications discussed by Hamilton in his logic lectures.

29 JCM, 'Whewell's Writings and Correspondence', *Nature,* Vol. 14/349 (6 July 1876), 206–208 (Also NVN1, 528–532): 'Dr Whewell's opinion with respect to the evidence of the fundamental doctrines of mechanics is repeatedly inculcated in his writings. He considered that experiment was necessary in order to suggest these truths to the mind, but that the doctrine, when once fairly set before the mind, is apprehended by it as strictly true; the accuracy of the doctrine being in no way dependent on the accuracy of observation of the result of the experiment . . . Thus, then, we are led by experiments which are not only liable to error, but which are to a certain extent erroneous in principle, to a statement which is universally acknowledged to be strictly true. Our conviction of its truth must therefore rest on some deeper foundation than the experiments which suggested it to our minds. The belief in and the search for such foundations is, I think, the most characteristic feature of all Dr Whewell's work.'

30 Hamilton, *Lectures on Logic and Metaphysics,* Vols 1 &2, (Edinburgh: Blackwood, 1859).

31 *Scotsman,* 5 Apr 1859. The review was anonymous, with Cairns' authorship revealed by MacEwen, *Life and Letters of John Cairns,* 67.

unconditioned, and it was Hamilton's implacable conviction on this point which lay behind his agnosticism. The mind could know *effects* which were the result of *causes*—thus the mind could know things which were *conditioned*—but the mind was incapable of knowing the *unconditioned,* that is, any reality which was not conditioned by, or caused by, known finite causes.[32] This followed Kant, for whom the *conditioned* was anything which depended on something else for its being.[33] For Hamilton, this ruled out any possibility of knowing God, who is the *uncaused* or *unconditioned* Creator, though Hamilton did offer an important caveat, to which we return later.

Significant for James as a budding scientist, Hamilton taught that the mind *could* perceive external objects immediately. He emphasised the validity of connections between cause and effect. And he emphasised that such connections were not impositions by the mind on reality but were embedded in reality itself. He tried to combine this with aspects of Kant's thinking, despite Kant proposing a radical disjunction between the *noumena* (what a thing is in itself) and the *phenomena* (how a thing is perceived by us). Hamilton held that things could be truly known, at least in their *relation* to the knower and other objects. At the same time, his philosophy tilted towards Kant's questioning of whether we can ever truly know the nature of external objects.

For Hamilton, philosophy was important in affirming morality though he personally struggled to establish solid foundations for ethics from a purely philosophical basis. He wrote to John Cairns: 'The foundation of ethics, connected as it is with moral freedom, has always appeared to me the most difficult, as the most important problem in philosophy.'[34] Following his teacher, James sought similarly to weave together science, philosophy, and ethics; but whereas Hamilton only worked within the parameters of philosophy, James included faith propositions derived from Scripture, as he alluded to in an 1850 letter to Lewis Campbell which we discuss later in this chapter.[35]

32 Kelly, 'Sir William Hamilton', *DSCHT,* 392.

33 Veitch, *Memoir of Sir William Hamilton,* 409.

34 Hamilton, *Letter to John Cairns,* 15 Apr 1851, (MacEwen, 347).

35 JCM, *Letter to Campbell,* 26 Apr 1850, (C&G1, 134–136; Harman, I, 191–192).

'Where Thought weds Fact'

At the heart of philosophy are the questions: How can we know things? Can the mind have knowledge of anything outside of itself? Do theories about reality correspond to what reality is in itself? Hamilton opened James' mind to these issues and to his own blend of Kantianism, Common Sense philosophy, and the knowability of the *relation* between things if not the things in themselves. All of this fed James' later thinking, though not without amendment.

As a young scientist, but with the invaluable ability of being able to reflect on his discipline, James sensed the importance of thinking through whether theories represent true knowledge of reality, or whether theories merely reflect subjective mental constructions. On this point, whereas Hamilton lived in the world of pure thought, James had experimented with physical objects from boyhood, predisposing him to believe that the mind could know things as they really are.

When Lewis Campbell commented that the ideas which James received from Hamilton were his habitual vantage ground, he insightfully added that one great difference was: 'While Sir William remained for the most part within the sphere of abstract logic, Maxwell ever sought to bring each concept to the test of fact.'[36] James' penchant was to combine the theoretical with the empirical, a tendency fostered by his love of physical experiments: no idea was true if determined by thought alone. He expounded this in his 1856 Inaugural Professorial Lecture at Aberdeen, emphasising that true knowledge required thought to be wedded to reality.[37] And in 1870, when President of the Mathematics and Physics Section of the British Association, he said:

> Who will lead me into that still more hidden and dimmer region where Thought weds Fact, where the mental operation of the mathematician and the physical action of the molecules are seen in their true relation? Does not the way to it pass through the very den of the metaphysician, strewed with the remains of former explorers and abhorred by every man of science?[38]

---

36 C&G1, 108.

37 Cf. Maxwell, *Aberdeen Inaugural Lecture,* 1856.

38 JCM, 'Mathematics and Physics: Presidential Address', *Report of the 40th Meeting of the BAAS,* (1870), 2. Cf. C&G1, 326.

By 1870, James had become convinced that an enquirer could indeed know reality as it is in itself, through the use of mathematics and physics, whatever philosophical and practical difficulties lay in the way. In the interface between enquirer and physical reality, the mind of the enquirer could be conformed to the nature of that reality. The task was difficult. But it was possible. With that conviction, he added:

> We are led up to questions the same in kind with those of metaphysics; and we approach them, not trusting to the native penetrating power of our own minds, but trained by a long-continued *adjustment of our modes of thought to the facts of external nature.* (Italics ours)

The closing phrase is important: '*adjustment of our modes of thought to the facts of external nature.*' This expressed his mature conviction that, despite many blind alleys, the mind of an enquirer could apprehend reality as it truly is. It was possible for the nature of a thing to be known by an enquirer. The pressure of external reality on the mind could reform, mould, and alter the very way we think and force the mind to think *kata phusin,* as Torrance would express it, i.e. in accordance with the nature of that which is being examined. If nature were examined properly, then nature itself would determine how we know it.

Natural Realism

That position was well in the future when James first sat under Hamilton, behind whose metaphysics lay generations of theorising on how truth about external reality can ever enter the mind. *Rene Descartes* (1596–1650) favoured the notion of innate or pre-existing ideas. *John Locke* (1632–1704) regarded the mind as a blank slate (*tabula rasa*) at birth, and taught that we are born without innate ideas, with knowledge only coming through experiences fuelled by sense perception. *Bishop George Berkeley* (1685–1753) questioned whether the mind could ever know anything outside itself or outside the realm of ideas. Was it even possible to prove that what we know through the senses is information from outside the mind at all?[39] Could pure

39 Berkeley argued that there is no material world to know, and that the sole objects of knowledge are ideas (hence: Idealism) and that these exist only in the act of being known: though, when these pass from *our* mind they do not cease to exist, for they are known by God.

thought ever prove that anything exists outside of thought? *David Hume* (1711–1766) wondered whether we can obtain knowledge even from empirical data.

Hume took Berkeley's thinking to its extreme and argued that the patterns which science discovers in the natural world—such as the sun rising every morning—are merely summaries of what an observer is *accustomed* to experience, but which the same observer can never *prove* to be part of the nature of things. For Hume, *custom* was the great guide of human life. What Hume meant was that theories were posited because of repeated experiences, and not because of a true understanding of objective reality.[40] In Hume's opinion, we know ideas or impressions: but where these come from, what they are, and what they may become, is inherently unknowable.

Immanuel Kant (1724–1804) expanded on the apparent impossibility of ever knowing anything outside of the mind. As already mentioned, Kant tried to overcome this impasse by creating the categories of *noumena* (what an object is in itself) and *phenomena* (our experience of an object in relation to us). This resulted in the so-called Kantian dualism between *things-in-themselves* and *things-for-us*, with Kant regarding things-in-themselves as unknowable.[41]

What all of this meant was that even the empirically based experimental method—science's apparently solid and impregnable foundation—was chipped away by philosophers sceptical of ever attaining true knowledge of anything apart from mental ideas. This affected how scientists viewed their work. What did a scientific theory mean? What did an equation mean? A theory might work. An equation might work. But did they work because they described what was out there? Or did they work only because they created models

---

40 Hume ignored concepts such as Newton's gravitational principle, which did in fact offer an underlying logic as to why the sun 'rose' every day: but Hume was weak on science, and adopted a simplistic mechanistic view of nature to argue against the possibility of miracles.

41 Torrance, *Theological Science,* 88f, wrote that for Kant, the 'thing in itself' was entirely independent of our experience of it, and led Kant to suppose that the truly fundamental laws of nature, similar to the truths of mathematics, are knowable only because they make no effort to describe the world as it really is, but *prescribe* the structure of the world as we experience it. For Kant, the reason *imposed* order, it did not *discover* order, since the mind could never know the *noumena.*

which accidentally fitted the data? As a realist, the Maxwell of 1870 had come to believe that theories could correspond to reality due to: '*the adjustment of our modes of thought to the facts of external nature.*' In other words, nature could alter the way we think. Nature could take hold of our minds, forcing them to think in accordance with the way things are.

Thomas Reid (1710–1796), a son of the manse and professor of philosophy in Aberdeen and later Glasgow, challenged the Locke, Berkeley, Hume *cul-de-sac*. Reid exchanged courteous correspondence with Hume; and the genial and avuncular Hume was appreciative of this thoughtful and friendly interchange, especially after having been damned as the 'Great Infidel' from numerous Scottish pulpits!

To escape from what he saw as a philosophical dead-end, Reid developed what is now known as Scottish Common Sense philosophy. He held that the universal common sense experience of humanity is itself a basis for objective knowledge, since a benevolent Deity would never give the whole of humanity a misleading common consciousness or an irresistible belief in falsehoods![42] Thus, humanity's universal awareness of things is a valid starting-point for the mind, from which it can begin to construct a proper representation of the world.[43] On this basis, Reid argued that the mind has an immediate perception of external objects; that these sensations of external objects are real; and that from this intuitive belief there is, and there need be, no appeal. All the reasonings of philosophy are futile against the common sense experience of humanity, which affirms an external world, free-will, and true causation.

Reid's thinking was modified by Hamilton, giving Hamilton a version of common-sense natural realism undergirding his metaphysics. But because Hamilton also affirmed aspects of Kant's thinking, this resulted in Hamilton's insistence that an object can only be known in its *relationship* to the knower and in virtue of its *relations* to other objects. This has been highlighted by some Maxwell scholars as a possible root of Maxwell's later 'relational' view of physics in which he viewed matter not so much in terms of discrete, individual particles, but in terms of a continuum of being, or being-in-relationship.[44]

42 Broadie, *The Tradition of Scottish Philosophy,* 158.

43 Gaal, *The Faith of a Scientist,* 14.

44 Francis Everitt, *James Clerk Maxwell: Physicist and Natural Philosopher,* 102.

There is a further possibility: did Maxwell detect elements of things only being known in their *relationships*—especially in their *relationship to the knower*—in Dean Ramsay's *Catechism for Young People*? Ramsay's *Catechism* states that the mystery of the Trinity is only known in its threefold *relation to us.*[45] Concerning the Trinity, Ramsay posed two questions. First, what do we learn practically from the great doctrine of the Trinity? Answer: We learn to consider the threefold *relation* that God bears to us in the great work of our redemption. Second, what are those relations? Answer: That God the Father stands to us in the *relation* of Creator; God the Son as Redeemer; God the Holy Spirit as Sanctifier. Ramsay himself italicised '*relation*'. We cannot prove that James made a connection between Hamilton's philosophy and Ramsay's doctrine of the Trinity, but he may have done.

In April 1850, James informed Campbell that he had been reading Kant's *Critique of Pure Reason* in German, determined to make it agree with Hamilton.[46] Gaal has interpreted this as James disagreeing with Hamilton; but James was actually trying to hold Kant and Hamilton together.[47] Whether Kant's fundamental distinction between 'what a thing is in itself' (*noumena*) and 'what a thing is as we perceive it' (*phenomena*) can ever be fully harmonised with any version of Scottish Common Sense philosophy is debatable, though Hamilton attempted it. Overall, James was aware that the instinctive common sense approach had to reckon with Kant's strictures concerning the inherent subjectivism of all human enquiry, which ever after made him sensitive to philosophical issues in the search for objective knowledge.[48]

The Unconditioned

Hamilton taught his students that the infinite (limitless time, space, power, etc.) is unable to be conceived of adequately because of

45 Ramsay, *Catechism for Young People,* Appendix, III: 9, 10.

46 JCM, *Letter to Campbell,* 26 Apr 1850, (C&G1, 135; Harman, I, 191).

47 Gaal, *The Faith of a Scientist,* 20.

48 On Maxwell's philosophy see: Harman, 'Edinburgh Philosophy and Cambridge Physics: The Natural Philosophy of James Clerk Maxwell', in: *Wranglers and Physicists,* 202–224. Also, George Elder Davie, *The Democratic Intellect: Scotland and her Universities in the Nineteenth Century,* 138–145 & 192–197.

the finite nature of mind and its experiences. This lay behind his *Philosophy of the Unconditioned*, which was the principle that there could be no knowledge of the infinite for the human mind. This made belief in God difficult for Hamilton, creating the impasse from which John Cairns tried to rescue him.

Years later, when Maxwell wrote on Michael Faraday (1791–1867) for the *Encyclopaedia Britannica,* he cited Faraday's similar insistence that human reasoning could never, by itself, know God.[49] Echoing the biblical text Job 11:7 ('*Can man by searching find out God?*'), Faraday had stated that no one 'by reasoning could find out God', which Maxwell quoted with approval. Faraday's conclusion would have been approved by Hamilton, though their reasoning diverged wildly. Though Cairns, Faraday, and Maxwell agreed with Hamilton that neither reason (the use of pure logic) nor experiment (the investigation of causal sequences in nature) could discover God—since God's eternal being and his free action of the will are both unconditioned—they believed that God had supplied another channel of knowledge: namely, the testimony of Holy Scripture.

The infinite gap between the finite and the infinite—unbridgeable by the finite mind—leaves the question of the possibility of knowing God with three options: agnosticism; belief that God has chosen to communicate himself; or the pursuit of a theological rationalism which maintains that God *can* be known deductively from clues in nature. Maxwell's option was the second of these: a faith that God had revealed himself through his Word. That bridged the otherwise unbridgeable, with Maxwell's simple trust in the Christian revelation lasting his whole life.

Hamilton was aware that full-blooded agnosticism was unwelcome in Edinburgh. That may have been why he offered a caveat which allowed for some knowledge of God. As a Scottish Common Sense philosopher, he knew of the notion that since all human thought appears to demand the idea of the infinite or the absolute (God), then there must be something to it. Consequently, he developed a 'philosophy of the unconditioned' to provide a

49 JCM, 'Michael Faraday', *Encyclopaedia Britannica,* (9th ed.), Vol. 9, 29–31. Maxwell cited: Faraday's, 'Lecture on Mental Education', 6 May 1854, which had been delivered before Prince Albert and members of the Royal Institution.

sphere for faith, with the role of theology being that of unpacking this human instinct. But he made a distinction between 'knowledge' and 'belief'—a distinction which James adopted. The 'philosophy of the unconditioned' meant that, within certain parameters, 'faith' was not in opposition to science or rationality. Rather, 'faith' was a type of common sense knowledge of which the human mind was aware, which it consciously demanded, and for which it reached out (as in Reid's philosophy): but it was *unconditioned.* That is to say, it was not conditioned by anything outside of itself. It was not based on deductive reasoning. Nor was it the end result of empirical causes. It was beyond empirical sense data. It stood on its own merits. It was grounded in the universal consciousness of humanity.[50]

The Question of Beginnings

Hamilton held that the human mind could not conceive of any phenomenon which had no beginning. This was in obvious tension with Christian faith which claimed knowledge of a God who was eternal. He offered two grounds for his stance. First, 'no beginning' implied an infinite existence—which he disallowed. Second, 'no beginning' implied the existence of a reality which was not itself the end product of a series of causes—which he also disallowed. Any entity with no identifiable natural cause was outside the range of valid human knowledge.

James' faith gave him different assumptions. Not only did he accept the validity of a knowledge of an uncaused God, but he accepted, from faith, the freedom of God to create out of nothing (*creatio ex nihilo*), and that there was an absolute beginning of space, time, matter, and energy. At the same time, he was influenced enough by Hamilton to permanently adopt the position that a natural cause could and should always be found for any phenomenon in

---

50 Maxwell refers to the Unconditioned in his 1878 poem, 'A Paradoxical Ode' (C&G1, 651) in the context of rejecting an infinite series of causes. He uses the notion to argue that science can only operate within given bounds. Unlike Hamilton, Maxwell's starting-point was faith, and he never considered that knowledge of God was impossible. In stating that science was incompetent to judge on questions of origin etc., Maxwell was not erecting barriers to protect faith, but was working out the implications of Hamilton's philosophy from a faith perspective.

nature—apart from the ultimate origin of matter and its laws. This conviction lay behind his later and much renowned 1873 'Discourse on Molecules'.[51] But his dominant conviction was that any entity or event in nature had a natural cause discoverable by science; hence there would never be a gap into which God could be introduced as the only possible solution! God had made a perfect created order, not one with loose ends. In James' mind, the concept of God was greater than simply being a solution for occasions when science came to an interruption in its chain of causality.[52] Contrary to a 'gaps' approach, if God were to be known through the work of science then it was by inductive reasoning. The 'whole' was to be grasped, and with it an awareness that the overall existence, structure, and logic of the created order were inexplicable without a higher form of reality, with nature 'crying out' for a Creator.[53]

For some Christians, Darwin's later explanation of the origin of humanity was uncomfortable since it removed a 'gap' in nature. Up to this point that gap was thought to be bridgeable only by a direct divine action, and therefore, if this 'gap' was removed, it destroyed an important 'proof' for the existence of God. But that was never a problem for James. His faith was not based on there being gaps which were only bridgeable by the God-hypothesis. For James, God had made a universe explicable within itself and from itself. It was the very explicability of nature which, for him, pointed to the Creator.

51 JCM, 'Discourse on Molecules', 1873, (C&G1, 359–360).

52 The 'God of the gaps' argument assumes there are gaps in scientific knowledge which can only be filled by introducing God as the cause. C.A. Coulson (1910–1974), professor of mathematics at Oxford and Methodist lay-preacher, argued in *Science and Christian Belief*, 32: 'There is no 'God of the gaps' to take over at those strategic places where science fails; and the reason is that gaps of this sort have the unpreventable habit of shrinking . . . either God is in the whole of Nature, with no gaps, or He's not there at all.'

53 Torrance rejected proofs of God from nature but did write of nature being inexplicable without God. Cf. Torrance, 'The Significance of John Philoponos', 53: 'The human genome . . . is laden with more information than would fill a vast encyclopaedia, and . . . by its astonishing complex nature could not have arisen in some sort of accidental or self-organisational way'.

The Question of Method

Sir William introduced his students to probability arguments, which may have aided James in later years as he incorporated probabilistic models into his scientific theories.[54] On this issue, he sent a letter to Lewis Campbell, in either June or July 1850, in which he discussed questions of methodology which might be common to all areas of enquiry, as he understood these at the time. He knew the importance of working within the principles of deductive logic, and that 'understanding ought to work by the rules of right reason'.[55] But he also knew that simple deduction with a unique solution was often impossible:

> The actual science of Logic is conversant at present only with things either certain, impossible, or entirely doubtful, none of which (fortunately) we have to reason on. Therefore, the true Logic for this world is the Calculus of Probabilities, which takes account of the magnitude of the probability (which is, or which ought to be in a reasonable man's mind). This branch of Mathematics, which is generally thought to favour gambling, dicing, and wagering, and therefore highly immoral, is the only 'Mathematics for Practical Men' as we ought to be.

The optimistic teenager reckoned that probability factors would decrease as overall knowledge grew.[56] In later life he realised that in some sciences a probabilistic model was a permanent feature. What he tried to do in 1850 was to transfer analogically some of his thinking in science over to moral philosophy and faith, aware of probabilistic thinking there also:

---

54 Elizabeth Garber, 'Maxwell's Kinetic Theory', in Flood (ed.), *JCM: Perspectives on his Life and Work,* 141.

55 JCM, *Letter to Campbell,* June 1850, (C&G1, 143; Harman, I, 197).

56 Ibid.: 'As human knowledge comes by the senses in such a way that the existence of things external is only inferred from the harmonious (not similar) testimony of the different senses, [our] understanding . . . will assign to different truths (or facts, or testimonies, or what shall I call them) different degrees of probability. Now, as the senses give new testimonies continually, and as no man ever detected in them any real inconsistency, it follows that the probability and credibility of their testimony is increasing day by day; and the more a man uses them the more he believes them.'

> What is believing? When the probability (there is no better word found) in a man's mind of a certain proposition being true is greater than that of its being false, he believes it with a proportion of faith corresponding to the probability, and this probability may be increased or diminished by new facts. This is faith in general. When a man thinks he has enough of evidence for some notion of his he sometimes refuses to listen to any additional evidence pro or con, saying: 'It is a settled question, probatis prolata; it needs no evidence; it is certain.' This is knowledge as distinguished from faith. He says: 'I do not believe; I know.' 'If any man thinketh that he knoweth, he knoweth yet nothing as he ought to know.'[57]

To support his argument, James alluded to the biblical text, *'And if any man thinks that he knoweth any thing, he knoweth nothing yet as he ought to know.'* (1 Corinthians 8:2). He interpreted this as a warning to anyone claiming absolute knowledge by implicit faith, with 'a shutting of one's ears to all arguments.' 'Shutting ears' was not what Christianity meant by a childlike faith! Faith had to be based on evidence and reason. Childlike faith was not to be confused with naïveté: 'Children are not credulous but find out sooner than some think that many men are liars.'

Having reflected on method, in which he emphasised the incompleteness, contingency, and necessary reviseability of all human knowledge—especially in the uncertain realm of moral philosophy—he concluded his long letter to Campbell by teasingly adding that his friend might write and state the *probability* of his coming to visit him.

GEOLOGY

Edinburgh has a small mountain within its bounds, known as Arthur's Seat, and generations of its citizenry have climbed it on the first day of May to bathe their faces in the morning dew. When James was a boy, his father took him there on several excursions, though not for the May Day jollities. They went to explore the local geological formations, and James later introduced his Edinburgh Academy school friends to the same places. Lewis Campbell recalled a walk in February 1845 during which James pointed out glacier-markings

57 JCM, *Letter to Campbell*, June 1850, (C&G1, 144; Harman, I, 198).

on the rocks and 'discoursed volubly on this subject, which was then quite recent and known to comparatively few.'[58]

Other boys at the Academy had similar interests. Leonard Horner, one of the school's founders, had a love of geology through reading Playfair's *Illustrations of the Huttonian Theory of the Earth* (1802). Horner's son-in-law was the geologist Charles Lyell (1797–1875), and, with Horner twice President of the Geological Society of London, it was natural for the school to encourage the science.

Alexander Fergusson, in his *Chronicles of the Cumming Club,* recalled that his class, one year behind James' cohort, formed an Association for the prosecution of scientific study, with geology the favourite topic.[59] The Association had Saturday excursions in and around Edinburgh. Fergusson remembered these as times of supreme enjoyment with the expeditions enhanced by banquets of oatmeal cakes, cheese, and ginger-beer in village inns. There was a particular interest in fossils, with a rich supply in a quarry five miles to the south of Edinburgh. Whether James joined the Association's excursions, or whether, being in a different school year, he had his own trips, we do not know: what is evident is the popularity of geology among James and his contemporaries. The boys of the High School formed similar clubs, with geological expeditions a popular activity.[60]

By the 1840s there had been significant advances in geological theory. The highly readable *Principles of Geology*, written by the Kirriemuir-born Charles Lyell and first published in three volumes from 1830 to 1833, had popularised new discoveries, reinforcing awareness that the earth was much older than once thought. Lyell focussed on slow and gradual changes, caused by ordinary natural processes of a type still operating day by day. His eloquently written book persuaded many lay men and women that the earth needed deep-time for it to become what it was now.

This approach was taken up in the popular press, which regularly published articles on geology. Charles Maclaren, co-editor of the *Scotsman* and friend of Louis Agassiz, was instrumental in spreading the science. For example, just after James started at Edinburgh Academy, a lead article in the *Scotsman,* entitled

58 C&G1, 73.

59 Fergusson, *Chronicle of the Cumming Club,* 56f.

60 Geike, *A Long Life's Work,* 15f.

'American Geology, Erratic Blocks, Glacial Action', focussed on new geology in the United States and Scotland.[61]

In the twenty-first century it is frequently assumed that most nineteenth-century Christian believers held to a young-earth theory with a six-day-creationist interpretation of the book of Genesis. This was not so. It is equally common for modern writers to assume that the seventeenth-century Archbishop James Usher's date of 4004 BC for the creation of the earth was what church-goers believed. Again, not so. Usher's dating had long been abandoned by the vast majority.[62]

By the mid-nineteenth century, and after considerable debate in the eighteenth century, old-earth theories had been absorbed into the worldview of most church-goers, with corresponding re-interpretations of Genesis. The crisis occasioned by Darwin's theories concerning the origin of species should not be confused with the response to geology. The factor which disturbed many Christians after Darwin published *On the Origin of Species* (1859) was not the great age of the earth which evolutionary theory demanded, but the possible undermining of the unique status of humanity if the theory were true.

The theological implications of modern geology were open questions among believers, and many evangelical Christians who were strongly committed to the Bible accepted the new ideas. To take one example, Thomas Chalmers, the emerging leader of the evangelical party in the Church of Scotland and future prime mover behind the formation of the Free Church, explained in his 1814 review of Cuvier's *Essay on the Theory of the Earth* why Genesis chapter one might not imply a literal six-day creation.[63] Chalmers argued that between the initial act of divine creation—which brought out of nothing the *matter* of the heavens and earth—and the initial act of the First Day's work as recorded in Genesis, periods of vast duration may have intervened. Chalmers wrote this review

---

61 *Scotsman,* 1 Dec 1841.

62 Cf. Ellicott, *Modern Unbelief: its Principles and Characteristics* (1877), 38: 'Much needless disquietude has been caused . . . by the unconscious assumption that the chronology of Archbishop Usher is as inspired as the [Bible] on which its results are placed.'

63 Cf. Miller, *The Testimony of the Rocks,* 117f. Chalmers considered the 'days' of Genesis to be twenty-four-hour days, but with a vast expanse of time existing before these specific days occurred.

after he was committed to the evangelical party, adhering to concepts on which he had lectured at St Andrews a decade earlier. Other evangelicals who absorbed the science into their faith, included: Robert Moffat (1795–1883), Thomas Dick (1774–1857), John Pye-Smith (1774–1851), and Hugh Miller (1802–1856).

It was not just the middle-class intelligentsia who took to the new ideas: an artisan gardener such as Robert Moffat did so also. Moffat was a leading Scottish missionary to southern Africa and father-in-law to David Livingstone. He had a copy of Lyell's *Principles of Geology* in his Kuruman home in Africa, and he wrote in his *Journal* about the amazing age of the rocks which he observed on his wagon journeys.[64]

Thomas Dick was another who was upbeat about recent geological findings, interpreting them as supportive of the Mosaic record: 'We have reason to believe, that as Geologists proceed in their researches and investigations, still more sensible proofs of the authenticity of Revelation will be brought to light.'[65] Dick made these assertions because, like Chalmers, he allowed for the earth existing for a lengthy period of time before the crucial six-day period commenced: 'Moses nowhere affirms that the materials or substance of our world, were created or brought from nothing into existence, at the time when his history commences.'[66] John Pye-Smith, who was tutor at the robustly evangelical Homerton Independent College where Maxwell's future father-in-law, Daniel Dewar, was trained, even suggested that some life-forms may have existed before the critical six days![67] Hugh Miller, the internationally known activist, author, journalist, geologist, and champion of all evangelical causes, interpreted the 'days' of Genesis

64 Moffat, *Journal,* 1 June 1854, (*The Matabele Journals of Robert Moffat,* Vol. I, 150). Following Chalmers' theories, Moffat referred to great periods of time prior to the emergence of humanity.

65 Dick, *The Christian Philosopher* (1843), 213.

66 Ibid., 222, and *The Discoveries of Modern Geology not Inconsistent with Revelation,* (1843), 21f. Like Chalmers, Dick taught that the phrase 'In the beginning God created the heaven and the earth' was a preface to the six days, covering an enormous era of time. For Dick, the purpose of Genesis 1:1 was to underline that the universe was not self-existent.: 'No specific period is here stated. . . . even millions of years may have elapsed.' (22).

67 Pye-Smith, *The Relation between the Holy Scriptures and some parts of Geological Science,*35, 146.

chapter one as six days in which Moses was given successive visions of creation, rather than as six days during which the earth was formed.[68]

What all of this meant was that though new ideas were accepted at varying speeds by different sections of the Christian community, the age of the earth was not in itself a stumbling-block for believers in the way that the origin of species would become. But, post-Darwin, the disquiet created by evolutionary theory fed back into views on geology, creating polarization and resistance where previously there had been growing acceptance.

The Clerk Maxwell family had direct connection with geology. James' great-grandfather, George Clerk Maxwell (1715–1784), had been a friend of James Hutton (1726–1797), accompanying Hutton on several geological expeditions around Scotland.[69] At the time there were several competing theories, each of which required the earth to have existed for an extended timespan. Hutton gathered evidence which he interpreted as consistent with slow, gradual, uniform, long-term processes, rather than by quick or sudden action, and formed his own ideas concerning the mechanism involved.

Hutton's publications, especially after their promotion in a more popular form by the Edinburgh mathematician John Playfair (1748–1819), contributed to a growing realisation among the general public that mountain ranges and shapes of rocks were not now as they had come from the hand of the Creator on the day of their creation. Instead, they had their current appearance through a series of processes. Some geological phenomena may have come by sudden catastrophic action, but not all. Geology was committed to the earth requiring an exceptionally long history for rock-formations to be what they are today. A famous line from Hutton was: 'We find no vestige of a beginning, no prospect of an end.'[70] The phrase was picked up by the publisher, amateur scientist, and prominent Edinburgh citizen

---

68 Hugh Miller gave a London lecture in 1854, '*The Two Records, Mosaic and Geological*'. This was followed by publication of his, *The Testimony of the Rocks*, which contained the earlier lecture plus a sequel, *The Mosaic Vision of Creation*.

69 Arthur, *Brilliant Lives*, 145.

70 What Hutton meant by his famous phrase is discussed by Ray Perman, *James Hutton: The Genius of Time*. Perman argues (237) that Hutton was not describing an eternal Earth, but was referring to the state of his present enquiry.

Robert Chambers (1802–1871), and adapted for the title of his book *Vestiges of the Natural History of Creation* (1844) which James read with fascination. Chambers' book also floated an early theory of the development of species but received a hostile reception.[71]

Was James a creationist? If by 'creationist' is meant someone who believes that the world was created over a literal six-day period, then he was not. But if by 'creationist' is meant someone who believes that God created the universe out of nothing and that the mind of God lies behind the structures and rationality of the created order, then he was. As a student, he wrote a paper for Sir William Hamilton's logic class, stating that there were two possibilities for matter: it was either created or uncreated, but if uncreated 'we are led into impiety.'[72]

ETHICS

Among the authors read whilst at Edinburgh were Descartes (1596–1650) and Leibniz (1646–1716). He also read Hobbes (1588–1679), 'chiefly on the ethical side.'[73] The moral philosophy classes at Edinburgh were given by Professor John Wilson, who, under his pseudonym of Christopher North, contributed to the literary journals of the day. Wilson was less renowned as a serious scholar, though his lectures were enjoyed by his students. His classes, plus Hamilton's philosophy (which included consideration of ethics), plus James' general reading, prompted him to reflect on a question which occupied his thoughts throughout his life: What is the Good?

In an 1850 letter to Campbell, James signalled his intention to study the moral philosophy of Hobbes' *Leviathan,* comparing it with Scripture, especially the first part of chapter seven of the Gospel of Matthew. His aim was to analyse the chapter's moral principles, and relate it to other biblical passages.[74] As already

---

71 The contemporary reaction to Chambers' *Vestiges* is discussed in David Secord's *Victorian Sensation* (2000). Secord notes Maxwell's interest in the book, but misleadingly implies that Maxwell regarded *Vestiges* in particular as an example of dark or false science (259). Maxwell nowhere mentions *Vestiges* when discussing 'dark science'. Our own discussion of what Maxwell regarded as 'dark science' comes in Chapter 9: The Apostles' Club.

72 JCM, *On the Properties of Matter,* (C&G1, 110; Harman, I, 111).

73 C&G1, 114.

74 JCM, *Letter to Campbell,* 26 Apr 1850, (C&G1, 134; Harman, I, 192).

noted, Sir William Hamilton had struggled to establish a basis for ethics from philosophy alone, but James' method was different. He accepted divine revelation, and his selected Scripture passage (Matthew 7) includes several critical sayings which James placed at the heart of Christian ethics and which he related to systems of moral philosophy. In the Authorised Version the sayings include:

- Judge not, that ye be not judged. (v.1)
- For with what judgment ye judge, ye shall be judged: and with what measure ye mete, it shall be measured to you again. (v.2)
- Why beholdest thou the mote that is in thy brother's eye, but considerest not the beam that is in thine own eye? (v.3)
- Give not that which is holy unto the dogs, neither cast ye your pearls before swine. (v.6)
- Ask, and it shall be given you; seek, and ye shall find; knock, and it shall be opened unto you. (v.7)
- What man is there of you, whom if his son ask bread, will he give him a stone? (v.9)
- All things whatsoever ye would that men should do to you, do ye even so to them: for this is the law and the prophets. (v.12)

A few months later James wrote another long letter to Campbell, in which he reflected further on the question of the Good.[75] It was important to him for his faith to be meaningful practically, and he explained to Campbell that he wanted to know what makes the good. He was asking the same question of moral philosophy as he had asked of natural philosophy since childhood: 'What's the go' o' it?', 'What makes it so?', 'How does it work like that?'

He noted that different philosophers used different criteria in applying the value 'good'. Some deemed an action 'good' because the action sought to be virtuous; others because the action intended benevolence; others because the action obeyed lawful rulers; others because the action was dictated by individual conscience; others because the action adhered to 'unalterable laws of right and wrong';

75 Ibid., June/July 1850, (C&G1, 138–144; Harman, I, 193–198). Campbell gives June; Harman prefers July.

still others because the action was consistent with an intuitive feeling about the 'fitness of things.'[76] Musing on such a diverse array of criteria, the nineteen-year-old posed ten questions which for him touched the root of the matter:

1 What is man?

2 What are the laws of human action? (Including: thoughts, words, deeds).

3 What are the motives of human actions?

4 What actions do men perform in preference to others, and why?

5 What is the principle by which men judge some actions right, others wrong?

6 What do particular men think of this principle? What are their doctrines?

7 What is the best criticism of right and wrong, or what (to us) is absolute right?

8 What are the best motives of human actions?

9 How are these motives to be implanted without violating the laws of human action?

10 What might, or rather, what will mankind become after this has been effected?

James was well aware of differences of method between moral philosophy (the ethical sciences) and natural philosophy (the physical sciences). He noted that in moral philosophy old ideas were often considered to be superior to new ideas, whereas the opposite was generally the case in natural philosophy. Similarly, moral philosophy was characterised by increasing diversity of opinion over concepts such as selfishness, self-love, appetites, desires, affections, disinterestedness; whereas in natural philosophy the thinking of experts tended to

76 Something had 'fitness' if it fitted; if it made sense; if it had the right feel to it; if it made things hang together; if it elevated something morally. 'Fitness' was grasped intuitively in situations where it might be difficult or impossible to deduce a truth from axioms. The concept is still used in science, when, for example, a theory is held onto because of its beauty and simplicity, indicating resonance with some deep truth as yet unveiled. Maxwell used the same term in a later paper: 'Is Ethical Truth obtainable from an Individual Point of View?', May 1855, (C&G1, 234; Harman, I, 294).

come together and coalesce. He had yet to experience the violent arguments which scientists can have with one another!

Faced with this divergence of opinions, he looked for a foundation for moral philosophy which might be equivalent to Newton's laws in natural philosophy. Sir William Hamilton had found this difficult to establish after side-lining Scripture and revelation. In James' quest for an axiomatic basis to ethics, he regarded his first Three Fundamental Questions as akin to Newton's Three Laws of Motion: (1) What is man? (2) What are the laws of human action? (3) What are the motives of human actions?

In James' view, these three questions cropped up in all systems of moral philosophy, even if for different philosophers they were expressed differently and were 'continually changing their shape, so that it is not easy to catch them in their best shape.'

'*What is man?*' Taking his cue from Hamilton, he argued that humanity has three faculties: cognition, feeling, and conation. *Cognition* is the same as understanding. *Feelings* are pleasures, pains, appetites, desires, aversions, approval and disapproval, love, hate, and all affections. *Conations* are acts of will.[77] The notion of a free act of the will became vitally important in James' future thinking.[78]

'*What are the laws of human action?*' He considered that the impulse of the will to moral conduct was primarily determined by a person's feelings, i.e. their affections. The will was moved by *affections* more readily than by *reason*: 'For a man does what he likes to do, not what he believes to be best for himself or others.' At the same time, 'feelings can only be moved by notions coming through the understanding, for cognition is the only inlet of thoughts.' He regarded the principle of self-love to be no proper basis for moral action: 'Although it can be proved [by some] that self-love leads to all goodness, or, in other words, that goodness is happiness—and

77 JCM, *Draft of Essay on Moral Philosophy,* Mar 1850, (Cf. Harman, III/2, 861f): 'This division of the faculties was made by Kant, and it is adopted by Sir William Hamilton, and it seems to me to be more perfect than that into Will and Understanding.'

78 JCM, *Letter to Litchfield,* 5 Mar 1858, (C&G1, 306; Harman, I, 588f). See Chapter 27: 'Automata or Free?'

self loves happiness—yet it can also be proved that men are not able to act rightly from pure self-love; so that, though self-love is a very fine theoretical principle, yet no man can keep it always in view, or act reasonably upon it.'[79]

'*What are the motives of human action?*' He compared two schools of thought. One focussed on the principle of *self-love*, while the other focussed on the principle of *benevolence*. He saw Thomas Hobbes and Adam Smith (1723–1790) as advocating self-love, with Frances Hutcheson (1694–1746) and Thomas Brown (1778–1820) promoting benevolence. The first group were 'selfish philosophers and the selfish school' who taught that a person acts morally only because he or she knows that in due course they will benefit from the action.

Having branded self-love as an inadequate basis for moral conduct, James considered benevolence: 'Others say that benevolence is the only virtue, and that any action not done expressly for the good of others is titled to no praise.' Benevolence appealed to James' natural generosity of spirit, and he noted that every work on moral philosophy seemed to make a reference to Jesus' summary of God's Law: '*All things whatsoever ye would that men should do to you, do ye even so to them: for this is the law and the prophets.*' (Matthew 7:12). He saw this as the most elevated mode of thinking.

He then returned to Hobbes and Smith in more detail, describing Hobbes' emphasis on selfishness, with men and women acting primarily for their survival pursuing what was good 'for them.' Adam Smith's *Theory of Moral Sentiments* was, in his view, similar to Hobbes, with Smith famously stating: 'It is not from the benevolence of the butcher, the brewer, or the baker that we expect our dinner, but from their regard to their own interest.'[80]

In later years, James viewed Adam Smith more positively. By 1855, when he wrote to his father from Cambridge, telling him that there was to be a meeting in his rooms that evening to discuss Smith's *Theory of Moral Sentiments,* he had changed his mind:

---

79 JCM, *Letter to Campbell,* June 1850, (C&G1, 142; Harman, I, 196).

80 Adam Smith, *An Inquiry into the Nature and Causes of the Wealth of Nations,* preface.

> The repeated action of what Smith calls sympathy calls forth various moral principles . . . Instead of supposing the moral action of the mind to be a speculation on fitness, a calculation of happiness, or an effort towards freedom, he makes it depend on our relation to others like ourselves.[81]

He now saw in Smith's *Theory of Moral Sentiments* more traces of Hutcheson's emphasis on benevolence than he had first thought: unsurprising, since Smith was taught by Hutcheson at Glasgow University. Hutcheson had reacted against a gloomy view of human nature taught by some theologians of his day, preferring to see humanity capable of a true generosity of spirit motivated by a genuine altruism.

James Clerk Maxwell returned to ethical themes throughout his life. His initial approach to the questions 'What is the Good?' and 'What is the basis for ethical conduct?', formulated whilst at Edinburgh University, endured in his thinking. However, unlike his admired teacher Sir William Hamilton, he accepted the Bible as a solid basis for faith and conduct.

---

81 JCM, *Letter to John Clerk Maxwell,* 5 May 1855, (C&G1, 212; Harman, I, 294). JCM, 'Is Ethical Truth obtainable from and Individual Point of View?' (Fragment), (C&G1, 234; Harman, I, 294).

Peterhouse and Trinity, Cambridge (1850–1856)

Chapter 6
Cambridge Scholar

Maxwell experienced two major spiritual developments at Cambridge. One involved practical Christianity. The other a deeper personal piety. He arrived at Peterhouse College in October 1850, but transferred to Trinity College that December after one term. Garnett wrote that Maxwell was unsettled at Peterhouse since most Peterhouse men were either classicists or pure mathematicians, with few interested in his empirical work, and therefore the possibility of a fellowship there, after graduation, seemed slim.[1] He went to Trinity with a reputation as an accomplished, if disorderly trained, mathematician, but with the advantage of knowing several fellow students, such as the Mackenzies.

Although he avoided sports such as cricket and rowing, he was physically active. When bathing in the river, he could take a running header from the bank and turn a complete somersault before touching the water![2] He also had the unfortunate routine of taking nocturnal exercise from 2 to 2.30 am, running along the upper corridor of his rooms, down the stairs, along the lower corridor, then back up the stairs. That lasted until the inhabitants of the other rooms set an ambush and threw boots and hair-brushes as he passed.[3]

Some years later, he had to clarify to his wife that another story about him at Trinity was quite untrue. It was said that he had discovered a method of throwing a cat which made it unable to land on its feet, and that he threw cats out of windows! 'I had to explain that the proper object of research was to find how quick the cat would turn round; and that the proper method was to let the cat drop on a table or bed *from about two inches*, and that even then the cat lights on her feet.'[4]

---

1 Garnett, 'James Clerk Maxwell', *Nature,* Vol. 21/524 (13 Nov 1879), 44.

2 Charles Hope Robertson, *Letter to Campbell,* (undated), (C&G1, 164).

3 C&G1, 153.

4 JCM, *Letter to Katherine,* 3 Jan 1870, (C&G1, 499; Harman, II/2, 528).

Garnett also recounted a later incident at Cambridge, which occurred when Maxwell was Professor in Aberdeen. Maxwell had invented a dynamical top whose properties could be varied to increase or decrease its moment of inertia, and he brought it with him to Cambridge in 1857 when he added the M.A. to his B.A. of 1854. One evening he showed the top to some friends who left while it was still spinning. Next morning, he sighted one of them coming across the courtyard; so he restarted the top and retired between the bed-sheets. The dawn visitor reported that Maxwell's amazing top had spun all night and was still going strong![5]

Cambridge and Religion

Cambridge University was founded in the medieval era on religious principles, and until the later nineteenth century its dons had to be men ordained into holy orders in the Church of England. Consequently, the overwhelming mass of tutors and masters had the designation 'Reverend'. The title was a qualification, and not necessarily a mark of piety.[6] Not every reverend professor holding a chair in the natural sciences had deep faith convictions. Some did. But by no means all.

Charles Darwin's granddaughter, Gwen Raverat (1885–1857), observed that the college people rarely went to any of the churches, and if they had to attend worship they preferred the College Chapel which had a short service, no sermon, and good music. In her view they showed up mainly for the music.[7] The *Universities Tests Act* of 1871 changed the culture by abolishing religious tests, and by allowing Roman Catholics, non-conformists, and non-Christians to take up professorships, fellowships, studentships, and other lay offices at the Universities of Oxford, Cambridge, and Durham.

In Maxwell's time each college had a distinctive religious ethos, and Brenda Colloms has written of how Trinity was regarded by some as still imbued with 'Cromwellian fervour' and low church

---

5 Garnett, 'James Clerk Maxwell', *Nature,* Vol. 21/524 (13 Nov 1879), 46.

6 Colloms, *Victorian Country Parsons,* 225, refers to William Kingsley (mathematics wrangler in 1838) who, simply in order to teach, was ordained in 1842.

7 Raverat, *Period Piece: A Cambridge Childhood,* 26.

evangelical ideas.[8] All students had compulsory examinations in religious subjects, regardless of their course of study. One of the trials, faced by Maxwell in March 1852, involved examination on the Greek text of the Gospel of Matthew; tests on Paley's *Evidences*; and assessments of his knowledge of the Old Testament from Genesis to Esther. This was in addition to questions on Homer's *Iliad,* Livy's *History,* and Euclid's *Elements.*[9] Maxwell knew about these before he went to Cambridge, referring to them in his April 1850 letter to Campbell: 'I suppose I must go through Wrigley's problems, and Paley's *Evidences* in the same sort of way, and be able to translate when required Euripides' *Iphigenia at Aulis*.'[10]

He lodged initially in the same house as Charles Hope Robertson (1831–1909) who had been an Edinburgh Academy schoolmate and who was studying theology in preparation for ordination in the Church of England. Robertson was the second son of Lord and Lady Benholme, with his father one of Scotland's Law Lords and Senator of the Scottish College of Justice. The Robertsons were Episcopalian and their Edinburgh house was at 76 Great King Street, only a short stroll from James' base in Heriot Row. Reunited at Cambridge, Maxwell gave up some of his free time to read textbooks to Robertson after Robertson hurt his eyes.[11]

Robertson was frequently quizzed by Maxwell on theological problems. Maxwell was aware that the reliability of Scripture, especially parts of the Old Testament, was questioned by contemporary biblical criticism; and on one occasion, when an attack on Mosaic history puzzled Maxwell, he was helped by an idea of Robertson's to the effect that some accounts were unlikely to occur in a book of purely *human* origin, 'where a leader would rather be apt to *magnify* than to *diminish* his own and his people's prowess.'[12] Robertson also recalled Maxwell delighting in what he saw as a 'beautiful mathematical illustration of St Paul's closing view of his ministry, in 2 Timothy 4:6-8.' Maxwell declared that, since Paul was looking backward, forward, and

8 Colloms, *Victorian Country Parsons,* 251.

9 Cf. Harman, I, 8.

10 JCM, *Letter to Campbell,* 26 Apr 1850, (C&G1, 134; Harman, I, 191).

11 Robertson, *Letter to Campbell,* (C&G1, 163).

12 Ibid., Undated, (C&G1, 164).

downwards then the resultant was upwards! In later years, Robertson stated that Maxwell had, 'an innate reverence for sacred things, which I do not think was ever much disturbed by the scepticism fashionable among shallow scientific men.'[13] He viewed Maxwell as too deep a thinker to be sceptical, but also saw him as too well read not to feel for others' difficulties. His verdict was that Maxwell's science led him to 'greater reverence for the Great First Cause' and that Maxwell was in tune with the sentiment expressed by Edward Young (1683–1765) in his *Night Thoughts,* that an undevout astronomer is mad.

Maxwell's confidence in the Scriptures lasted his whole life. He never thought that miraculous events in the Bible had to be deleted in order to accommodate intellectual sophistications. Karl Pearson (1857–1936), who studied under Maxwell at Cambridge in the 1870s, recollected an incident, during an examination diet, when he and other students wrote their papers in Maxwell's dining room, with an adjournment for lunch at which Maxwell himself presided. The conversation turned to Darwinian evolution, during which Pearson questioned Noah's flood. This annoyed Maxwell, who reproved Pearson for lack of faith in the Bible. Pearson commented: 'I had no idea at the time that he had retained the rigid faith of his childhood; and was, if possible, a firmer believer than Gladstone in the accuracy of Genesis.'[14]

It was in November 1852 that Maxwell delivered his undergraduate Declamation on the Scottish Covenanters.[15] Around the same time he wrote a long letter to his father detailing the various groups in the Church of England.[16] His father replied that this went far beyond his knowledge, and that he would need an explanatory lecture before he could follow the High, Broad, and Low, through their ramifications![17] Despite his protestations, John Clerk Maxwell was well versed in church affairs south of the Border, with various Episcopalian/Anglican connections in his family.

---

13 Robertson, *Letter to Campbell,* (C&G1, 163f).

14 Pearson, 'Old Tripos Days at Cambridge: As seen from Another Viewpoint', *Mathematical Gazette,* 20, No. 37 (1936), 32. Pearson became a biostatistician who was an advocate of eugenics and social Darwinism.

15 C&G1, 161.

16 Ibid., 171.

17 John Clerk Maxwell, *Letter to JCM,* 16 Dec 1853, (C&G1, 194).

Harvey Goodwin (1818–1891) and St Edward's Church

Maxwell settled easily into the Anglican ethos of Cambridge life with its college prayers and Sunday services. His experience of Scottish Episcopalianism made it a comfortable environment, and he soon discovered and enjoyed the ministry of Harvey Goodwin at St Edward's church.[18] In the 1850s Cambridge still retained vestiges of the evangelical tradition which had been championed by Charles Simeon (1759–1836). Though Goodwin was no Simeonite evangelical, his orthodoxy and his emphasis on practical Christianity struck a chord with Maxwell.

After a high-flying undergraduate career, Goodwin gained a lectureship at Cambridge, becoming a fellow of Caius in 1841. In 1848 he was made Rector of Cambridge St Edward: King and Martyr, better known as simply St Edward's, with the church a short walk from Trinity. It was there that he enhanced his reputation as a popular and persuasive preacher He remained at St Edward's until he became Dean of Ely Cathedral in 1858, before his elevation as Bishop of Carlisle in 1869. It was Goodwin who preached in Westminster Abbey in commemoration of Charles Darwin on the Sunday after Darwin's funeral in 1882.[19]

Goodwin was not only a preacher but a mathematician, and so Maxwell met at Cambridge what he had already encountered in Edinburgh at the RSE—clergymen with high competency in the natural sciences. This fusion reinforced Maxwell's life-long conviction that science and faith were comfortable companions. Goodwin was a prolific author, penning books both religious and scientific, including: *An Elementary Course of Mathematics for Students of the University of Cambridge* (1846), *Walks in the Regions of Science and Faith* (1883), and *The Foundations of the Creed* (1889).

Though several were written after Goodwin left St Edward's, their content indicates his cast of mind, especially in relation to questions of science and faith, and they exhibit significant continuity with Maxwell's own thinking. For example, in his 1880 work, *The Origin of the World according to Revelation and Science*, Goodwin emphasised that science cannot account for the ultimate origin of

18 C&G1, 153.

19 *English Lakes Visitor*, 25 Nov 1882.

the material universe but deals with the laws which govern matter *once it exists*, and that neither science nor theology impinges on the other if properly understood. Goodwin argued, against Ernst Haeckel of Geneva, that the biblical account was never meant to be taken literally, but was a poetical presentation of fundamental truths. He cited Hugh Miller's interpretation of Genesis 1 as a series of visions presented to the Mosaic writer over a period of several days.[20] Goodwin regarded the Genesis creation narrative as essentially moral and didactic, climaxing in humanity being in the image of God. He later saw in the Mosaic account the concept of progressive development, the implication being that it could harmonise with some evolutionary thought.[21]

St Edward's was crowded every Sunday during Goodwin's ministry. Each seat was occupied. Every square yard of standing room was filled. As many as a hundred persons crowded the porches or stood out in the street.[22] So what was it that attracted students such as Maxwell? In terms of style, Goodwin's biographer stressed simplicity and practicality.[23] Whenever Goodwin himself was asked how he prepared for preaching to so many young intellectuals, he replied that he never, or very seldom, directed his preaching to them at all, but 'preached plain, practical sermons to my own people.' Goodwin's view was that it was better for undergraduates to hear an

20 Miller, *Testimony of the Rocks,* Lecture IV.

21 Goodwin, *The Origin of the World according to Revelation and Science.* Writing in a post *Origin of Species* environment, Goodwin noted that Genesis 1 was criticised for being geocentric and anthropocentric (23). He dismissed this, arguing that such expressions were common in normal human speech: we commonly say 'the sun rises' without rejecting heliocentricity (25). Concerning the inference that Darwin, in demonstrating that man is descended from animals also demonstrated that man is no more special than other species (29f), Goodwin argued that the purpose of the opening chapters of Genesis was to 'assure us that we were made in the image of God.' (50). In line with the Victorian emphasis on the perfection of creation, Goodwin did not dwell on the fallenness of the created order. Though not all seemed beneficent; he believed that all things would ultimately prove to be so (55). George Wilson (Maxwell's mentor at Edinburgh University), in his *Religio Chemici,* 26–50, regretted the tendency of natural theology to ignore pain and suffering in the world when using science to prove the benevolence of God.

22 Rawnsley, *Harvey Goodwin, Bishop of Carlisle,* 71.

23 Ibid., 70–92.

ordinary parish sermon than an address geared to students: 'Many of them would have resented the one; they listened gladly to the other.' This worked for Maxwell who, throughout his life, preferred plain and unadorned preaching.

Goodwin never strayed from official Church of England theology.[24] His main contribution to Maxwell's spiritual development was to present Christianity as an applied faith. Maxwell wanted a preacher who addressed ethical questions such as: 'What is the Good?', 'How can I do the Good?' This longing for a practical Christianity lay alongside his acceptance of traditional doctrine. Though Goodwin's sermons were simple, they were well constructed and delivered eloquently. This was perfect for Maxwell.

In 1851 Maxwell read Thomas Carlyle's translation of Goethe's *Wilhelm Meister,* which chimed with Goodwin's emphasis. Carlyle's home at Craigenputtock was only a few miles from Glenlair, and a destination to which Maxwell occasionally took guests on a country drive. The novel was about a young man named Wilhelm who attempted to disassociate himself from the bourgeois socio-economic class into which he had been born, and who sought to discover an authentic life away from that environment. Goethe's conclusion was that all life is an apprenticeship and is to be lived with that realisation.

In some respects—especially in striving to escape from a bourgeois culture masquerading as Christian—Goethe anticipated Kierkegaard's desire for a meaningful, authentic, existentialist philosophy of life, with a religion to match. Maxwell was sympathetic to these issues, especially the need to live a genuine Christian faith grounded in the realities of human life and human need. Maxwell recommended *Wilhelm Meister* to Campbell, though he added that the book should be read with discretion.[25]

The Mackenzies

Maxwell's distant relatives and fellow-students, Henry and Frank Mackenzie from Peeblesshire, were also attracted to Goodwin's ministry. Goodwin was a particular friend to their older sibling,

24 E.g. Goodwin, *The Foundations of the Creed.*

25 C&G1, 154.

Charles Mackenzie, who lectured in mathematics at Caius and had been one of John Clerk Maxwell's advisors when the possibility of James going to Cambridge was being considered. Years later, it was Goodwin who wrote Charles Mackenzie's biography after he died of blackwater-fever on an island in the Shire River in southern Africa.[26]

Charles Mackenzie was ordained a priest of the Church of England in 1852. Three years later he went to Natal with Bishop John Colenso (1814–1883), serving as Archdeacon of Natal, before coming back to England in 1859. On his return, he found that many young men had been moved and enthused by David Livingstone's 1857 University Speeches and their commanding call to mission. Mackenzie was then appointed head of the Oxford and Cambridge Mission to Central Africa (renamed the Universities' Mission to Central Africa) which was set up in the aftermath of Livingstone's challenge. After consecration in St George's Cathedral, Cape Town, as Missionary Bishop of Central Africa, he proceeded towards what is now Malawi, with the twin aims of establishing a mission station near Zomba and opposing the slave trade. Then came his tragic death. A year later, David Livingstone reached the spot and erected a cross above his grave.

Cousin Jemima knew and liked Charles Mackenzie. She wrote: 'One of my friends was Charles Mackenzie, brother of my brother-in-law James. I knew him well since he was twelve years old, he being a little younger than I. He was the most Christian man I ever knew.'[27] She commented that, whilst Charles was at university, he had heard a sermon preached by Bishop Selwyn, pleading for young men to go as missionaries to South Africa, and had volunteered immediately.[28]

Charles Mackenzie's African work was still in the future when Maxwell arrived at Cambridge. He remained a presence in Maxwell's life, and when he went to Africa Maxwell was an enthusiastic supporter. In 1860, Maxwell attended meetings in Aberdeen which were held to support the Universities' Mission, with the *Aberdeen Press & Journal*

26 Goodwin, *Memoir of Bishop Mackenzie,* (1864).

27 Fairley, *Jemima,* 115.

28 Probably Bishop George Augustus Selwyn (1809–1878) of New Zealand, who was in England in 1854.

listing him as a member of the Aberdeen District Committee chaired by the Bishop of Aberdeen, whose aim was to raise money in the north-east districts of Scotland for Mackenzie's Mission.[29]

The Academic Journey

In Maxwell's final year (1854) he was Second Wrangler in the Cambridge Tripos examinations, plus first-equal Smith's prize-winner. The Wrangler concept was daunting, with only outstanding students having any hope of making their mark. Pipping him as Senior Wrangler was Edward Routh (1831–1907) who went on to have a notable career, both as an original mathematical theorist and as a successful coach to gifted students, with twenty-seven Senior Wranglers to his credit. Passing exams with high marks was Maxwell's bread and butter, though not without cost in terms of nervous and physical exertion. Despite the academic pressures, he contributed papers to mathematical journals, and made time to explore topics outside the prescribed course.[30]

As an undergraduate, he was tutored by William Hopkins (1793–1866) from Peterhouse, under whose guidance his methods became more systematic. Hopkins had been Seventh Wrangler in 1827, after which he specialised in private tuition. From 1829 until 1854, Hopkins tutored twenty of the twenty-six Senior Wranglers plus nineteen of the Second Wranglers; and when Maxwell graduated in 1854, seven of the top nine Wranglers were Hopkins' students.[31] Like Maxwell, Hopkins was interested in geology and a few years later, whilst at

29 *Aberdeen Press & Journal*, 6 June 1860: 'The above Committee has been organised at Aberdeen for the purpose of receiving contributions from parties in the North-East Districts of Scotland. . . . The Committee feel they have only to call attention to it, in order to ensure the support of all who feel interest in the Christianising and civilisation of the heathen, and the suppression of that most atrocious of all traffics—the slave Trade; to say nothing of the national advantages which must accrue by following out, energetically, the path which has been opened up by the discoveries of our distinguished countryman, Dr Livingstone.'

30 C&G1, 161, (NVN1, xiii) Maxwell wrote on fish-eye lenses in the *Cambridge & Dublin Mathematical Journal* (1853). For other papers, see: Harman, I, 230–235.

31 McCartney, 'I Remember Years', in Flood (ed), *JCM: Perspectives on his Life and Work*, 298.

Aberdeen, Maxwell read with great interest a paper by Hopkins on the mathematics of geology.[32]

At Trinity it was Hopkins' coaching, rather than the formal lectures, which trained Maxwell. A weakness of the system, which Maxwell found to his cost when he became a teacher himself, was that students spent most of their time with their coaches, cramming for the exams, only turning up at lectures because attendance was compulsory for undergraduates.[33] Maxwell sought more from study than cramming. He wrote to Campbell: 'Man requires more. He starves while being crammed. He wants man's meat, not college pudding.'[34]

His friends noted that his natural way of thinking involved visual diagrams (often geometrical), either forming them mentally in his head or scribbling them on paper.[35] He used this method wherever possible, even though others might solve the question by abstract analysis.[36] Throughout his career he found geometry and analogy to be indispensable aids, and it was his go-to way of thinking.[37] J.J. Thomson, one of Maxwell's distinguished successors at the Cavendish Laboratory, remarked that Maxwell's preference for geometrical methods over analytical ones was an example of his general method of thought, which was 'to proceed step by step, from one definite idea to another until he had reached the goal, instead of getting there by means of symbols and equations without any visualisation of the intermediate stages.'[38] In Maxwell's mind, each stage of a mathematical

32 JCM, *Letter to Litchfield,* 7 Feb 1858 (C&G1, 302; Harman, I, 583). Cf. Hopkins, *An Abstract on Physical Geology,* (1836).

33 JCM, *Letter to Tait,* 3 Dec 1856, (Harman, I, 480). See also, Forsyth, 'The Old Tripos Days at Cambridge', 168: 'In the teaching, the coaches were all-dominant, almost the sole agents.' A more positive view is given by Pearson, 'Old Tripos Days at Cambridge as seen from Another Viewpoint.'

34 JCM, *Letter to Campbell,* 9 Nov 1851, (C&G1, 158; Harman, I, 208).

35 NVN1, xiii.

36 Lawson, *Letter to Campbell,* 6 Jan 1882, (C&G1, 175).

37 Cf. Gaal, 'The Faith of a Scientist, 15, re Olson, *Scottish Philosophy and British Physics: 1750–1880,* 21. Gaal and Olson (22) refer to Sir William Hamilton insisting on the importance of the geometrical method compared with the algebraic one.

38 Thomson, J.J, 'James Clerk Maxwell', 11

description of nature was a portrayal of an aspect of reality, and he strove to visualise what each meant in real terms.

Maxwell's preference for pictorial and geometric thinking introduces the issue of how he viewed space and time. It has been argued by some that his later work encouraged physicists to break free from a Newtonian container-box notion of space and time, which viewed space and time as a 'box' independent of events occurring within it. Yet Maxwell appears to have worked all his life with an essentially container-box view of the universe. This can be detected not only in his early paper 'Notes on Space and Position', in which he discoursed on Absolute Space, but also in his highly acclaimed *Matter and Motion* (1877).

In *Matter and Motion* he stated that it was conducive to scientific progress to recognise with Newton that the ideas of space and time were: 'distinct, at least in thought, from that of the material system whose relations these ideas serve to coordinate.'[39] His theories and equations would eventually push the container-box model beyond breaking point: he then realised that abstract mathematical descriptions, rather than physical visualisations, were sometimes the only way to portray reality when at the boundaries of knowledge. But for Maxwell himself, his geometric way of thinking was rooted in the traditional Newtonian view of space and time.

This means that we need to qualify Torrance's assertion that Maxwell consciously moved away from a Newtonian container-box conception of space and time. Torrance has argued that there is 'no evidence in Clerk Maxwell's papers and books for Newtonian container notions of space and time.'[40] The Finnish Lutheran theologian Tapio Luoma (b.1962) questions this generalisation. In his PhD thesis on 'Natural Science in the Theology of T.F. Torrance', Luoma argues that Torrance did not pay proper attention to the 'archaic concept of ether which was still to be found in Maxwell's thinking.'[41] Given the ether notion, a receptacle notion of space had to be presupposed, with

39 JCM, 'Notes on Space and Position', (Harman, I, 515–516). Also, JCM, *Matter and Motion,* 11.

40 Torrance, 'Christian Faith and Physical Science in the Thought of JCM', *Transformation and Convergence in the Frame of Knowledge,* 235.

41 Luoma, *Incarnation and Physics: Natural Science in the Theology of Thomas F. Torrance*, 107.

space 'containing' the ether. Luoma suggests that Torrance was so determined to present Maxwell's idea of space and time in a certain light, that he over-stated his conviction that there were no hints of Newtonian receptacle space in Maxwell's literary productions. As we have seen, that was not entirely the case. Luoma also points out that Maxwell himself, when defining the electromagnetic field, maintained that the field was 'that *part of space which contains and surrounds* bodies in electric or magnetic conditions.'

Torrance's statements require refinement. Luoma's point is valid: for Maxwell, space was something that contained bodies, and was something that could be *filled* by the ether. In that respect, Maxwell's space was a receptacle. As Luoma puts it: 'Maxwell was not, after all, as free from the container notion of space as Torrance would like us to believe.'[42]

When Maxwell became twenty-one in June 1852, his father sent a letter expressing the hope and prayer that, in the years following this milestone, his son would be as careful and sensible as he had been before it: 'Be as discreet when *major,* as you have been while *minor.*'[43] He directed his son to the Scripture text: *'A wise son brings joy to his father, but a foolish son brings grief to his mother.'"* (Proverbs 10:1). Children of Scottish Presbyterian parents were well familiar with the book of Proverbs, with selections from it included in one-volume educational primers, alongside the *Shorter Catechism,* the *Lord's Prayer*, the *Apostles' Creed*, prayers for before and after a meal, an *ABC* of the letters of the Alphabet, plus basic multiplication tables. These were regarded as the essential basics for any child's well-rounded Christian education.

There was little cause for John Clerk Maxwell to be anxious about his son's moral and spiritual character. James, who, according to Campbell, never even drank wine, had no tendency towards rebellion or bohemianism.[44]

42 Ibid., 155.

43 John Clerk Maxwell, *Letter to JCM,* 12 Jun 1853, (C&G1, 166).

44 C&G1, 105.

Chapter 7
Maurice and Christian Socialism

On the tenth of February 1852, Cambridge was abuzz. A recital of Handel's Oratorio, *The Messiah,* was to be performed by the Cambridge Choral Society and would take place the next evening in the Town Hall.[1] Celebrated soloists, such as Miss Clara Henderson and Miss Sophy Law from the Royal Academy of Music, had been secured for the occasion, though the main choir was local with members from both town and university. Whilst the concert organisers were busy with last-minute preparations, Maxwell and some of his friends gathered for conversation in the rooms of Isaac Taylor (1829–1901), a fellow student at Trinity.

Taylor came from a family which had a strong commitment to Christian Socialism, and the movement was striking a chord with Maxwell. Though instinctively conservative, he was left-of-centre in some of his politics. In 1855 for example, when his friend, Robert Henry Pomeroy (1830–1857), spoke on the British in India, Maxwell expressed reservations about the whole concept of Britain attempting to manage India at a period, 'when all the old institutions and religions must break up' and when it was 'by no means plain how new civilisation and self-government among people so different from us is to be introduced.'[2] Maxwell's disquiet with British policy was based not only on cultural and governance issues but on fears concerning the economic exploitation of the sub-continent. He added: 'One thing is clear: that if we neglect them, or turn them adrift again, or simply make money of them, then we must look to Spain and the Americas for our examples of wicked management and consequent ruin.'

---

1 *The Cambridge Independent Press,* 7 Feb 1852.

2 JCM, *Letter to his Father,* 1 Nov 1855, (C&G1, 220; Harman, I, 332). This mitigates Stella Pratt-Smith's concern in, 'Boundaries of Perception: James Clerk Maxwell's Poetry of Self, Senses, and Science', in Flood (ed.), *JCM: Perspectives on his Life and Work,* 256, that Maxwell's poems did not address politico-social issues such as the Crimean War or expansion of the British Empire. We bear in mind that we only have Campbell's selection of Maxwell's letters. There is also Maxwell's commitment to Christian Socialism.

These sentiments on Empire ran parallel to his interest in Christian Socialism.

At Cambridge, three of the principal stimuli in the direction of Christian Socialism were: the Isaac Taylors (father and son), Richard Litchfield, and Harvey Goodwin. Behind them lay the considerable influence of F.D. Maurice (1805–1872) whom Maxwell admired, first through reading Maurice's books and later through personal association.

The Isaac Taylors

When Maxwell arrived at Taylor's apartment for an evening of convivial discussion, he met Isaac Taylor senior (1787–1865) who was visiting his son. Maxwell later described the elder Taylor as a cold little man with a tremulous voice, and who talked about the weather with precise exactitude as if he were on oath.[3] Appearances aside, Maxwell was impressed.

The Taylors discoursed at length about the state of society, associations of workmen, and systems in which each workman had a share in the firm. Maxwell saw merit in these ideas. They represented a practical Christianity applicable to the day-to-day working life of society. He wrote to Aunt Jane:

> I met old Isaac Taylor in his son's rooms some time ago. He began by speaking of the weather in a serious way, and went on to his Manchester concerns, effective motives to work, actual methods adopted, and so got into the merits of socialism, joint-stock workmen's associations, and so forth; appearing all the while to say nothing, but quietly feeding on the wisdom of the undergraduates, as they enounced their opinions.[4]

The elder Isaac Taylor had trained in the family business as draughtsman, engraver, and book illustrator, before becoming a prolific author, mainly on literary and philosophical subjects. In 1836 he contested the Chair of Logic at Edinburgh University, losing out narrowly to Sir William Hamilton. The *Scotsman,* which devoted several columns to the election of the new professor, reported that, out of four candidates, the choice came down to Hamilton or Taylor. Hamilton's religion was suspect, and Thomas Chalmers lobbied for Taylor; this annoyed some of the

3 JCM, *Letter to Campbell,* 10 Feb 1852, (C&G1, 177).

4 JCM, *Letter to Jane Cay,* 23 March 1852, (C&G1, 180).

appointing committee who felt that Chalmers had 'interfered enough in this election already.' Though Taylor missed the Edinburgh post, he was later influential in persuading Maxwell to make a commitment to Christian Socialism and give his services as a tutor for working-men. Others who dined with Maxwell at the scholar's table in college were of the same mind, and the idea of self-improvement in society through education took a firm hold on his thinking.

Richard B. Litchfield (1831–1903)

A significant proportion of Maxwell's letters in Campbell and Garnett's *Life of James Clerk Maxwell* are addressed to Richard Litchfield, and of all of Maxwell's Cambridge companions it was Litchfield who had the earliest, deepest, and most enduring commitment to F.D. Maurice's ideas on Christian Socialism. Litchfield and Maxwell were part of the circle who met in each other's rooms to discuss topics scientific, theological, and political. Maxwell once described spending a whole evening, from eight until midnight, in Litchfield's rooms, along with Farrar (later the distinguished Dean Farrar), Pomeroy, and D.Y. Blakiston (1832–1914), discussing the topic of eternal punishment. That issue had been brought to the fore by some of F.D. Maurice's other publications.

After graduation, Litchfield taught mathematics at Trinity before starting a legal career. He entered the Inner Temple in 1854, before being called to the Bar nine years later. In 1854 he was co-founder of Maurice's Working Men's College in Red Lion Square, London, to which he gave service for over fifty years as tutor, college bursar, vice-principal, and editor of the college's magazine. Among his fellow-tutors and friends were the art critic and social thinker John Ruskin (1819–1900), and the artist William Morris (1834–1896). According to Litchfield's niece Gwen Raverat (Charles Darwin's grand-daughter), Uncle Richard adored Ruskin and worshipped Morris, and had slept for years with a copy of Tennyson's *In Memoriam* under his pillow. She wrote: 'Uncle Richard had done everything that an enlightened person, flourishing in the middle of the nineteenth century, ought to do: taught at the Working Men's College, organised great country walks, admired Nature, and all the rest of it.'[5] She described him affectionately as a nice funny little man whose socks were always

5 Raverat, *Period Piece: A Cambridge Childhood*,126.

coming down, who had an egg-shaped waistcoat, and who sported a fuzzy, waggly, whitey-brown beard, which was quite indistinguishable in colour and texture from the Shetland shawl which Aunt Etty made him wear round his neck.

In 1871 Litchfield married Henrietta (Etty) Emma Darwin (1843–1927), the daughter of Charles Darwin. Though formidable Aunt Etty was not a churchgoer, she did make frequent recourse to the Darwin family bible, if only to record illnesses and vaccinations.[6]

When James and Katherine Clerk Maxwell moved to London in 1860, James revived his links with Litchfield and other former Cambridge companions.[7] Whilst in Aberdeen, Maxwell had asked Litchfield about the College in London and the working-men.[8] As soon as Maxwell arrived in the metropolis, Litchfield had him tutoring, enabling Maxwell to carry on a commitment which started in Cambridge and had continued in Aberdeen. Litchfield died in France in 1903, and was buried in Cannes' cemetery, with his gravestone highlighting his fifty-year connection with the London Working Men's College.

Harvey Goodwin

The cause of Christian Socialism was turbo-boosted in Cambridge by the support of Harvey Goodwin of St Edward's through his status, leadership, popularity, and public-speaking talents. Despite heavy traces of paternalism, Goodwin had a genuine concern for working-men and believed that education was key to releasing latent abilities. He was convinced that if untapped talents were given opportunity to flourish then whole families could be moved out of poverty and into secure and comfortable lives.

Significantly, for Goodwin as with Maurice, the matter was theological as much as sociological. He saw the promotion of the further education of working-class men as a Christian duty, expounding his vision at a meeting in Cambridge Town Hall on the twenty-ninth of October 1855. Goodwin's speech was billed as the 'Inaugural Lecture of the Working Men's College in Cambridge', whose programme of lectures was starting the next evening. Goodwin addressed a crowded audience of working-men, students,

6 Ibid., 133.

7 C&G1, 319.

8 JCM, *Letter to Litchfield,* 29 May 1857, (C&G1, 268f; Harman, I, 508f).

university staff, and members of the public, with his lecture receiving 'long and continuous cheering.'[9]

In a pulsating and energetic speech, Goodwin described Working Men's Institutions—avoiding the term 'college' since in Cambridge a 'college' was part of the university—as places where education was not limited to utilitarian or practical subjects. His conviction was that a complete general education should be made available and he extolled the virtues of history, mathematics, and Latin, alongside classes in so-called 'useful knowledge.'

Halfway through his speech he mentioned F.D. Maurice, the nationally recognised champion of the movement. Loud cheers went up from the audience. Goodwin responded: 'I am glad to perceive that the name of Mr Maurice is met by you with a cheer: for indeed I believe that there are few persons at present in England to whom we owe more.'[10] Goodwin traced the inspiration behind Cambridge's nascent Institution to the college founded by Maurice in London the previous year. He then asked why clergymen should be involved in such a project. His answer was that as a Christian he saw it his religious duty to be aligned with the movement:

> Everything which tends to improve the condition of men in this world—to raise their views in regard to that condition, and their hopes in regard to another—ought to be considered strictly a religious thing. I say, therefore, that we ought all to look on this work as a religious one. I have considered it so; and that, as a clergyman, I am doing the work of a clergyman in taking part in it.[11]

Goodwin told his audience that 28 men had already signed up for English Grammar, 35 for French, 6 for German, 26 for Latin, 54 for Mathematics, 8 for English History, 20 for Music, and 31 for Drawing.[12] There was also a Bible Class on Sunday evenings for which no fee was required, starting with a critical examination of the text and meaning of Matthew's Gospel.[13]

9 *Cambridge Independent Press,* 3 Nov 1855.

10 Goodwin, *Education for Working-Men,* 5.

11 Ibid., 46.

12 Ibid., 13.

13 Ibid., 35.

Maxwell was already committed. A month before Goodwin's tone-setting lecture, he wrote to his father expressing enthusiasm for the scheme, and explaining that during the coming term 'I intend to addict myself to the working-men who are getting up classes, rather than to pupils who are in the main a vexation.'[14] The classes for working-men would take precedence over private pupils at Trinity: 'I have refused to take pupils this term, as I wanted to get some time for reading and doing private mathematics; and then I can bestow some time on the men who attend lectures.'[15] He kept his word. In the preface to the published version of Goodwin's lecture, the name 'James Clerk Maxwell, Fellow of Trinity' is registered as one of the tutors for the new Institution. His friend, R.H. Pomeroy, was also listed.

The classes were soon in full swing and after a few weeks F.D. Maurice came to see how matters were progressing. Maxwell wrote to his father: 'Maurice was here from Friday to Monday, inspecting the working-men's education. He was at Goodwin's on Friday night, where we met him and the teachers of the Cambridge affair. He talked of the history of the foundation of the old colleges, and how they were mostly intended to counteract the monastic system and allow of work and study without retirement from the world.'[16]

The following March, another letter to his father spoke of Maxwell's continuing enthusiasm, and noted efforts to persuade employers to give men time off during the week to attend classes: 'I was at the Working College to-day, working at decimal fractions. We are getting up a preparatory school for biggish boys to get up their preliminaries. We are also agitating in favour of early closing of shops. We have got the whole of the ironmongers, and all the shoemakers but one. The booksellers have done it sometime. The Pitt Press keeps late hours and is to be petitioned to shut up.'[17]

From 1854 to 1856, when he was a fellow at Trinity, several of his lectures for workingmen were delivered in Barnwell.[18] Barnwell was a Cambridge ward which had the reputation of being populated by

---

14 JCM, *Letter to John Clerk Maxwell,* 27 Sep 1855, (C&G1, 217; Harman, I, 327). Also, JCM, *Letter to Campbell,* 17 Oct 1855, (C&G1, 218).

15 Ibid., 25 Oct 1855, (C&G1, 219; Harman, I, 331).

16 JCM, *Letter to his Father,* 3 Dec 1855, (C&G1, 222; Harman, I, 336).

17 Ibid., 12 March, 1856, (C&G1, 252; Harman, I, 404).

18 Garnett, *Heroes of Science: Physicists,* 284.

itinerant workers and prostitutes, and was described by some as 'poor degraded Barnwell'.[19] Establishing classes there was lauded in the local press: 'Great praise is due to these gentlemen who have delivered lectures to the workingmen of Barnwell'.[20] The newspapers earnestly 'recommended all the workingmen of Barnwell to attend the delivery of each lecture.' Maxwell was one of the lecturers.

John Frederick Denison Maurice (1805–1872)

F.D. Maurice was born into a Unitarian family but was baptised in 1831 in the Church of England, and later became an Anglican clergyman. It was within his calling as an ordained priest that he sought to stimulate the social conscience of the Church. Maxwell admired his thinking, with Maurice becoming a significant factor in the development of Maxwell's practical application of his Christian faith. But if Maurice's Christian Socialism was inspirational, other aspects of his theology were problematic.

After studies at Cambridge, Maurice read for the Bar before being ordained and placed as chaplain to London's Guy's Hospital in 1836.[21] He was then appointed professor of English literature at King's College, London, before becoming its professor of theology in 1846.

Living and working in London, he was keenly aware of appalling social deprivation and widespread poverty. Faced with these, he believed that profound change was necessary, and that religion—and not a secular movement such as Chartism—should be the catalyst for progress. It ought to be the values of the Kingdom of God, activated by and through the Church, which should instigate improvement in all parts of life, including education, the workplace, and every practical need of the population. He believed that a true Christianity automatically created a true socialism. Christianity, not a secular philosophy, should be the natural driving-force for social change. Henry Goodwin agreed with this wholeheartedly, and Maxwell found the vision immensely attractive.

Maurice argued that Christ did not come to establish a religious sect or a new society but a kingdom. That kingdom should embrace

---

19 *Cambridge Independent Press*, 29 Sep 1855.

20 *Ibid.*, 10 Nov 1855.

21 Moorman, *A History of the Church in England*, 355-356.

all humanity, and the whole world should be brought under the rule of Christ.[22] This chimed with the prevailing optimistic post-millennial view of history, in which history was viewed as moving inexorably towards a period of a thousand years in which Christ's kingdom of peace and harmony would rule on earth before Christ came again. Maurice believed that in this kingdom there would be no class distinctions, no rich and poor, no oppressor and oppressed: there would be a world in which the love of Christ governed everyone. He saw the free market as merciless and fundamentally unchristian. Over against individualistic capitalism he set the power of Christian brotherhood. Economics and politics needed to have a ground beneath themselves apart from a focus on profits. For Maurice, that foundation was a society regenerated by finding its order and harmony in God. It was this theological underpinning which made him found a string of co-operative societies, twelve in London alone. Several failed, but eight in London, and fourteen across the country, prospered for decades.

However, alongside Maurice's faith-rooted socio-political programme, he held his own theological views. These precipitated the 'Maurice Controversy'. This exploded in early 1853 when publication of his *Theological Essays* brought matters to a head, resulting in his dismissal by King's College. This actually freed Maurice to pursue his vision further. It was then that he co-founded the Working Men's College in Red Lion Square which offered classes every evening. Maurice lectured on St John, John Ruskin on Art, J.S. Brewer (1809–1879) on History, and other scholars on various subjects. Some of Maurice's former colleagues at King's helped out, showing their solidarity with his ideals.[23] In 1866 he gained a chair at Cambridge whilst continuing as Principal of the Working Men's College, and he remained an activist and author until his death in 1872.

Why did Maurice's *Theological Essays* precipitate such a reaction? Much of the book is conventional doctrine, with Maurice defending Anglican beliefs and trinitarian faith over against the Unitarianism of his childhood. However, in one chapter he doubted the notion of eternal punishment for the unsaved. He refused to interpret the concept of *eternal damnation* as implying *everlasting punishment*. This

---

22 Maurice, *The Kingdom of Christ, or Letters to a Quaker concerning the Principles, Conceptions, and Ordinances of the Catholic Church,* (1837).

23 Moorman, *A History of the Church in England,* 356.

was the issue which Maxwell and his friends discussed at length one evening in February 1853.[24]

For Richard William Jelf (1798–1871), the Principal of King's College, the matter was more than a late-night discussion topic! Jelf regarded Maurice's view as close to heresy since Maurice threw doubt on the meaning of the word *eternal* and conveyed the impression that everyone would ultimately be saved. In Jelf's opinion this contradicted the Church of England's *Thirty-Nine Articles*. Jelf wrote to Maurice, seeking clarification. The issue was relevant to Jelf because King's College had been founded largely to counterbalance the University of London (UCL) which was regarded as a secular institution. Jelf saw it his duty to safeguard Anglican orthodoxy within its walls. Maurice replied to Jelf, but the principal was unconvinced by his explanation. The matter escalated. Maurice was dismissed.

The sequence of events was closely followed by Maxwell, for whom Maurice's Christian Socialism and the question of universal salvation were both hot topics of debate and discussion.[25] He wrote to Aunt Jane: 'You will have heard how the Council of King's College have sat upon Professor Maurice and intend to turn him out of the college. So, there are pamphlets and replies on the meaning of the word "Eternal", and broadsides of the Record on the side of the attack.'[26] Three weeks later, in a letter to Campbell, Maxwell elaborated:

> We have the usual amount of discussion here on labour parliaments, multiplicity of votes, Eternity, and Maurice and Jelf; or the contest between those who think that there is a real depth to which thought must go, though words cannot well follow it, and those who maintain that that which is not obvious to a man of sense, cannot be really connected with a religion which is not confined to deep thinkers, but professes to afford the highest principles to the simple. That is what most men discuss. Maurice has settled it for himself, believing that the things of which he treats do actually form the necessary thoughts of all men whether learned or not.[27]

---

24 JCM, *Letter to Campbell,* 20 Feb 1853, (C&G1, 182).

25 Ibid., 20 Feb 1853, (C&G1, 182f).

26 JCM, *Letter to Jane Cay,* 12 Nov 1853, (C&G1, 194).

27 JCM, *Letter to Campbell,* 3 Dec 1853, (C&G1, 194). See also, C&G1, 172.

Maxwell was hesitant about Maurice's theology. He felt that Maurice, in his *Theological Essays,* sat lightly to some Christian doctrine. Though admiring Maurice, he was troubled that Maurice seemed ready to side-line any doctrines which did not, for Maurice, 'form the necessary thoughts of men whether learned or not.' Maurice was in fact appealing to a principle which became increasingly popular in the second half of the nineteenth century, namely: *Anything strictly essential to true religion must be intuitively obvious to a man or woman of sense.* The German theologian Adolf von Harnack (1851–1930) would put this principle at the heart of his presentation of the essence of Christianity in his celebrated Berlin lecture series of 1899–1900, arguing that nothing over-complex could be of the *essence* of the faith. This principle, as employed by both Maurice and Harnack, was used to excise any doctrine which seemed speculative, uncertain, or undesirable.

Was Maurice revealing his rationalistic Unitarian origins? Was he dismissing from Christianity anything apparently too complicated? Was he dumbing-down religion? Or was he only trying to simplify faith? In his letter to Campbell, Maxwell indicated that, though he (Maxwell) was prepared to leave some theological issues as undecided, he could not go as far as Maurice.[28] His problem with Maurice was expressed more strongly in another letter:

> Maurice is a man I am loath to say nay to, or to accuse of wilful perversion of facts; but in some matters I think he is in great error, especially in his estimate of respectable ordinary Christians, as far as regards their creed. He cannot go too far in enforcing practice and work on people who were bound to it before, and theoretically confess it; but he is too hard upon the theories, and totally misrepresents them. *I would rather be taken for a Yezidi than for one of Maurice's popular religionists.*[29] (Italics ours)

Maxwell respected Maurice. He was inspired by Maurice. But he did not idolise him. He regretted that Maurice went too far in devaluing truths which were precious to many ordinary Christians. In Maxwell's

28 Ibid., 15 Sep 1853, (C&G1, 192f), 'An attempt must be made to find what is requisite to a Christian system, and whether the variables of such a system ought to remain constant, as they were at some arbitrary epoch (that of sect-founders, Fathers, General Councils, Reformers, etc.),

29 Ibid., 17 Oct 1855, (C&G1, 218).

assessment, Maurice edged towards an (unintended) perversion of facts in some of his statements about traditional Christian doctrine. Nevertheless, Maurice was a great man, through whose vision Maxwell had seen how his own longing to 'do the good' and express his Christian faith in practical terms could be realised.

Maxwell was committed for life to the principles which Goodwin and Maurice set out. He taught at Working Men's Colleges in Cambridge and London. He lectured at a roughly equivalent institution in Aberdeen. When a similar society was founded in Belfast, Thomas Andrews (1813–1885), the Queens College professor of chemistry, invited Maxwell to come and speak, though sometimes pressure of work made that impossible: 'I am sorry that I cannot give a lecture to the [Belfast] Working Men's Institute this winter, and that I must decline your hospitable invitation to stay with you, but I have avoided all engagements other than my regular work at Cambridge, and must continue to do so for some time.'[30]

Maxwell loved that part of his life. Not least because the working-men who came to learn were serious, whereas undergraduates could be frivolous! Even at Glenlair, Maxwell found an opportunity to further the same principles, supporting the Corsock Mutual Improvement Society.

30 JCM, *Letter to Thomas Andrews,* 15 July 1875, (Harman, III/1, 236).

Chapter 8
Tayler and Evangelical Piety

In June 1853, Maxwell experienced what has been described by some biographers as his conversion to a truly evangelical faith. That may be too dramatic a claim, but living with the family of the Revd Charles B. Tayler during the early summer did have a profound effect on his spiritual life. It may not have been conversion—as understood in the evangelical sense of regeneration—but it certainly resulted in a deepening of his existing Christian life.

One of Maxwell's friends at Trinity, G.W.H. ('Freshman') Tayler (1832–1890), invited him to Otley in Suffolk, to stay at the rectory of his uncle, the Revd Charles Benjamin Tayler (1797–1875). The intention was for Maxwell to visit for a few days or for a week or so at the most.[1] To begin with, all went according to plan. He had a relaxing time. He enjoyed walks and expeditions. He chatted about science to Charlie, the Rector's young son.[2] He took part in discussions about parish life; and he involved himself in household domesticities.[3] Suddenly, on the seventeenth of June, he was taken dangerously ill. The cause was possibly exhaustion, brought on by concentrated study in preparing for the formidable Tripos examinations which were looming in January 1854, as well as by the summer exams he had just sat. Tripos preparation involved candidates being in contact with their tutor day after day for month after month, in an intense, rigorous, unrelenting programme of exercises and instruction.[4]

Tayler described the illness as a type of brain fever. In Victorian times, brain fever referred to symptoms considered to be consistent with an inflamed brain and was characterised by headaches, flushed skin, delirium, and over-sensitivity to light and sound. Maxwell's father was kept informed of his son's illness, and, when he discovered

1 C&G1, 169–170.

2 JCM, *Letter to C.B. Tayler,* 8 July 1853, (C&G1, 188; Harman, I, 221).

3 JCM, *Letter to Campbell,* 14 July 1853, (C&G1, 190).

4 Tripos preparation is described by Forsyth, 'Old Tripos Days at Cambridge', 173–175.

how serious it had been, his gratitude to the Taylers was almost beyond expression. This was not the first time Maxwell had suffered this way. His father recalled an earlier episode during schooldays:

> I have just received your letter and Mr Tayler's. You may be sure I am thankful to hear of your recovering, although not previously made anxious about the illness. I cannot but think of the fever fit you had in Edinburgh after an Academy exam., when we had settled to go to Melrose: that was in 1846. Nothing can exceed the kindness of Mr and Mrs Tayler, and I hope you will not need long nursing. If you are well and not much hindered, you can let me know more fully how you are getting on.[5]

The illness left Maxwell not at all well. He stayed on with the Taylers, and through their care was tended back to full health. It was during this period of illness and convalescence, that he was exposed to the impressive piety of Charles Tayler and his family. For two weeks the Taylers nursed him as they would have their own son. He was deeply moved by this. As was his father.[6]

The Otley Evangelical

The Reverend C.B. Tayler was an evangelical Anglican who took a serious and solemn view of a clergyman's calling. He decried the attitude of some who, in his opinion, regarded settlement in a parish through the patronage of a favourite relative as a passport to a life of ease.[7] He grieved over any preacher who was himself 'unconverted' yet attempted to 'preach the gospel of converting grace.'[8] He may have had in mind the Revd John Mitford of Benhall parish, only fourteen

---

5 John Clerk Maxwell, *Letter to JCM,* 24 June 1853, (C&G1, 187).

6 Ibid., 28 June 1853, (C&G1, 187): 'I am most thankful and happy to hear of your convalescence through Mr Tayler's most kind and daily bulletins. I know not how sufficiently to thank Mr and Mrs Tayler.'

7 Tayler, *Facts in a Clergyman's Life,* 9, described such a clergyman: 'His uncle has promised him the family-living … [He is] an idle trifling young man, who has already given sufficient proof, that he has neither energy nor application enough to succeed in a secular profession, and whose only distinct perception of so sacred a calling, whose only reason for a decided preference towards a clergyman's life, are, that he shall thus secure to himself a comfortable income, a pleasant parsonage, and an easy idle life.'

8 Ibid., 15.

miles from Otley, described by Brenda Colloms as the prime example of a scholarly parson whose 'lack of religious conviction in no way interfered with the overall supervision of his parish responsibilities.'[9]

Belonging to the Church of England's evangelical wing, Tayler was a member of several bodies, including the London Missionary Society. One of his correspondents was Robert Moffat, the renowned Scottish missionary to southern Africa. In 2020, a letter from Moffat to Tayler was sold at Bonhams, London.[10] Addressed by Moffat to 'My dear Friend and Brother', the eight-page letter reflected on the East African slave trade: 'The astonishing explorations of Dr Livingstone & others have laid open to Christian philanthropy lands teeming with inhabitants, whose degradation has been increased a thousand-fold by the demon of slavery.'

Tayler believed firmly in the doctrines of the Reformation, especially its core principle of justification solely through God's grace received by faith. For Tayler, this meant that a clergyman's personal experience of the redeeming grace of God in Christ was an essential precondition of ministry. In all the known sermons he preached, and in all the books he wrote, Tayler's unswerving aim was to convert men and women to Christ. That was his consistent theme, and that evangelical intensity pervaded all his work, with the Reformed doctrine of justification the active principle at the heart of his faith and ministry. The doctrine was not simply another theological dogma on a check-list of orthodoxy but permeated his entire calling, both as preacher and pastor. His writings reveal him as pro-active in introducing the doctrine into pastoral situations, and as pro-active in challenging his flock to regard it as the door to a new life both in this world and the world to come. For Tayler, the doctrine was a practical and pastoral doctrine. As Maxwell stayed in Tayler's Suffolk Rectory, and as he was ministered to by the Rector daily, he would almost certainly have had deep discussions with him on this theme.

Long afterwards, Maxwell said that the care shown to him by the Taylers gave him a new perception of the love of God. There is little doubt that the family's affection, as they brought him back to full health, made a deep impression. William Garnett's tribute article

9 Colloms, *Victorian Country Parsons*, 107.

10 Robert Moffat, *Letter to C. B. Tayler*, 9 July 1869 (from Kuruman Mission Station). Autograph letter signed by Robert Moffat.

for *Nature* stated that the care and kindness of Mr and Mrs Tayler's nursing never faded from Maxwell's memory. This was written and published only days after Maxwell's death, and therefore based on what Maxwell must have said to Garnett directly rather than gleaned from a third-party some months later. The Taylers so impressed Maxwell with the power of love that their example was a critical factor in the future formation of his own Christian character.[11]

The Otley Effect

When Maxwell did recover, his father suggested he should go to Glenlair to convalesce.[12] He actually chose to return to Cambridge and resume his studies, though he was careful 'not to read inordinately hard.' Hopkins recommended light work.[13] Concerning the experience he had at Otley, we have no direct record of the bedside discussions between Tayler and Maxwell. Nor is it easy to work out what the 'Otley Effect' was, though there are clues in three sources: 'Freshman' Tayler, Lewis Campbell, and Maxwell himself.

'Freshman' Tayler believed that the Otley experience clarified Maxwell's faith and made it more decisively evangelical. This was what he wrote to Lewis Campbell after Maxwell's death:

> My uncle's conversation seemed to make such a deep impression on his mind. [Maxwell] had always been a regular attendant at the services of God's house, and a regular communicant in our college chapel. Also, he had thought and read much on religious subjects. But at this time (as it appears from [Maxwell's] own account of the matter) his religious views were greatly deepened and strengthened.[14]

The letter concluded:

> Maxwell has indeed left a very bright memory and example. We, his contemporaries at college, have seen in him high powers of mind and great capacity and original views, conjoined with deep humility before his God, reverent submission to His will, and hearty belief

11 Garnett, 'James Clerk Maxwell', *Nature,* Vol. 21/524 (13 Nov 1879), 44.

12 John Clerk Maxwell, *Letter to JCM,* 28 June 1853, (C&G1, 187).

13 JCM, *Letter to C.B. Tayler,* 8 July 1853, (C&G1, 188; Harman, I, 220).

14 G.W.H. Tayler, *Letter to Campbell,* 4 March 1882, (C&G1, 173).

> in the love and the atonement of that Divine Saviour who was his Portion and Comforter in trouble and sickness, and his exceeding great reward.

'Freshman' Tayler emphasised Maxwell's belief in the atonement, which was a key principle of evangelical Protestantism. But, in his specific mention of the atonement, was Tayler projecting his own piety onto Maxwell? It is true that in Maxwell's prayers—as quoted in the opening pages of this book—the atonement is mentioned, and Gaal has interpreted this as incontrovertible proof that the atonement had a central and active role in Maxwell's faith.[15] But such expressions were conventional in nineteenth-century religious language across the theological spectrum. At the same time, given the evangelical ministries with which Maxwell associated himself, it is not unreasonable to assume that he did accept the central role of the atonement in Christian faith, though in his extant writings there are few direct references to it.

Lewis Campbell's interpretation of the Otley experience was a little different. Campbell was less evangelically inclined, and anxious to emphasise Maxwell's feeling for something akin to Campbell's own broad-churchmanship. Consequently, though Campbell accepted that Otley had been significant for Maxwell, he was more guarded in his evaluation. For Campbell, Maxwell's strongest conviction following this experience was: '*Love abideth, though knowledge vanish away*' (I Corinthians 13:8), with this coming to Maxwell 'at the very height of the intellectual struggle.'[16]

What was that 'intellectual struggle'? Was it Maxwell's intense concentration on mathematics in preparation for examinations? Some considered that too much mathematics was harmful to health. Did Campbell have that in mind? Or was Maxwell struggling with faith? We know he was aware of the impact of liberal scholarship on the veracity

15 Gaal, *The Faith of a Scientist*, 46. Torrance, 'Christian Faith and Physical Science in the Thought of JCM', 221, argued similarly that Maxwell never doubted God's self-revelation in Christ, or the full sufficiency of atonement that flowed from the love of God. Torrance added that Maxwell's base theology was the love of God; and that he sought to live a daily life directed and shaped by that love through his union with Christ.

16 C&G1, 170.

of Scripture, discussing this with his friend Robertson. Was that his intellectual struggle? Did the love and care of this strongly evangelical family enable him to feel more assured in his faith? As Campbell reflected on Maxwell's experience, he wrote a cautionary footnote:

> At the same time, it is not to be supposed that Maxwell was ever completely identified with any particular school of religious opinion. He was too much 'the heir of all the ages', and, as he himself expressed it: 'his faith was too deep to be in bondage to any set of opinions.' Scottish Calvinism was the theological system which had most historical interest for him, and most claim on his hereditary piety. He was learned in the writings of Owen and Jonathan Edwards.[17]

Campbell continued:

> But that which his latest pastor [William Guillemard] has called 'his deep though simple faith,' was not enclosed in any system. Even his youthful training (which in the case of one so loyal is not to be disregarded) was favourable to a comprehensive view of Christianity. Beginning with the Bible, which, he knew by heart from a child, Presbyterian and Episcopalian influence had been blended as we have already seen. Hence, when he went to Cambridge, there was nothing strange to him in the service of the English Church. His mind was always moving . . . above and 'beyond these voices' yet they were not indifferent to him.

Campbell was reluctant to pigeon-hole his friend theologically. He preferred to see him as 'the heir of all the ages', citing Maxwell's own view that Maxwell's faith 'was too deep to be in bondage to any set of opinions.' But was Campbell claiming Maxwell for his own standpoint? And if Maxwell appreciated the writings of the great Puritan theologians John Owen (1616–1683) and Jonathan Edwards (1703–1758)—as he did—then, despite his ecumenical instincts, he was no broad churchman in Campbell's sense.

What is not disputed is that Maxwell's own abiding recollection of the Otley Rectory was of the immense kindness which he received from the Tayler family. Shortly after returning to Cambridge, he wrote to Campbell:

---

17 C&G1, 170, fn.

> After examination I went to visit the Rev. C.B. Tayler . . . I intended to return on the 18th June, but on the 17th I felt unwell, and took measures accordingly to be well again, i.e. went to bed, and made up my mind to recover. But it lasted more than a fortnight, during which time I was taken care of beyond expectation (not that I did not expect much before). When I was perfectly useless, and could not sit up without fainting, Mr Tayler did everything for me in such a way that I had no fear of giving trouble. So did Mrs Tayler; and the two nephews did all they could. So, they kept me in great happiness all the time, and detained me till I was able to walk about, and got back strength. I returned on the 4th July. The consequence of all this is that I correspond with Mr Tayler, and have entered into bonds with the nephews, of all of whom more hereafter.[18]

The letter gives the bare facts. But in another missive—a long and affectionate letter of thanks to C.B. Tayler, written *before* he communicated with Campbell—Maxwell referred to a change. He saw the experience as having taken him out of an *inward-looking*, slightly self-obsessed, spoilt-child attitude, to one which made him more *open* and aware of others. In a spirit of confession, he blamed only himself for having been the person he had been. This was expressed in his response to a message from C.B. Tayler, which he received one morning at Cambridge after early chapel:

> You dimly allude to the process of spoiling which has gone on during the last two years. I admit that people have been kind to me, and also that I have seen more variety than in other years; but I maintain that all the evil influences that I can trace have been internal and not external, you know what I mean—that I have the capacity of being more wicked than any example that man could set me, and that if I escape, it is only by God's grace helping me to get rid of myself, partially in science, more completely in society, but not perfectly except by committing myself to God as the instrument of His will, not doubtfully, but in the certain hope that that 'Will' will be plain enough at the proper time. Nevertheless, you see things from the outside directly, and I only by reflexion, so I hope that you will not tell me you have little fault to find with me, without finding that little and communicating it.[19]

18 JCM, *Letter to Campbell,* 14 July 1853, (C&G1, 190).

19 JCM, *Letter to C.B. Tayler,* 8 July 1853, (C&G1, 188f; Harman, I, 220f).

What did Maxwell mean by 'spoiling'? What was the 'capacity' for wickedness? What had Tayler seen 'from the outside directly'? Had he emphasised the sinful condition of every man? Had he brought that home to Maxwell personally? Had Maxwell, as he lay ill, reflected on his life and the state of his soul? But, if so, and if there had been a standard evangelical 'conversion', why did Maxwell not refer to the cross or to his Saviour? What did he mean when he wrote that God's grace helped to 'get rid of myself, partially in science, more completely in society?' Did this refer to conversion? Or was it the phrase of a believer who, following a crisis, reflected on his life, became more aware of his selfishness and self-preoccupation, and was now giving his life a reconsecration and new direction?

Had Maxwell's Christian faith been *spoiled* during the previous two years at Cambridge? Was Maxwell admitting that the causes for *spoiling* came from within himself rather than from people or circumstances? Was he saying that he alone had been to blame? Had he become conceited?

Something happened at Otley. At the very least it brought about a re-evaluation of faith and attitudes. Definitive analysis is difficult, given that few of Maxwell's personal letters survive, and the piety revealed in these letters can support either 'Freshman' Tayler's interpretation, or the one by Lewis Campbell. What may be a stronger clue is the type of preaching ministry which Maxwell subsequently chose to sit under, most of which were of a decidedly evangelical persuasion.

The Taylers' example, as they nursed him through illness, stayed with him. He saw that nothing was too much trouble: 'I had got into habits with you of expecting things to happen, and if I wake at night I think the gruel is coming.'[20] That model, of how to care for someone and of how to do it wholeheartedly and cheerfully, was his template when he became support and aid to others: 'I remember everything about you and Otley much better than most things before and after that time. I got advantage from your nursing when my father was ill, and many other things have since brought you and Mrs Tayler to mind.'[21]

20 JCM, *Letter to C.B. Tayler*, 2 Feb 1866, (C&G1, 344).

21 Ibid.

At Otley, Maxwell had a wake-up call concerning his attitudes as a Christian, making him reflect deeply on what constituted a good and meaningful Christian life. A year later he wrote to F.W. Farrar expressing his wish to make the most of life in a religious sense.[22] He told Farrar that the secret to true enjoyment was to 'have the work of the day continually before his eyes' but in a liberating way. Focus should not be on *yesterday's* work, as despair would take over since that was always incomplete. Nor should it be on *tomorrow's* work, as that might result in unrealistic visionary day-dreaming. Nor should it be on something which '*ends with the day*' and has no lasting significance. Nor should it be on something which '*remains to eternity*' (that is: only starts in eternity) since that cannot shape actions in the here and now. Instead, Maxwell regarded real happiness as being linked to an ability to recognise in the work of *today* an embodiment of the work of *eternity*, realising that in doing the work of *today* he was a partaker of *infinity*.[23]

In other words, what is done today, if done well, has eternal significance. Moreover, through such work being done well, humanity reflects the 'divine process of nature' and exhibits 'the union of the infinite with the finite.'[24]

Maxwell's choice of words is intriguing, given that he had been schooled by Sir William Hamilton to allow no link between the finite and the infinite! But as a Christian, Maxwell thought differently. Moreover, he believed that anyone who lived and worked with a perspective moulded by awareness of eternity, would never be in danger of underestimating, or devaluing, or 'slighting' his temporal existence. Instead, such a person would know that the experience of temporal physical existence is what actually makes individual action possible and meaningful, yet without losing view of the eternal. He believed that his work in this life—and indeed the work of anyone, if approached in the proper manner—was part of the creative work of God himself. As such, it was rooted in the infinite, and had infinite value.

22 JCM, *Letter to F.W. Farrar*, 1854, (C&G1, 200).

23 Maxwell may have been thinking of 2 Peter 1:4: '*Ye might be partakers of the divine nature.*'

24 JCM, *Letter to F.W. Farrar*, 1854, (C&G1, 200).

Chapter 9
The Apostles' Club

In March 1852, a year before Otley, Maxwell wrote an extended letter to Lewis Campbell, reflecting on the foundations of truth, the nature of faith, and the search for objective knowledge. He told Campbell that his stated intention was to subject every truth, sacred or secular, to thorough examination. There were to be no off-limits. There was to be no holy ground. Nothing was untouchable. Any restriction would only lead to truncated science and barren faith. Faith should never fear truth:

> My great plan . . . is to let nothing be wilfully left unexamined. Nothing is to be holy ground consecrated to stationary faith, whether positive or negative. All fallow land is to be ploughed up and a regular system of rotation followed. . . . Never hide anything, be it weed or no, nor seem to wish it hidden. . . . Again, I assert the right of trespass on any plot of holy ground which any man has set apart . . . to the power of Darkness. Such places must be exorcised and desecrated till they become fruitful fields. Again, if the holder of such property refuse admission to the exorcist, he *ipso facto* admits that it is consecrated, and that he fears the power of Darkness. It may be that no such darkness really broods over the place, and that the man has got a habit of shutting his eyes in that field, which makes him think so. . . . But a candle is coming to drive out all ghosts and bugbears. Let us all follow the Light.[1]

When Torrance commented on this passage, he made the observation that both Michael Faraday and James Clerk Maxwell were overwhelmed by a profound sense of the integrity and orderliness of nature as God had created it.[2] They had confidence in scientific research precisely because they had a faith-based confidence in a God-given inherent orderliness of creation. And so Maxwell saw his faith as *emancipating* him to investigate. In no way did he see his faith as *limiting* his intellectual curiosity or independence of mind.

1 JCM, *Letter to Campbell,* 7 March 1852, (C&G1, 178).

2 Torrance, 'Christian Faith and Physical Science', 216.

A month previous to his 'no holy ground' letter, he had emphasised to Campbell that 'objectivity alone is favourable to the free circulation of the soul', adding: 'but let the object be real and not an image of the mind's own creating, for idolatry is subjectivity with respect to gods.'[3] His March 'no holy ground' message reaffirmed his comfort with radical openness. Openness was not *despite* being a Christian believer. Nor was it in tension with faith. Nor was it divorced from his religion. It was an openness precisely *because* he was a believer. As a Christian, he was liberated for free enquiry:

> I am convinced that no one but a Christian can actually purge his land of these holy spots. . . . I do not say that no Christians have enclosed places of this sort. Many have a great deal, and everyone has some. But there are extensive and important tracts in the territory of the Scoffer, the Pantheist, the Quietist, Formalist, Dogmatist, Sensualist, and the rest, which are openly and solemnly tabooed. *Christianity—that is, the religion of the Bible—is the only scheme or form of belief which disavows any possessions on such a tenure. Here alone all is free.* You may fly to the ends of the world and find no God but the Author of Salvation. You may search the Scriptures and not find a text to stop you in your explorations. (Italics ours)[4]

The extent to which Maxwell put his Christian beliefs to such severe intellectual examination can be judged only incompletely from his writings; but his desire for open, objective enquiry was clearly expressed. This was why, later in life, he was reluctant to join the Victoria Institute which was seen—though not always fairly—as having predetermined that certain scientific theories were wrong. In his March 1852 epistle, Maxwell used bold words to express audacious intentions: 'Let nothing be wilfully left unexamined' and 'Never hide anything, be it weed or no, nor seem to wish it hidden.' These were the words of a young man assured in his faith and confident in his science.

The Select Essay Club

By 1852, his second year at Trinity, Maxwell was known not only as a fine mathematician, but as a thinker. It was then that he was

---

3 JCM, *Letter to Campbell*, 10 Feb 1852, (C&G1, 177).

4 Ibid., 7 March 1852, (C&G1, 179).

elected to a scholarship, joining the scholars who dined together at table. This brought him into close contact with other high-flyers, resulting in his election to the *Select Essay Club*. The Club was also called the *Cambridge Conversazione Society*, but was best known as the *Cambridge Apostles*, or simply the *Apostles*.

An air of mystery surrounded its activities. Its records were stored in a wooden trunk known as the Ark. It met at late hours on a Saturday evening. The host read an essay or in default gave a supper. The members drank coffee, ate sardines on toast, and debated issues into the small hours of Sunday morning. Membership was by invitation only, and so secret that newcomers might only know after their election that they had been proposed. Numbers were originally limited to twelve, hence the name *Apostles*. Until the 1850s membership was almost exclusively for men interested in the arts and philosophy, with Maxwell perhaps the first scientist to be elected.[5]

After Maxwell's era the *Apostles* became an avant-garde group with libertine tendencies. But in the 1850s its culture was far less extreme, though cutting-edge matters were discussed. The Society had religious roots with an emphasis on debating faith and morals, and it tended to take a progressive stance on social and ethical questions. Samuel Taylor Coleridge was an early star, not as a poet but as a proponent of social reform. Elements of social radicalism characterised the group from the start, preparing Maxwell for F.D. Maurice's brand of Christian Socialism. This environment reinforced his conviction that his Christianity should be practical, should address the needs of society, and should help ordinary working people have a better life. All of this lay alongside Maxwell's robust doctrinal orthodoxy in matters of faith.

'Idiotic Imps' and 'Dark Science'

Maxwell presented several papers to the *Apostles*, taking the opportunity to reflect further on the relationship between science, philosophy, and faith. One tendency of the period—which Maxwell abhorred—was the attempt by popular writers to swaddle their ideas

5 Deacon, *Cambridge Apostles: A History of Cambridge University Elite Intellectual Society*, 33.

with the vocabulary of science in order to give their speculations an appearance of solid thought. In his 'Idiotic Imps' 1853 paper to the *Apostles*, he deplored this fashion of clothing superstition with scientific terminology. He labelled it 'dark science', since it attempted to link true science with spiritualism and séances.[6] 'Dark science' aped the language of physics and tried to give occult phenomena some kind of validation.

Elsewhere in 'dark science' the phraseology of physics was exchanged for that of psychology, with a determination to prove that things could be done by the power of will alone, and suggesting that two or more minds could have communion through mesmerism or a spiritual medium, even though separated by space. Maxwell regarded it all as nonsense. In his opinion, the real 'spiritual' forces at work in these farces were simply 'the familiar spirits of money-making media' which, through a Ouija board, could 'rap out lies for hours together.'[7]

Later in life, in his pencilled comments on Stewart and Tait's *Paradoxical Philosophy*, he reaffirmed his disapproval of the Ouija board craze which had gathered momentum in the intervening years.[8] He was anxious to distance his own faith and science from spurious phenomena which were increasing in popularity. Ironically, his vocabulary of 'fields' and 'lines of force' was later misappropriated for extravagant theories of spiritualism and ley-lines.

Apostles' Papers: William Paley and Design

At an *Apostles*' meeting in 1853, Maxwell reflected on Paley's 'Argument from Design'. As a boy he had read some of Paley's works which he had found in Aunt Isabella's library, and he was familiar with the famous Watchmaker Argument. His 1853 paper was titled: 'What is the nature of Evidence of Design.' Unfortunately, only part of it survives.[9] Colebrooke in his monograph on Maxwell's

6 JCM, 'Idiotic Imps', 1853, (C&G1, 227–230; Harman, I, 224–226).

7 Ibid., (C&G1, 229; Harman, I, 226). See also: JCM, *Letter to Campbell*, 14 July 1853, (C&G1, 191; Harman, I, 222).

8 NLS: Acc.10000/399, 'Balfour Stewart & P.G. Tait, *Paradoxical Philosophy* with annotations by JCM.', 208, 226.

9 JCM, 'What is the nature of Evidence of Design?', (C&G1, 225–227; Harman, I, 227–228).

spirituality has argued that Maxwell was never impressed by Paley's design argument. But that is a hasty judgement, with Colebrooke over-anxious to distance Maxwell from some modern versions of Intelligent Design theory.[10] In fact, Maxwell saw himself as making Paley's argument more robust.

For Maxwell, the notion of design required two essentials. First, an entity which appeared to have design. Second, a presupposed personality to do the designing. Maxwell applied these criteria to Paley's illustration of a watch, but made the important point that, when considering a watch, we become aware of *two* levels of design and not just *one*.

One level was the physical complexity of its mechanism (springs, wheels, dials, etc.), which was what Paley focused on. The other level included the fact that a functioning watch faithfully indicates the unit of civil time which is a construct of society. Thus, for Maxwell, evidence of purposeful design was not simply in the *complexity* of the mechanism alone (the first level), but in the watch having apparent *purpose* and function in society (the second level) indicating conscious intention. He then asked whether the *complexity* and the *purpose* were independent of each other, or if they were connected. As a careful logician, he made a key point:

> It is the business of science to investigate these causal chains. If they are found not to be independent but to meet in some ascertained point, we must transfer the evidence of design from the ultimate fact to the existence of the chain.[11]

In the fragment of Maxwell's talk available to us, he did not state explicitly—in the case of the watch—that the two chains of design did meet, but that was implied. Given such a conjunction, what could

---

10 Colebrooke, *Science and Religion: The Spirituality of James Clerk Maxwell*, 17: 'Maxwell had little time for William Paley's watchmaker's analogy, where the argument for God was based on the complexity of nature . . . Maxwell's argument for design is not based on complexity, but on unity, which avoids all 'God of the gaps' criticisms. He would not have supported intelligent design; for him there was no point at which science must stop looking, and there was always more to discover.' Our view is that Maxwell did not reject the 'complexity' argument, but augmented it with a 'purpose' argument.

11 JCM, 'What is the nature of Evidence of Design?' (C&G1, 225; Harman, I, 227).

cause design to exhibit both *complexity* and *purpose*? He stated: 'The only subordinate centres of causation which I have seen formally investigated are men and animals.' By 'subordinate centre' he meant any cause 'other than a divine agent'. In other words, if we put the Deity aside, any possible 'designer' had to be human or animal. It certainly had to be a *being*. But what were the necessary attributes of such a *being*? Did it have to be *personal*? Did it have to be *intelligent*? He wrote: 'Whether that being be *personal* is a question which may be determined by induction.' In other words, inductive logic allows the reasonable inference that the central cause must be *personal being*. He added: 'The less difficult question whether the *being* be *intelligent* is more practicable, and should be kept in view in the investigation of organised beings.'

Maxwell then had a brief excursus (as a mathematician he had to cover all hypothetical options) in which he rejected the idea of any other 'supernatural' agents, such as witches or demons, being causes of phenomena. By this stage, he had argued that if an entity exhibited strong evidence of both *complexity* and *purpose*, then it had to have a designer who was both *personal* and *intelligent*.

There was a warning. Acceptance of such an ultimate cause (God) should never make the scientist abandon his or her search for natural causes. In Maxwell's mind—and at the core of all his scientific work—human science never comes to a point where a further natural link in the chain of natural causes cannot be discovered. Within nature there is always another natural cause around the corner. There is no 'end of the road' in causations. That was why, for Maxwell, God could not be discovered *deductively* as the only remaining possibility, after the chain of all natural causes had been exhausted. Consequently, Maxwell's argument was based on *inductive* rather than *deductive* reasoning. A truly first cause or prime mover cannot be *deduced* from the chain of natural causes: but, by *inductive* reasoning—parallel with the conviction that all things have a beginning—a first and ultimate cause (which is both *personal* and *intelligent*) can be posited as necessary.

Following his mentor Sir William Hamilton, Maxwell held that purely deductive reasoning could never make the leap from the material to the spiritual, or from immediate natural causes to an ultimate divine cause, or from the finite to the infinite. Inductive reasoning however—in which one began with the whole and moved to the parts, rather than starting with the parts and attempting to move to the whole—did provide rational grounds for stating the nature and necessity of a first and ultimate cause.

In his paper Maxwell was anxious to avoid three common fallacies. First, the fallacy of putting the final cause in the place of a physical connection. Second, the fallacy of the erroneous assertion of a physical relation. Third, the fallacy of applying an argument from final causes to misunderstood phenomena.[12] The third concerned sloppy and illogical thinking. The second was about scientists starting with a false or non-existent relation between two physical entities. But it is the first of these fallacies which is of most interest. For Maxwell, it concerned a situation in which a scientist, at some stage in his or her investigation of nature, abandoned trying to find a natural cause and reverted to simply stating: 'God must have done it!' But in no way did this cut across his core belief that in considering the 'whole', there had to be a designer with intelligence.[13]

As stated above, Maxwell believed that any natural phenomenon was capable of never-ending examination and discovery, and that scientists should always work with the assumption that yet *another* link in the chain of natural causation could always be found. That was because God had made the universe in such a way that, as part of its divinely-gifted integrity, any natural phenomenon is causally connected to other natural phenomena.

In working with this assumption—which dovetailed with Hamilton's philosophy of the *conditioned*—Maxwell shared considerable common ground with out-and-out materialists, at least in methodology if not in conclusions. But, unlike them, he asserted the necessity of an intelligent designer, without whom both the existence and the logic of the universe were impossible. This established two fundamentals in his faith and science. First, in science there is never a point where a brick wall absolutely precludes detecting another natural cause: in that sense, a 'material explanation' can always be found. Second, the actuality and structure of the universe cries out

12 JCM, 'What is the nature of Evidence of Design?', (C&G1, 226; Harman, I, 228).

13 Cf. JCM, 'Has Everything Beautiful in Art it's Original in Nature', Essay for the Apostles, Spring 1854, (C&G1,230–232; Harman, I, 248).

for, and demands, the mind of a Deity. Though a simplistic 'God of the gaps' argument was unacceptable to Maxwell, the universe made no logical sense without God.

Apostles' Papers: Ethical Truth

Whilst a student in Edinburgh, Maxwell had written long letters to Campbell on ethics in which he contrasted self-interest and altruism as stimuli for ethical behaviour. He returned to the topic in May 1855, in another paper for the Apostles' Club: 'Is Ethical Truth obtainable, from an Individual Point of View.' In it, he reflected on some of the core questions of moral philosophy, having considered them further over the intervening five years. Previously, he had interpreted Adam Smith as teaching that human behaviour was only motivated by self-interest. Now he realised that Smith's thinking was more nuanced, allowing unselfish factors to direct actions, and he accepted that Smith saw men and women as influenced by positive forces and not just by selfishness.

As Maxwell dug into possible motivations for human behaviour, he listed three criteria which were to be considered in relation to human actions: fittingness, pleasure, and freedom. Freedom was the preferred choice, yet freedom was an incomplete notion by itself, and needed to be driven by values. It was this which had made him look more closely at Adam Smith's use of the principle of *sympathy* or *fellow-feeling*: 'Instead of supposing the moral action of the mind to be a speculation on fitness, a calculation of happiness, or an effort towards freedom, *he makes it depend on a recognition of our relation to others like ourselves*.'[14] (Italics ours). Having reread Smith, Maxwell concluded that it was a person's ability to project beyond the self and self-interest which was the 'only true one' (i.e. the only one which carried integrity), and which should be extended to 'embrace other relations than that of mere similarity'. What Maxwell meant was that there was a serious need to embrace persons who were not like-minded or belonged to the same group. Outward-altruism rather than an inward-selfishness was all-important.

---

14 JCM, 'Is Ethical Truth obtainable from an Individual Point of View?', Apostles' Club, May 1855, (C&G1, 234f; Harman, I, 294).

The Foundations of Truth

In November 1851, before his election to the *Apostles*, the twenty-year-old Maxwell wrote to Campbell, musing on the foundations of knowledge.[15] This complemented his long letter of June 1850 from Edinburgh on the foundations of ethics.[16] His question, 'Where does truth lie?', was prompted by awareness that there seemed to be few academic disciplines whose conclusions were complete and assured. The only approximation to certainty seemed to be in mathematics:

> Is Truth nowhere but in mathematics? Is 'Beauty' developed only in men's elegant words, or 'Right' in Whewell's *Morality*? Must Nature as well as Revelation be examined through canonical spectacles by the dark lantern of Tradition, and measured out by the learned to the unlearned, all second-hand?

In believing that mathematics, if nothing else, could provide absolute truth and certainty, Maxwell echoed Sir William Hamilton who stated: 'Logic, within its proper sphere, is of such irrefragable certainty, that, in the midst of all the revolutions of philosophical doctrines, it has stood not only unshattered but unshaken. In this respect, logic and mathematics stand alone among the sciences.'[17]

At that time, it was assumed that mathematics could give certainty. Eighty years later, Kurt Gödel (1906–1978) demonstrated that the provability of such certainty, even for mathematics, was impossible. But in 1851 Hamilton and Maxwell were blissfully unaware of this philosophical conundrum, seeing mathematics as a solid starting-point for truth.

Maxwell alluded to the abstract notion of *beauty*. Whilst in Edinburgh he had met the artist R.D. Hay who tried to explain *beauty* in terms of mathematical principles, especially in shapes such as the oval or ellipse.[18] For some, Art and Science were viewed as the two critical mechanisms which were capable of raising humanity to the 'level of the gods', thus elevating humanity beyond brute existence.

---

15 JCM, *Letter to Campbell*, 9 Nov 1851, (C&G1, 157–160; Harman, I, 208f).

16 Ibid., June 1850, (C&G1, 138–144; Harman, I, 193–198).

17 Hamilton, *Lectures in Logic and Metaphysics*, Vol. 3, 43.

18 C&G1, 73.

Art was based on the *aesthetic appreciation of* beauty; science was based on the *logic of mathematics.* Both disciplines were regarded as powerful levers for human progress. The poet John Keats (1791–1821), in his *Ode on a Grecian Urn,* wrote: 'Beauty is truth, truth beauty; that is all ye know on earth, and all ye need to know'. Keats drew on a tradition which linked reasoning with aesthetics, and whose origins lay in the Platonist notion of the Ideal. This tradition expressed the conviction that the most truthful statements in any and every discipline would also be the most beautiful and the most elegant. Maxwell was tuning into the thought that beauty would be characteristic of any essential truth of nature, especially any fundamental mathematical equation.[19]

In the same letter, he referred to the *Shorter Catechism*: 'I believe, with the Westminster Divines and their predecessors *ad infinitum* that *'Man's chief end* [purpose] *is to glorify God and to enjoy him for ever.*"[20] Maxwell interpreted this as giving a theological justification for scientific work, since science enabled humanity to glorify God even more intensely through understanding the wonders of God's world in ever greater depth. He continued:

> It is natural to suppose that the soul, if not clothed with a body, and so put in relation with the creatures, would run on in an unprogressive circle of barren meditation. In order to advance, the soul must converse with things external to itself. In every branch of knowledge, the progress is proportional to the amount of facts on which to build, and therefore to the facility of obtaining data.[21]

Glorifying God included knowing God's world, and it was in relation to scientific work that Maxwell wrote: 'In order to advance, the soul must converse with things external to itself.' Science dealt with external things: with what was 'out there'. It had to do with *realities in nature,*

19 On Maxwell and the relationship between beauty and truth see: McCartney, 'I Remember Years', 299, who quotes Hardy (1877–1947) and Dirac (1902–1984). Maxwell referred to science as dealing with 'the highest principles of order and beauty' in his Aberdeen Inaugural Address of November 1856.

20 *The Shorter Catechism,* Q&A 1.

21 JCM, *Letter to Campbell,* 9 Nov 1851, (C&G1, 159; Harman, I, 209).

and not just with *ideas in the mind*. For Maxwell, this conversation with external reality increased the number and quality of proper 'facts' known to humanity. In turn, these 'facts' resulted in an increase of 'actual truth'.

Maxwell opined that truth was 'proportional to the amount of facts on which to build.' This prompted him to rank academic disciplines in relation to the material facts at their disposal. In decreasing order his list was: mathematics, chemistry, natural history sciences, medicine, metaphysics, law, and ethics, with pneumatology and theology last of all. It was not that he disbelieved in the truths of Christian theology: the opposite was the case. But theology was not built on the same type of material facts as were the physical sciences. Following Sir William Hamilton, he was aware that the philosophical validity of theology lay on a different foundation from that of material phenomena. Material phenomena were part of the *conditioned*. Spiritual phenomena belonged to the *unconditioned*, with its own data and its own methods.

The natural order spoke eloquently to Maxwell. During his 1853 Easter break he expressed this in poetry. He stayed on at Cambridge to work at high intensity, but found time to compose *The Student's Evening Hymn*.[22] This was not one of his comic verses, but a serious poem expressing his conviction that the purpose of science was to celebrate the wonder and mystery of what God had made. Knowledge was to be pursued, not *despite* faith in God' but *because* of it. It is too long to quote in full, but the stanzas which follow give a flavour:[23]

Through the creatures Thou hast made
Show the brightness of Thy glory,
Be eternal Truth displayed
In their substance transitory,
Till green Earth and Ocean hoary,
Massy rock and tender blade
Tell the same unending story
"We are Truth in Form arrayed."

---

22 C&G1, 167.

23 Ibid., 594–596, gives the full text.

Teach me so Thy works to read
That my faith, new strength accruing,
May from world to world proceed,
Wisdom's fruitful search pursuing;
Till, thy truth my mind imbuing,
I proclaim the Eternal Creed,
Oft the glorious theme renewing
God our Lord is God indeed.

Give me love aright to trace
Thine to everything created,
Preaching to a ransomed race
By Thy mercy renovated,
Till with all thy fulness sated
I behold thee face to face
And with Ardour unabated
Sing the glories of thy grace.

Chapter 10

Books! Books! Books!

Maxwell loved books! He immersed himself in a vast range of literature: from poetry to the latest novels, and from theology to metaphysics. Book-worming was habitual, and increased significantly once the Tripos examinations were in the past. During 1855 his lighter reading included Charlotte Brontë's *Shirley*. He had already read *Jane Eyre,* and had devoured *Villette* as soon as it appeared in 1853.[1] Other 1855 books included William Thackeray's *The Newcomers,* and Charles Kingsley's *Westward Ho!*[2] These, plus his reading of George Eliot's *Middlemarch* in 1873, soon after its publication, amply contradict the hasty statement made by Sir John Randall (1905–1984) that Maxwell did not read modern nineteenth-century novels.[3] On the serious side, he enjoyed pulpit discourses, especially from the older Divines, with Campbell commenting that his friend's interest in old sermons was like Macaulay's interest in old novels or Charles Lamb's in old plays.[4]

The process of reading and thinking intrigued him. He wondered 'whether words mould thought, or thought brews words', concluding that one theory may be as good as another![5] What his letters reveal is that—both before and after the Otley experience—his reading encompassed a wide range of theological opinion. As well as delighting in John Owen and Jonathan Edwards, he felt impelled to understand those who were not necessarily likeminded, viewing that as a Christian imperative and not just as an exercise in intellectual curiosity. For his faith to be credible, his mind had to be informed. Among the authors read at that time

---

1 JCM, *Letter to Campbell,* 14 July 1853, (C&G1, 190).

2 JCM, *Letter to John Clerk Maxwell,* 5 Oct 1855, (C&G1, 217; Harman, I, 328).

3 Randall, 'Aspects of Maxwell's Life and Work', *Clerk Maxwell and Modern Science,* 23.

4 C&G1, 171.

5 JCM, *Letter to Campbell,* 17 Oct 1855, (C&G1, 218).

were: Julius Hare, Robert Chambers, Erskine of Linlathen, and Thomas Browne.

Archdeacon Julius Hare (1795–1855)

In the spring of 1853, Maxwell was deep into Julius Hare's works. Many of Hare's books, and those by F.D. Maurice, were published in Cambridge by the MacMillan brothers from Arran whom Maxwell knew personally. Julius Hare—married to F.D. Maurice's sister, Esther (1814–1864)—belonged to the broad-church party in the Church of England. He was an enthusiast for German liberal scholarship, having helped to introduce it into Britain. Though Maxwell was uncomfortable with some of the conclusions drawn by such scholarship—as seen in his reaction to D.F. Strauss, and in his conversation with Charles Robertson on the reliability of Mosaic history—his acquaintance with Hare's work meant he was well-informed on the subject, and he enjoyed Hare's pulpit discourses: 'I have been reading Archdeacon Hare's sermons, which are good.'[6]

Robert Chambers (1802–1871)

In August 1853, Maxwell and a group of friends holidayed at Coniston in the English Lake District, where they studied Robert Chambers' *Vestiges of Creation,* and Maurice's *Theological Essays.*[7] The book 'excited thought and talk'. This was when Robert Chambers was still reluctant to acknowledge his authorship of *Vestiges*. The book had caused controversy when first published in the 1840s, mainly because of its sympathy towards evolutionary concepts. Maxwell was familiar with the ideas behind Chambers' work, but for several of his companions these were new.

Thomas Erskine of Linlathen (1788–1870)

In a December 1853 letter to his father, Maxwell referred to Thomas Erskine of Linlathen. Erskine, who influenced F.D. Maurice, was part laird and part theologian. Though raised Episcopalian, in adult life he became involved with the Church of Scotland and Scottish

---

6 JCM, *Letter to Miss Cay,* 11 March 1853, (C&G1, 184).

7 JCM, *Letter to Litchfield,* 23 Aug 1853, (C&G1, 191).

Congregationalism. Maxwell's father knew the Erskine family in Edinburgh through the Clerks of Penicuik and had visited them socially.[8] He recalled that Thomas and his mother had lived at 30 Heriot Row, and that Thomas had come to the Edinburgh Bar a year before himself. Personal connections between families, such as the Clerks and the Erskines, are to be kept in mind when considering Maxwell's readiness to read books which some modern evangelical Christians might regard as liberal. With the likes of Erskine, Maxwell was not reading the works of total strangers. They were books written by friends, by friends of family, or by friends of friends. Personal connections created an extensive circle of authors.

Thomas Erskine and John MacLeod Campbell tried to change Scottish Calvinism to something gentler from what it was perceived to be at the time. Though it was not until 1856 that MacLeod Campbell published his influential work *The Nature of the Atonement*, by 1853 his views and those of Thomas Erskine were well known, especially their rejection of the Limited/Particular Atonement doctrine that on the cross Christ only died for the elect. Erskine's conviction that Christ died for all, and not only for the elect, eventually led him to believe in the universal salvation of all humanity.

Erskine and MacLeod Campbell concentrated on the *filial* rather than the *legal* as the primary relationship between God and humanity. That was what Thomas Crawford, Maxwell's Edinburgh minister, also emphasised, and which Crawford later debated publicly with Robert Candlish. Crawford's emphasis on the *filial* may have been expressed, directly or indirectly, in sermons which Maxwell heard as schoolboy and student in Edinburgh.

John MacLeod Campbell (1800–1872)

We know that Maxwell read John MacLeod Campbell's *The Nature of the Atonement*, though we do not have a full account of his reaction. All we have is a footnote by Lewis Campbell: 'About the year 1860 I remember discussing with him J. Macleod Campbell's book on the Atonement, which had lately appeared.' [9] It may have been Campbell himself who prompted Maxwell to read the book, since

8 John Clerk Maxwell, *Letter to JCM*, 16 Dec 1853, (C&G1, 194).

9 C&G1, 431.

Campbell was close to Benjamin Jowett who in turn was firm friend to both Erskine of Linlathen and MacLeod Campbell.[10] Concerning Maxwell's thoughts on MacLeod Campbell's book, Lewis Campbell's comment was: 'He made some criticisms, which I have forgotten; but I remember the emphatic tone in which he said, "We want light."' Maxwell's lukewarm reaction may indicate discomfort with some of MacLeod Campbell's theology.

MacLeod Campbell was being read by other scientists. At the 1870 Meeting of the British Association, when Maxwell was President of the Mathematics and Physics Section, Professor George Rolleston (1829–1881) referred admiringly to *The Nature of the Atonement,* though mainly touching on aspects of scientific and theological methodology.[11]

Thomas Browne (1605–1682)

After Maxwell died in 1879, 'Freshman' Tayler told Lewis Campbell that Sir Thomas Browne's *Religio Medici* (*Religion of a Doctor*) had been one of his favourite books: 'Any such author, who propounded his speculations in a quaint, original manner was sure to be a favourite with him. But I particularly remember his attraction to Sir Thomas Browne during the long vacation, when he was laid up with severe illness (a brain fever) in my dear uncle's house in June 1853.'[12]

---

10 Lewis Campbell, *Life of Benjamin Jowett,* Vol. 1, 266, wrote: '[Jowett] repeated his visits to . . . friends in Scotland, amongst whom was Thomas Erskine of Linlathen, a friend of Bishop Ewing's. He was a thoughtful mystic, of great liberality of mind, and after retiring from the Scottish Bar, had written several theological essays, which had a powerful influence in promoting the reaction against Calvinism. He was an admirer of William Law, and F. D. Maurice acknowledged himself to have derived something from him. . . . [Erskine] was the sworn brother in matters spiritual of J. Macleod Campbell of Row . . . Campbell's book on the *Atonement*, in which he sought by metaphysical argument to eliminate the crudities of current theology, and justify the ways of God, appeared in 1856, a year after Jowett's *St. Paul*, and about the same time Jowett's annual visits to Linlathen began.'

11 Rolleston, 'Presidential Address: Biological Section', *Report of the 40th Meeting of the BAAS* (1870), 96.

12 G.W.H. Tayler, *Letter to Campbell,* 4 March 1882, (C&G1, 173).

Maxwell's interest in Browne may have been sparked by Professor George Wilson of Edinburgh. We know that Wilson gave a copy of *Religio Medici* to Archibald Geike, another of his protégés; and the title of one of Wilson's own books, the posthumously published *Religio Chemici*, was a play on the title of Browne's work.[13]

Browne's *Religio Medici* had been a best-seller across seventeenth-century Europe. He composed his book around the virtues of *Faith* and *Hope*, emphasising traditional doctrines such as: salvation by faith alone, the existence of hell, the last judgment, and the resurrection. When Archibald Geike wrote about *Religio Medici,* he focussed on Browne's stress on the kindliness of a guiding providence. Though we do not know what Maxwell enjoyed in Browne, his interest may have been sparked by Browne's use of scientific imagery to illustrate religious truths, or by Browne's treatment of the relationship of science to religion as understood in his day.

Around 1854, Maxwell read George Berkeley's *Minute Philosopher*, commenting that it had 'some very bad Political Economy and much very good thinking on more interesting subjects. Paradox is still sought for and exaggerated. We live in an age of wonder still.'[14] After 1854, his more serious authors included the afore-mentioned Benjamin Jowett (Lewis Campbell's mentor), Charles Ellicott, Thomas Carlyle, Geoffrey Chaucer, Francis Bacon, Alexander Pope, Oliver Goldsmith, William Cowper, Bishop Berkeley, Robert Burns, Isaac Taylor, John Ruskin, F.D. Maurice, and Richard C. Trench among others.[15] Several were poets. Ruskin and Maurice were advocates of Christian Socialism. Others wrote specifically religious works.

Benjamin Jowett (1817–1893)

Jowett was Anglican theologian, cleric, and translator of Plato and Thucydides. As Master of Balliol College, Oxford, he was a major figure, and we have already noted his influence on Lewis Campbell. At one time Jowett was linked romantically to Florence Nightingale

13 Geike, *A Long Life's Work,* 41.

14 JCM, *Letter to Litchfield,* 25 Mar 1854, (C&G1, 208; Harman, I, 247).

15 JCM, *Letter to Campbell,* 17 Oct 1855, (C&G1, 218).

of Crimea fame, but without the relationship progressing to marriage. As with Julius Hare, Jowett was a follower of liberal German theologians, especially Tubingen's F.C. Baur. Some of Jowett's views of the Bible were regarded within Church of England circles as near heresy, though these were 'balanced' by his other religious convictions being conservative and orthodox. Jowett was not the only scholar with a liberal attitude towards Scripture with whom Maxwell was acquainted. He never restricted himself to safe, middle-of-the-road, theology.

William Robertson Smith (1846–1894)

William Robertson Smith, the Scottish Old Testament scholar, adopted a modernist approach to Scripture which resulted in him being removed from his teaching post at the Aberdeen Free Church College in 1881.[16] Ten years prior to his dismissal, and before he wrote the controversial articles on Scripture for the ninth edition of the *Encyclopaedia Britannica* which brought about his downfall, Robertson Smith had been acquainted with both Maxwell and Peter Tait. As an accomplished mathematician as well as a noted Hebraist, Robertson Smith assisted Tait in natural philosophy at Edinburgh before his Aberdeen appointment in 1870. He also provided a Hebrew title to one of Maxwell's humorous poems.[17]

Charles John Ellicott (1819–1905)

The Anglican Charles Ellicott was professor of divinity at King's College, London before being appointed Hulsean Professor of Divinity at Cambridge. After serving as Dean of Exeter, he became Bishop of Gloucester (and briefly Bristol) where he had a decidedly anti-ritualistic reputation.[18]

Ellicott authored numerous books, mainly on New Testament issues or apologetics. He also chaired a committee which aimed at producing a fresh translation of the New Testament. In middle-age he was involved in a railway accident, injuring both legs—a calamity from which his limbs never fully recovered—but was compensated

16 Cf. JCM, *Letter to Tait*, 14 Nov 1870, (Harman, II/2, 577).

17 Knott, *Life and Scientific Work of Peter Guthrie Tait*, 71, 171.

18 Cf. *Westminster Gazette*, 17 Oct 1905.

by the Great Eastern Railway Company with a free ticket between Cambridge and London for life. According to accident reports, though Ellicott was himself injured he tried to minister to the dying whilst still clasping in his hands the manuscript of a commentary on which he had been working![19]

In the 1870s, Maxwell's relationship with Ellicott progressed beyond reading Ellicott's books. At that point the bishop engaged in lengthy correspondence, to clarify what Maxwell had meant in his paper 'Discourse on Molecules' of September 1873. Maxwell had made remarks which some interpreted as stating that modern science had validated the argument from design for the existence of God. Ellicott asked if that were what Maxwell really meant, and Maxwell responded with a fuller exposition of his views.[20] Ellicott's subsequent book, *Modern Unbelief, its Principles and Characteristics* (1877), drew on their correspondence. We examine the Ellicott/Maxwell correspondence in Chapter 25, 'Orderliness and Design.' Unlike Hare and Jowett, Ellicott was biblically conservative.

John Henry Newman (1801–1890)

Of the many religious authors whom Maxwell read, John Henry Newman had arguably the highest profile. Newman had been leader of the High-Anglican Tractarian movement but in 1845 he left the Church of England for the Roman Catholic Church, in which he eventually became a Cardinal.

Lewis Campbell recalled that Maxwell studied Newman's seminal 1845 work, *The Development of Christian Doctrine,* with great concentration if not full agreement.[21] Newman argued that a weakness of Protestantism was in not allowing for development of doctrine and liturgical practice under Divine guidance. For Newman, such 'spirit-led' developments legitimized Roman Catholic teaching on the role of Mary, interpretation of the sacraments, and the authority of the Papacy. He saw Protestantism as rooted in a simplistic reading of the Bible.

---

19 *Eastern Daily Press*, 17 Oct 1905.

20 C&G1, 392–396.

21 Ibid., 232, 462.

Maxwell was unimpressed. Concerning purgatory, one of the 'developed' doctrines, he commented:

> Mr Newman and his predecessors take up some single practice of Christians, and by means of analogies derived from the practices of Egyptian priests, Roman emperors, or Jewish rabbis, they determine, most precisely, the situation, extent, and exposure of the place of purging by fire, together with all the technicalities, observances, and etiquette of that mysterious region. The convolutions of the brain are very wonderful.[22]

Maxwell returned to Newman in 1878, when annotating his draft copy of Stewart and Tait's *Paradoxical Philosophy*. In a section in which Stewart and Tait referred to the doctrine of purgatory, Maxwell commented that, historically, the notion of purgatory had been utilised as a 'source of revenue', but added that—if simply taken as a scholarly account of the evolution of the idea of purgatory as emerging from the problem of post-baptismal sin—Newman's explanation in his *Development* paper was 'first rate'.[23] Maxwell's problem with Newman was that Newman believed it was perfectly valid for the Church to move *beyond* Scripture in developing doctrine. In contrast, a Presbyterian such as Maxwell regarded the limit to be determined by the *given* content of Scripture, and though the Church's understanding of Scripture might deepen, no doctrine which was not supported by Scripture could be valid.

What is striking about Maxwell's reading, before and after Otley, is its width. As he stated in 1852, he set no boundaries to what may be examined or questioned, even in matters of faith. True to this principle, he made himself familiar with a wide range of theological opinion.

---

22 JCM, 'Envelopment: Can Ideas be developed without Reference to Things as their developing Authorities?', 1854, (C&G1, 232).

23 NLS: Acc.10000/399, 'Balfour Stewart & P.G. Tait, *Paradoxical Philosophy* with annotations by JCM.', 178. Maxwell alluded to Newman's treatment of the same issue in his 5 Feb 1878 paper 'Psychophysik', (C&G1, 462; Harman, III/2, 606).

Chapter 11
Fellow and Researcher

The Tripos examinations took place in January 1854. From distant Glenlair, John Clerk Maxwell assisted his son as much as he could, offering practical counsel for cold winter mornings: 'You will need to get muffetees for the Senate-Room. Take your plaid or rug to wrap round your feet and legs.'[1] Maxwell passed with flying colours, graduating BA later that year. He now had to consider the future. He loved Glenlair. At some point he would like to spend the bulk of his time there. But at present his father was in charge. So he stayed at Trinity, took pupils, studied the work of Michael Faraday, and dug deeper into optics and colour theory which were his favourite areas.[2]

He read Faraday's *Experimental Researches* in detail, thinking through how he might express Faraday's experimental insights in mathematical form.[3] William Garnett, in his tribute to Maxwell in *Nature,* wrote that Faraday found in Maxwell a mind constituted after the same plan as his own, but with the advantage of a mathematical training, and with the ability to interpret his ideas to the mathematical world.[4] Throughout his career, Maxwell had the knack of bringing together diverse scientific insights and expressing them mathematically.

---

1 John Clerk Maxwell, *Letter to JCM,* 30 Dec 1853, (C&G1, 195).

2 Cf. C&G1, 198: 'When actually emancipated he seems to have reverted principally at first to his beloved optics. He makes inquiries about a microscope manufactory at Zurich; reads Berkeley's *Theory of Vision* taking up Mill's *Logic* by the way, and finding there by no means the last word on the relation of sense to knowledge; looks up his stock of coloured papers furnished by D. R. Hay, and sets to work spinning and weaving the different rays; inquires [about] colour-blind persons . . . and invents an instrument for inspecting the living retina, especially of dogs. By and by it is the art of squinting which again has charms for him, and he combines it with the teaching of solid and spherical geometry, by drawing wonderful stereoscopic diagrams.'

3 J.J. Thomson, 'James Clerk Maxwell', *A Commemoration Volume,* 11. Cf. Tait, 'Maxwell', *Encyclopaedia Britannica,* (9th ed.), Vol. 15, 64, & Forsyth, 'The Old Tripos Days at Cambridge', 169.

4 Garnett, 'James Clerk Maxwell', *Nature,* Vol. 21/524 (13 Nov 1879), 45.

In October 1854, he made his first application to become a fellow of Trinity. This was unsuccessful, but he was elected a year later in October 1855. Though his second attempt, it was still sooner than the norm as it usually took at least two years for a fellowship to be granted. As a fellow he had to remain unmarried. Revised statutes came into force in 1878, allowing fellows of colleges to marry without losing their standing. Until then, with a few exceptions, only heads of houses and professors could marry. The restriction was still in place when Maxwell married Katherine Dewar in Aberdeen in 1858, at which point he surrendered his fellowship, though he was re-elected as an honorary fellow in 1872.

The fellowship carried teaching responsibilities. For Maxwell, these involved preparing lectures on hydrostatics and optics and setting examination papers. To make time for his own research he stopped taking private pupils, though he did continue with his commitment to the Working Men's Institution. It was suggested he might prepare a textbook on optics, and this progressed as far as making initial drafts. His friend Cecil James Monro (1832–1882) proposed that he might translate Newton's *Principia Mathematica*, with a suitable commentary.[5] Andrew Motte (1696-1734) had translated Newton's work into English from Latin, but it was thought that a fresh version would be useful. Maxwell was interested, but the project faded from sight.

Colour Theory

In March 1855, Maxwell presented a paper to the Cambridge Philosophical Society on the 'Transformation of Surfaces by Bending.'[6] The same month he travelled to Edinburgh, lecturing on experiments in colour to the RSE, during which he demonstrated, by means of his own carefully designed colour-top, that white light resulted from a mixture of red, green, and blue light. Two months later he showed the same 'colour trick' to the Cambridge Society.[7]

---

5 JCM, *Letter to C.J. Monro,* 7 Feb 1855, (Harman, I, 280; Cf. C&G1, 204).

6 JCM, 'On the Transformation of Surfaces by Bending': *TCPS*, Vol. IX. Part IV (IV). (Harman, I, 239–241).

7 JCM, *Letter to John Clerk Maxwell,* 1 May 1855, (C&G1, 211).

Maxwell's insistence that red, green, and blue are the primary components of white light is now accepted, but that was not always so. That was evident at the 1855 British Association gathering in Glasgow. At that meeting Maxwell was a rookie compared to the established scientific figures of the day, but he was present when Sir David Brewster poured scorn on William Whewell's optical theories. Whewell was the much-respected Master of Trinity—effectively Maxwell's boss—but Whewell and Brewster had clashed frequently since the late 1830s.[8]

Their most recent dispute came after Whewell published *On the Plurality of Worlds* (1853), in which he argued that there were no grounds for believing in higher intelligences on worlds other than this. Brewster, a licensed preacher and intense evangelical, had a slate of reasons for believing the contrary. He responded with *More Worlds than One: The Creed of the Philosopher and the Hope of the Christian* (1854), denouncing Whewell's book.

Though Charlotte Brontë had found Brewster charming when he accompanied her through the scientific section of the Great Exhibition of 1851, Brewster had a prickly personality and an explosive temper. Raised in the Scottish Border town of Jedburgh, with its strong connections to Mary Somerville (1780-1872),[9] his brilliance took him to international prominence. In temperament he was instinctively anti-establishment and seceded from the Church of Scotland at an early stage, joining the Free Church soon after its launch in 1843.

8 According to Brewster's daughter, Margaret Gordon, in her *The Home Life of Sir David Brewster,* 140f, Whewell and Brewster made up their differences. Gordon acknowledged her father's temper and intemperate speech. Whewell and Brewster's conflict was so well known that Trollope referred to it in *Barchester Towers* (1857): the character Charlotte Stanhope introduced the topic, citing Whewell and Brewster directly, in an attempt to initiate conversation between her impecunious brother and the wealthy young widow Eleanor Bold.

9 Maxwell knew of the mathematician Mary Somerville (1780-1872), referring to her in his 21st February 1873 Royal Institution paper 'On Action at a Distance' (*Proceedings of the Royal Institution of Great Britain,* Vol. VII, 49). It was in reviewing her 1834 book, *On the Connexion of the Physical Sciences,* that Whewell used the term 'scientist' which in time replaced the term 'natural philosopher'. In February 2016, Somerville was listed by the Royal Bank of Scotland, along with Maxwell and civil engineer Thomas Telford, as a person whose face might be on the bank's new £10 notes. She was chosen.

Personality traits, as much as intellectual reasons, lay behind his antipathy to the likes of Whewell.

Brewster adhered to the older view that the eye had receptors for red, blue, and yellow. In his Glasgow lecture, Brewster continued to expound these convictions, mockingly suggesting that Whewell might book himself into a sanatorium run by Professor Wartman of Geneva which specialised in colour-blindness! It was a cruel jibe, especially as Whewell was present. Though Maxwell disagreed with Brewster, he stayed quiet. He was a young graduate, whereas Brewster, fifty years older, had an international reputation and was the current doyen of optical theory. Their personalities were also different. Brewster was caustic, confrontational, and combative. Maxwell was the opposite. Moreover, five years earlier, at the 1850 British Association meeting in Edinburgh, Maxwell had found that he struggled to present his thoughts fluently in speech when he stood up to comment.

Maxwell wrote to his father, telling him that Whewell sat through Brewster's lecture but, when Brewster finished his talk, Whewell left the room to avoid a public quarrel.[10] On the same day, Maxwell demonstrated his colour-top at an afternoon fringe-event in the house of Professor William Ramsay (1806–1865), illustrating his conviction that white light came from red, blue, and green, rather than from Brewster's red, blue, and yellow. Brewster had also been invited to the occasion but did not appear. The events were recorded in a letter from James' cousin, William Dyce Cay, to James' father:

> Sir David Brewster was upon the triple spectrum. As far as I can understand, he believes the spectrum to be composed of three colours red, blue, and yellow; and that the intermediate colours are composed of mixtures of these, as, for example, the green from a mixture of blue and yellow, which, I think, is different from what James believes. James did not say anything in the controversy which followed [Brewster's] speech, as he was to meet Sir David Brewster at the Ramsay's afterwards, where he would have his top and other apparatus to show him.[11]

---

10 JCM, *Letter to his Father*, 24 Sep 1855, (C&G1, 216; Harman, I, 325) & JCM, *Letter to C.J. Monro*, 24 July 1858, (Harman, I, 596).

11 W.D. Cay, *Letter to John Clerk Maxwell*, 18 Sep 1855, (C&G1, 215).

Brewster's treatment of Whewell was uncalled for. Whewell was a man of standing, not only in science but in the wider academic community, and Maxwell disapproved of how his Master at Trinity had been treated. A few years later, when Maxwell wrote a paper on the composition of light, he carefully avoided any direct statement about Brewster's views.[12]

As well as differing from Brewster on the primary colours, Maxwell also diverged on what colour actually was. He wrote to George Stokes: 'I think [Brewster] regards the three elements of colour as objective realities or actual substances of which the light is compounded. I consider them as properties of homogeneous rays having reference to the human organ of vision whether dichromic or trichromic.'[13] For Maxwell, colour was an *effect* on the human eye caused by rays of light. To put it in later terminology: colour was an *effect* caused by specific wavelengths of the electromagnetic spectrum.

Faraday's Philosophy

Before we reflect on Maxwell's important 1855/56 double-paper, 'On Faraday's Lines of Force', we look at Michael Faraday, whose intuitions lay behind it. Maxwell's achievements cannot be separated from Faraday's insights, a fact which Maxwell himself always acknowledged. He had enormous respect for Faraday, and wrote the 'Faraday' entry for the ninth edition of the *Encyclopaedia Britannica*.[14]

Like Maxwell, Faraday was a devout Christian believer. Whereas Maxwell had roots in the Church of Scotland, Faraday belonged to the Sandemanian Church, which had grown from the Glassite movement, itself a secession from the Kirk in the first half of the eighteenth century. Among the tenets of their faith, Glassites believed that the Church should be visibly independent of the State, and this was one of the reasons why Faraday turned down some State honours and privileges later in life. He was twice elected an elder of the London branch of his denomination.

Faraday's science was inextricably rooted in his religious faith, especially his belief that the various forces observed in nature—such

12 JCM, *Letter to Stokes,* May/June 1860, (Harman, I, 657).

13 Ibid..

14 JCM, 'Michael Faraday', *Encyclopaedia Britannica* (9th ed.), Vol. 9, 29–31.

as electricity, magnetism, and gravity—were all manifestations of one single fundamental force, which came from one single Creator. Some of his most creative scientific thinking centred around that conviction. Though Faraday was not a trained mathematician, and hence unable to give his ideas a solid mathematical basis, he was an exceptional theoretician, whose ideas became Maxwell's seed-bed.

The Royal Institution in London, of which Faraday was Director from 1825 until 1861, hosted regular Friday-evening public lectures, and it was at one of these, on Friday the tenth of April 1846, that Faraday shared speculative thoughts on energy, matter, light, magnetism, and lines of force. The circumstances surrounding his lecture were unusual. On the Friday in question, Charles Wheatstone (1802–1875) was due to lecture on his electrical device, the Wheatstone Bridge, but the timid Wheatstone turned tail and fled when he learned that a well-known barracker was in the audience. This left a packed audience with no speaker. There was huge relief when Faraday took over. He first described to the audience what he thought his friend Wheatstone would have said about his Bridge. But a full hour had to be filled. So Faraday talked off-the-cuff about some of his ideas on light, the ether, magnetism, and gravity.

The audience thoroughly enjoyed the lecture, but Faraday worried about having shared personal conjectures without solid evidence. A few days later, he wrote to Richard Phillips (1778–1851), Secretary to the Royal Institution.[15] The overt purpose of his letter was to give Phillips an approximate record of what he had said extemporaneously, but he also apologised for having been so speculative. He explained why he had spoken so unguardedly: 'I do not think I should have allowed these notions to have escaped from me, had I not been led unawares, and without previous consideration, by the circumstances of the evening on which I had to appear suddenly and occupy the place of another.' Faraday knew he had expressed thoughts which, as a scientist, he still had to establish.

Faraday voiced notions which had been swirling in his mind for some time, and it was these creative insights which would feed

15 Faraday, 'Thoughts on Ray-Vibrations', *Letter to Richard Phillips, Secretary,* 15 Apr 1846 (published: *The London, Edinburgh, and Dublin Philosophical Magazine and Journal of Science,* Vol. 28/188, 345–350).

into Maxwell's thinking in the decades which followed. During his impromptu lecture, Faraday shared a list of mind-bending thoughts with the Royal Institution audience. For example, he was prepared to dispense with the concept of the ether—something which Maxwell never fully did—and he suggested that lines of force proceeded through space without needing particles of matter or ether to carry them: 'I do not perceive in any part of space, whether (to use the common phrase) vacant or filled with matter, anything but forces and the lines in which they are exerted.'[16] He went further:

- Atoms may not be 'stuff', but may simply be centres of force.
- Electricity may not be instantaneous, but may have an actual velocity as great as that of light.
- Through gravitation even the smallest atom on earth may act directly on atoms in the sun, through lines of force 'extending from and belonging to each', despite being 90 million miles apart.
- Lines of force associated with atoms may be the medium by which light waves are propagated.
- There may be several types of lines of force: gravitation, electrostatic-induction, magnetic-action, electricity etc. (Maxwell unified these—apart from gravitation—as deriving from the one electromagnetic spectrum.)
- Radiation may only be a species of vibration in the lines of force which connect all particles and masses of matter.

Faraday closed his lecture by setting the ether-model and the lines-of-force-model in opposition.[17] He now viewed space as full of gravitational, magnetic, electrical lines of force, and speculated that *matter* was in essence simply *force*.

Years later, Maxwell's set of differential equations would give scientific credibility to Faraday's ideas, but it was Faraday who first thought the thoughts. Maxwell's brilliance was in grasping Faraday's

16 Faraday, 'Thoughts on Ray-Vibrations', 349.

17 Cf. Tyndall, *Faraday as a Discoverer,* 119ff. Like the theist Maxwell, the materialist John Tyndall was a huge admirer of Faraday.

mind, in being open to his whole new way of seeing reality, and in being able to express everything mathematically.

The unscheduled lecture of April 1846 was Faraday's boldest presentation of his embryonic ideas, though there had been other occasions when he had hinted where his thoughts were going. In 1845 he had demonstrated that light which travelled through heavy glass had its plane of polarization rotated if a magnetic field were applied to the glass, proving a connection between light and magnetism.[18] And in 1844 he had already been contemplating new ways of understanding matter, suggesting that matter filled all space, or, at least, 'all space to which gravitation extends' since gravitation was a 'property of matter dependent on a certain force, and *it is this force which constitutes the matter*.'[19] (Italics ours).

Maxwell's 'On Faraday's Lines of Force'

Maxwell immersed himself in Faraday's work, as well as following up suggestions made by William Thomson of Glasgow.[20] After checking that he was not trespassing on Thomson's active research area, Maxwell then wrote his remarkable double-paper 'On Faraday's Lines of Force' over the winter of 1855/56.[21] In it he compared the behaviour

18 Faraday, 'The Magnetization of Light and the Illumination of the Lines of Magnetic Force' (read 20 Nov 1845), *Philosophical Transactions of the Royal Society,* Vol. 136, Dec 1846, 1–2: 'I have long held an opinion ... that the various forms under which the forces of matter are made manifest have one common origin; in other words, are so directly related and mutually dependent, that they are convertible, as it were, into one another, and possess equivalents of power in their action [including light].'

19 Faraday, 'A Speculation touching Electric Conduction and the Nature of Matter', *Letter to Richard Taylor,* 25 Jan 1844, 143. The paper is found in *The Philosophical Magazine,* Vol. 24/157, 136–144.

20 JCM, *Letter to Thomson,* 13 Sep 1855: (Harman, I, 319–324, esp. 322f.).

21 Siegel, 'Maxwell's Contributions to Electricity and Magnetism', [Flood (ed.), *JCM: Perspectives on his Life and Work*], 188: 'Maxwell had three goals: first, to develop a theory that would encompass all the known facts of electricity and magnetism; second, to accomplish this unification by using Michael Faraday's approach to electricity and magnetism–the *lines of force* or *field* approach; third, to have the new theory go beyond what was known from experiment.' The third references Maxwell's use of 'induction' rather than 'deduction': antithetical to positivism.

of lines of force to the flow of a liquid. He also constructed equations for known electric and magnetic effects, though he did not yet link these to variations in magnetic fields.[22]

Tait and Garnett claimed that Maxwell had in fact completed most of the work on this whilst still an undergraduate. Tait wrote that he had seen Maxwell's manuscript, in a form little different from the final version, a year before Maxwell took his degree.[23]

Whatever its history, the double-paper would lead Maxwell, in the 1860s when he was at London's King's College, to more advanced submissions on electromagnetic forces. But it was this early Cambridge work which prepared the way for four of his future permanent contributions to science. These were: his comprehensive theory of electricity and magnetism; his discovery that the electromagnetic spectrum *included* light as one segment of it; his more advanced equations on electromagnetism; and his move away from an atomistic view of physical reality to viewing it as a *continuum* or *universal plenum* involving *fields of force*.[24] At all times he acknowledged Faraday's legacy.

The underlying problem which Faraday and Maxwell sought to solve was that of *action at a distance*. One of the basic questions in science is: how can one object exert force on another object?[25] The simplest scenario is when one object touches another object

22 JCM, 'Faraday's Lines of Force': *TCPS*, Vol. X. Part I. (VIII). (Draft: Harman, I, 337–352; Abstract: Harman, I, 353–369, 371–375). Cf. Lindley, *Degrees Kelvin*, 200.

23 Tait, 'James Clerk Maxwell', *Encyclopaedia Britannica*, (9th ed.), Vol. 15, 647. Garnett, *Heroes of Science: Physicists*, 283.

24 In his Cambridge Inaugural Lecture (1871), Maxwell used the term 'universal plenum' to describe what might now be termed a 'continuum'. Cf. JCM, *Introductory Lecture on Experimental Physics*, 20: 'Two theories of the constitution of bodies have struggled for victory with various fortunes since the earliest ages of speculation: one is the theory of a universal plenum, the other is that of atoms and void. The theory of the plenum is associated with the doctrine of mathematical continuity, and its mathematical methods are those of the Differential Calculus, which is the appropriate expression of the relations of continuous quantity.'

25 For overview see: Hesse, *Forces and Fields. The concept of Action at a Distance in the history of Physics.*

and pushes it. That was the scenario envisaged by Newton's laws of motion.[26] Though in the 1870s, at another of the Royal Institution's Friday evening lectures, Maxwell would question whether any two objects *ever* touch![27] Putting that aside for the moment, what about objects which are clearly separate from one another? In these instances, how do bodies act on one another? Sun, earth, and moon exert the force of gravity on one another, even though they do not touch. Magnets exert a force on certain metals without touching them. Electricity has similar properties. These all involve *action at a distance.*[28]

As an aside, we note that, in the parlance of the time and as understood by Maxwell himself, to believe in *action at a distance* indicated belief that remote bodies also acted on one another *immediately*, without the intervention of anything else.[29] Additionally, *action at a distance* generally assumed that the space between a magnet and the object it attracted was empty space.

Faraday questioned the current assumptions. He argued that *action at a distance*, understood in that sense, was unable to explain electric and magnetic forces. Instead, he proposed that *lines of force* filled the space. Under this model, the intervening space was not empty but was occupied; and the action of a magnet, for example, was through lines of force, whatever these were! Faraday pondered the patterns made by iron filings when sprinkled on a sheet of paper and influenced by a magnet, and it was these patterns which he termed 'lines of force', and which made him think that any space containing such lines of force was in a special state.[30] That was what Maxwell picked up on, as he explained in his 1873 *Treatise on Electricity and Magnetism*:

---

26 Hesse, *Forces and Fields*, 189.

27 : JCM, 'On Action at a Distance', 47. (Royal Institution Lecture, 1873).

28 Until the seventeenth century, some theoreticians gave light an infinite velocity, impacting instantaneously from its source. Others argued its speed was finite. The belief that light had infinite speed encouraged the idea of instantaneous action at a distance.

29 JCM, 'On Action at a Distance', 44.

30 Randall, 'Aspects of Maxwell's Life and Work', *Clerk Maxwell and Modern Science*, 8. Cf. Peierls, 'Field Theory since Maxwell', *Clerk Maxwell and Modern Science*, 26.

> Faraday, in his mind's eye, saw lines of force traversing all space where the mathematicians saw centres of force attracting at a distance: Faraday saw a medium where they saw nothing but distance: Faraday sought the seat of the phenomena in real actions going on in the medium, they were satisfied that they had found it in a power of action at a distance impressed on the electric fluids.[31]

It was later mooted that *lines of force* indicated the existence of a *field of force* which had a continuous action, and which acted even if objects were not obviously contiguous. R.V. Jones (1911–1997), one of Maxwell's successors at Aberdeen, stressed that what Maxwell set out to do in his 1855/56 double-paper was to take Faraday's pictorial idea of lines of force and give them mathematical expression.[32] Maxwell succeeded in this, and in doing so he recognised a new force between bodies other than gravity. This was *electromagnetism*.

In 1855/56, Maxwell presented his thinking to the Cambridge Philosophical Society over two evenings. Part One was on Monday, the tenth of December 1855. Part Two was on Monday, the eleventh of February 1856.[33]

In the run up to the first lecture, Maxwell was exceptionally busy, with student exams taking up much of his time. Nevertheless, he was able to get his paper ready.[34] Two papers were usually read at each meeting, and Maxwell was preceded by Robert Leslie Ellis (1817–1859) who spoke on, 'Napier's solution of right-angled spherical Triangles.'[35] Maxwell completed his presentation at the February meeting.

Part One revealed Maxwell's love of analogies in nature, and his predilection for pictorial thinking. He saw analogies between lines of force on the one hand, and streamlines in an incompressible fluid on the other. He also reflected on the close parallel between electric currents and magnetic lines of force as observed by Faraday. He later

31 JCM, *A Treatise on Electricity and Magnetism,* Vol. 1, x.

32 Jones, 'The Complete Physicist: James Clerk Maxwell', 9.

33 NVN1, xix–xxiii, considers the implications of this paper. For the paper itself, see: NVN1, 155ff.

34 Cf. JCM, *Letter to his Father,* 3 Dec 1855, (C&G1, 221; Harman, I, 336).

35 *Cambridge Chronicle & Journal*, Saturday 15 Dec 1855. Cf. Walton (ed.), *Mathematical and other Writings of Robert Leslie Ellis,* 328–335.

realised that his use of analogies could be confusing, and needed to stress that the analogy between lines of force and incompressible fluids was meant to be *illustrative*: i.e. the similarity was fortuitous, with no essential connection between them.

In Part Two, Maxwell looked further at electromagnetic theory, setting out mathematical equations which he had developed. Part of his success came from his ingenious adaptation of mathematical techniques used by William Thomson at Glasgow in relation to the flow of heat, with Maxwell applying them to lines of force.[36]

Taken as a whole, the double-paper addressed a range of issues. It described aspects of Faraday's work. It discussed how electricity and magnetism were interconnected. And it presented an initial set of equations which could account for nearly all known theory at that time. His 1855/56 work was the catalyst for his future insights.

Faraday was grateful to Maxwell.[37] One of Maxwell's aims had been to show that Faraday's *lines of force* could be expressed mathematically, and that they were not just a pretty artistic model, as some had regarded them. John S. Reid (1942–2021) pointed out that Faraday's ideas had been dismissed by several researchers on the Continent as 'mere hand-waving metaphor, devoid of depth and, ultimately, reality.'[38] That was why Maxwell wrote:

> An impression still prevails that there is something vague and unmathematical about the idea of lines of force, and that they will inevitably lead astray any student who is not guided by that familiar acquaintance with the laws of nature which distinguishes [any] great electrical philosopher. To illustrate the mathematical character and scientific usefulness of lines of force I have developed certain of their properties in a series of propositions which I have proved in full without the introduction of technical mathematical expressions.[39]

---

36 Reid, 'Maxwell at King's College, London', [Flood (ed.), *JCM: Perspectives on his Life and Work*], 61. Flood himself in his 'Introduction' (13), points out that, in Part Two, Maxwell also made extensive use of Stokes' Theorem, which he studied at Cambridge as an undergraduate.

37 Faraday, *Letter to JCM*, 25 Mar 1857, (C&G1, 519f).

38 Reid, 'Maxwell at King's College, London', 60.

39 JCM, *'Abstract of Paper on Faraday's Lines of Force'*, 10 Dec 1855, (Harman, I, 353f).

Maxwell's double-paper kick-started a change in the way he thought about reality. As early as June 1856, Maxwell intimated that he was beginning to think less in terms of material objects in space, and more in terms of lines, or fields, of force—though he was not yet thinking in terms of a *force-field*.[40] Maxwell was following Faraday. At the same time, he had not yet broken free from atomistic thinking, and he did not yet see physical reality in terms of a *field-continuum*. Eventually he would conceive of a *field*, not as a collection of particles distributed throughout the *field*, but as a radically new type of reality, later termed by others as a *continuum*.

But there was a problem. Maxwell's favourite method of explaining an idea—by citing analogies from other areas of science—was creating confusion. Seeing similar patterns in nature had always appealed to Maxwell, and, as we have seen, in Part One he had posited an analogy between electrical/magnetic phenomena and fluids affected by heat. Some interpreted the analogy too literally despite Maxwell's warnings.[41] In later papers he set aside that method of illustrating the subject and appealed more to the purely abstract mathematical notion of a *field* in order to clarify the issues.

Maxwell had also not solved the dilemma of *why* objects affected one another through *action at a distance*, which gravity, magnetism, and electricity seemed to demand. He had produced equations which explained and predicted. But how and why? A favourite solution (accepted by Maxwell at the time) was that the 'ether' filled empty space, touched both objects, and was the medium through which one object could exert a force on another. There might even be more than one ether: a *gravitational ether* enabling gravitation to cross the gap from the earth to the moon; and a *luminiferous ether*, allowing light to travel through the vacuum of space. Yet, despite

40 Cf. JCM, *Letter to Henry Cook*, 6 June 1856, (Harman, III/2, 869): 'I think more of the distribution of lines of force in space, and less of any distinct centres to which they are supposed to tend. In fact, in many cases, including all cases of electromagnets, the lines of force return into themselves and do not run towards any centre whatever.'

41 Cf. Harman, I, 371. Maxwell stated in Part Two: 'It was shown in the first part of this paper, that electrical and magnetic phenomena present a mathematical analogy to the case of a fluid whose steady motion is affected by certain moving forces and resistances. (*the purely imaginary nature of this fluid has been insisted on.*)' (Italics ours).

retaining notions of the ether, Maxwell had laid the groundwork for a new way of understanding reality. After 1855/56, he thought through things even further, with Lewis Campbell recalling him expounding extraordinary ideas during a carriage drive at Glenlair in 1857:

> His interest in Faraday's investigations must have dated from a very early time, certainly before 1849. And now he sought to give to these speculations, at which his imaginative mind had long been working, precise mathematical expression. I wish I could recall the date (1857?) of a drive down the Vale of Urr, during which he described to me for the first time, with extraordinary volubility, the swift, invisible motions by which magnetic and galvanic phenomena were to be accounted for. It was like listening to a fairy-tale.[42]

Even with his forays into Faraday's work, he was still thinking *mechanically* rather than in terms of a *field.* [43] That shift came later, even though a radically new way of seeing things had already been suggested by the likes of John Herschel (1792–1871) who, in his presidential address to the British Association in 1845, had prophesied: 'The time seems to be approaching when a merely mechanical view of nature will become impossible.'[44]

Herschel's hint fell on unprepared soil in the 1840s. It was Maxwell's work which proved to be a game-changer, with his publication *On Physical Lines of Force* (1861) a major step towards a new conception of physical reality.[45]

---

42 C&G1, 199.

43 Harman, I, 19.

44 Herschel, 'Presidential Address', *Report of the 15th Meeting of the BAAS (1845),* xli. Rowlinson, 'Maxwell and the theory of liquids', in Flood (ed.), *JCM: Perspectives on his Life and Work,* 155, pointed out that scientists of the 1840s and 1850s *did* question a purely molecular way of explaining physics; however, the growing acceptance of the real existence of atoms and molecules militated against continuum theories.

45 John Tyndall suggested to Maxwell that a whole new way of thinking about reality might be necessary. Cf. Tyndall, *Letter to JCM,* 7 Nov 1857 (C&G1, 288): 'I never doubted the possibility of giving Faraday's notions a mathematical form, and you would probably be one of the last to deny the possibility of a totally different imagery by which the phenomena might be represented.'

The Use of Analogy

When the Revd George Walker (1842–1919) reminisced on Maxwell for his local newspaper, he wrote that Maxwell's mind was so subtle that he could see a 'host of hidden analogies', and that he used them so frequently in conversation that he could be hard to follow.[46]

Analogical thinking came naturally to Maxwell. As a boy he had loved to see similarities in nature. Added to that early inclination, Sir William Hamilton's emphasis on knowing things in their *relations* meant that he was as aware of similar patterns of relationship between entities, as of similarities between the entities themselves. Thus, in his posthumously published *Elementary Treatise on Electricity*, Maxwell wrote that when thinking analogically he worked with similarity between *relations*, rather than similarity between the *things related*: 'The similarity which constitutes the analogy is not between the phenomena themselves, but between the relations of these phenomena.'[47]

For Maxwell, analogy cross-fertilised diverse fields of enquiry, and served as a stepping-stone between pictures in the mind and abstract thought. Yet, precisely because analogy had to do with *partial resemblance* rather than *identity*, it had limits. So, what was an analogy? Was an analogy relevant if not ultimately rooted in an identity? Were analogies fortuitous? Did analogies work because they emerged from something common across all being?

Maxwell distinguished between three types of analogy: *illustrative* analogies, *objective* analogies, and *false* analogies. *Illustrative* analogies were when the similarity was purely accidental. *Objective* analogies arose when there was a common factor deeply embedded in reality. *False* analogies ensued when pseudo-scientists hijacked science to provide 'proof' for their wild theories.

To explain his methods to others—and perhaps also to clarify them to himself—he spoke on 'Analogies in Nature' to the *Apostles* in February 1856.[48] This was soon after he had presented Part Two

46 Walker, 'James Clerk Maxwell', *Galloway News & Kirkcudbrightshire Advertiser*, 14 May 1883.

47 JCM, *Elementary Treatise on Electricity*, Section 64 (p.51):

48 JCM, 'Analogies in Nature: Essay for the Apostles' Club', Feb 1856, (C&G1, 235–244; Harman, I, 15, 376–383).

of his double-paper to the Cambridge Philosophical Society, and after becoming aware of misunderstandings arising from his use of analogies in it.

The 'Analogies in Nature' paper is complex, not just because of its tight reasoning, but because Maxwell covered a wide range of analogous phenomena, including laws of nature, laws of thought, and laws of morality. What he argued was that there are only a limited number of types of relation at all logically possible in thought or in reality: therefore, because that number of possibilities is finite, patterns of relation are necessarily repeated, and objective analogies are inevitable.

Neither Sir William Hamilton nor Maxwell were so Kantian that they denied any possibility of knowing *things in themselves,* but both were aware that our primary knowledge of the natural world is of how things interact with ourselves and with one another.[49] Thus, though *objects* are seldom analogous to one another, *patterns of relationships* can be, and analogy usually emerged in that realm of interactions:

> Whenever [men] see a relation between two things they know well, and think they see [that] there must be a similar relation between things less known, they reason from the one to the other. This supposes that although pairs of things may differ widely from each other, the *relation* in the one pair may be the same as in the other. . . . The relation is the most important thing to know, a knowledge of the one thing leads us a long way towards a knowledge of the other. If all that we know is *relation,* and if all the relations of one pair of things correspond to those of another pair, it will be difficult to distinguish the one pair from the other, though [they are] not presenting a single point of resemblance, unless we have some difference of relation to something else, whereby to distinguish them.[50]

49 Cf. Torrance, 'The Significance of John Philoponos', 55: 'Clerk Maxwell became convinced that "in a scientific point of view the *relation* is the most important thing to know". The kind of relations he wanted to express and develop were not of a putative kind but real relations of an ontological kind inhering in reality, for the inter-relations of things are ontologically constitutive of what they really are.'

50 JCM, 'Analogies in Nature', February 1856, (C&G1, 242f; Harman, I, 381f.)

For Maxwell, thinking analogically was a creative tool which connected different fields of science, and which enabled insight gained in one area to become a tool of discovery in another, such as when he borrowed some of William Thomson's mathematics for fluids to apply to his work on electricity. Alongside *analogy* was Maxwell's enduring fascination with *geometrical* representation, which could help express an idea apart from algebraic formulae.[51]

But *analogy* did not point to *identity*. Or did it? It indicated a *partial* and not a *total* resemblance between two entities. It had limits. But where was that limit? And why? Why was the pattern of movements in fluids analogous to lines of force in electricity and magnetism? Was it because identical forces were at work? Or was the similarity fortuitous?

In answer to these questions, Maxwell posited three reasons why analogies may occur: (i) by pure accident; (ii) because the patterns in diverse areas are generated by something common to all existence; or (iii) because the human mind is conditioned to think in certain ways and imposes these ways on what it observes. In other words, analogical connections may be accidental, objective, or subjective. As noted, Maxwell's view was that *true* analogies are objectively rooted in nature, and that they exist because of the limited number of ways in which entities can relate to one another in any field of knowledge.

Maxwell had to reject the possibility that analogies appear to exist only because the mind imposes its structures of thought onto reality. He offered an ingenious proof based on space and time.[52] First, he argued that space and time were distinct from each other, proven by the fact that they had non-shared characteristics: for example, a movement in space could be reversed, but not a movement in time. For Maxwell, that also indicated that space and time were not mere products of the mind, otherwise they would have been given identical characteristics. Second, considering space on its own, he argued that it was not a projection of the mind because any point in space could be defined objectively by three variables (x, y, and z coordinates) *independent of who the observer might be,* meaning that position was not a subjective mental creation. Third, satisfied that space and time

51 Harman, II/1, 1.

52 JCM, 'Analogies in Nature', 1856, (C&G1, 236; Harman, I, 377).

were objective entities, he concluded that the type of *relations* they exhibit must also be objective.

For Maxwell, this meant that the mind could describe objective relations, and that the mind was able to reason in accordance with something exterior to it. This conviction lay behind his later speech to the 1870 British Association, in which he spoke of exploring regions where '*thought* weds *fact*', and where 'the mental operation of the mathematician and the physical action of the molecules are seen in their true relation'. He was convinced that scientists were able to make an 'adjustment of our modes of thought to the facts of external nature.'[53] This important aspect of Maxwell's thinking was highlighted by Torrance: 'Every science must have modes of thought by which the process of our minds is brought into complete harmony with the process of nature.'[54]

What Torrance saw in Maxwell was a realisation that knowledge of reality did not come from human ingenuity, but was 'gained and developed under the power of nature's own self-revelation and in harmony with its dynamic process.'[55] The ability of the mind to describe objective relations in nature lay behind what Maxwell regarded as the first and most basic analogy of all: an *analogy of thought* between mind and reality: 'Here, therefore, we have a *real* analogy between the constitution of the intellect and that of the external world.'[56] In other words, the mind could recognise and reproduce patterns which exist objectively in nature, precisely because it could reason in harmony with the structures of nature. The mind was able to conform itself to the logic of external reality, and was able to submit itself to the forms and structures of objective reality. It could think in parallel with—*analogically* in relation to—external reality.[57]

53 JCM, 'Mathematics and Physics: Presidential Address', *Report of the 40th Meeting of the BAAS,* (1870), 2. C&G1, 326.

54 Torrance, 'Christian Faith and Physical Science', 220.

55 Ibid., 229.

56 JCM, 'Analogies in Nature', (C&G1, 238; Harman, I, 378).

57 Cf. JCM, 'Analogies in Nature', 1856, (C&G1, 240; Harman, I, 380): 'The thing to be observed is, that the laws which regulate the RIGHT process of the intellect are identical with the most abstract of all laws, those which are found among the relations of necessary truths, and that though these are mixed up with, and modified by, the most complex systems of phenomena in physiology and physics, they must be recognised as supreme among the other laws of thought.'

As far as Maxwell was concerned, that ability existed because, under God's providence, the human mind and the world which God had created were correlated in a fundamental way with one another.[58] The mind was not correlated with the *mind of God* but with the structures and logic of *created being*. God had given this quality to humanity, enabling humanity to explore the cosmos in which humanity had been placed.

For Maxwell, scientific knowledge involved the object controlling the enquiring mind, rather than the mind dictating to the object. The nature of the object could mould thought, rather than thought imposing itself on the object: 'The only laws of matter are those which our minds must fabricate; and the only laws of mind are fabricated for it by matter.'[59] In the twenty-first century the word 'fabricate' implies something artificial, but that was not what Maxwell meant. The critical words are: 'The only laws of mind are fabricated *for it* by matter.'

In other words, in the encounter between mind and nature, human modes of thought are restructured by the form in which matter actually exists in nature; and ultimately, the nature of human logic is determined by the nature of matter in its inter-relations. Because matter has certain structures, then the mind—through its interaction with matter—restructures (fabricates) its thinking in a form analogous to these, with the logic of mind mirroring the logic of reality. That is why logic has the form it has. That is why the mind can recognise patterns in nature. That was what lay behind Maxwell's *analogy of thought*. Nature imprints its patterns on the mind. Nature

58 Cf. JCM, *Aberdeen Inaugural Lecture,* (1856). See also: Torrance, 'Christian Faith and Physical Science in the Thought of JCM', 220. Also: Torrance, 'The Significance of John Philoponos', 48: '[For both Maxwell and Einstein] there is and must be a fundamental harmony between the laws of the mind and the laws of nature; that is, an inherent harmony between how we think and how nature behaves independently of our minds. Thus, the more profoundly our understanding penetrates into the rationality of the universe of space and time, the more clearly and fully a pre-established harmony between mind and nature becomes manifest, between the way we think and what we think about.'

59 JCM, 'Analogies in Nature', (C&G1, 244; Harman, I, 383).

moulds the *form* of human thinking as well as its *content*.[60] On this point Maxwell echoed some of Thomas Chalmers' thinking.[61] Moreover, since the universe created by God was essentially one then, for Maxwell, there were fundamental affinities between all of its laws, and between the constitution of the human intellect and the structure of the external world.[62]

Maxwell's conviction that the human mind is correlated with the structures of nature had far-reaching implications. He rejected the continental positivism which cast doubt on the possibility of gaining any objective knowledge at all. In Maxwell's opinion, positivists regarded scientific theories simply as models which happened to work and happened to fit the data, and were sceptical of ever knowing what reality is truly like. In contrast, Maxwell believed that the mind was constituted with an ability to apprehend real truth, with such truth 'distinguished from artificial ideas or systems of thought.'[63] Given a correlation between the logic of nature and the logic of the mind, the task of the scientist was to re-present the pattern of God's creative operation.[64] The *analogy of thought* made this possible. Science was the task of thinking in faithful conformity to the inherent structure and dynamic configuration of the universe as it came from the wisdom and power of the Creator.

---

60 Campbell, (C&G1, 246), notes other material, including two unfinished essays (*Sensation*, and *Reason and Faith*) which may date from when Maxwell tended his father in Edinburgh in the Spring of 1855. Cf. JCM, *Letter to C.J. Monro,* 19 Feb 1855, (C&G1, 210).

61 Cf. W.G. Blaikie, *Thomas Chalmers,* Chapter 1: 'Under Professor Robison's instructions [Chalmers] was led to ponder the remarkable harmony between the human mind and the processes of nature—the wonderful adaptation of the one to the other; and the conclusion was irresistible that this must be due to an intelligent Divine Being who had framed these adaptations. In after years this was the theme of his Bridgewater treatise, and it was one of the corner-stones of his Natural Theology.'

62 Torrance, 'Christian Faith and Physical Science', 221.

63 Ibid., 220f.

64 Ibid., 227.

Reasons, Forces, and Causes

Looking deeper, Maxwell saw analogy between the notion of a *reason* in logic and the notion of a *force* in physics, since both 'made things happen.' As a *reason* made a sequence of thought happen, so a *force* made a sequence of events happen. As a reason was something in the mind which connected thoughts, so a force was something in nature which connected objects. The relationship of reasons to thoughts in the mind was analogous to the relationship of forces or causes to objects in nature, with each relationship having the 'power of predicting events not previously known.'[65]

What is a *reason*? Maxwell suggested: a *reason* or *argument* is a conductor by which the mind is led from a proposition to a necessary consequence of that proposition: 'In pure logic, reasons must all tend in the same direction.'[66] A 'reason' was what connected one thought with another, and what 'mind' did was to hold together several 'thoughts' in a matrix. 'Reasons' assisted by enabling a person to move one thought to another thought. When that occurred, 'reasons' created a dynamic mental matrix, with Maxwell seeing an analogy to the matrix of objects in nature where 'forces' or 'causes' connect objects to one another, and where forces move or transform one object to or into another. A matrix model was common to both the mind and the external world, with analogous laws of causation applying in both realms.

Yet, whereas in thought reasons tended to give unique solutions, that was not always so with forces in nature.[67] Maxwell considered *cause* to be a metaphysical word, implying something unchangeable and irresistible, and always producing its effect. In contrast, he regarded *force* as a practical scientific word which signified something which may well meet with opposition—sometimes significant opposition—'but yet never fails to do what it can [do] in its own favour.'[68] His description of *force* may have been coloured by Newton's Third Law which states that when two bodies interact, they apply forces of action and reaction on one another.

65 JCM, 'Analogies in Nature', (C&G1, 238; Harman, I, 379).

66 Ibid., (C&G1, 239; Harman, I, 379)

67 Ibid.

68 Ibid.

For Maxwell, *cause* was something which brought inevitable logical consequences; whereas a *force* was a blunt instrument, *forcing* objects in nature to do things, and obeying the rule 'might is right.' *Causes* were like *moulds*: they defined and pre-set what would be the inevitable outcome if the appropriate *force* were applied. *Causes* determined the result.[69] They shaped the end product of forces. Maxwell called these *Organic Laws.*[70]

Science was about studying varieties of movements and motions in nature, and, for Maxwell, because logic and the laws of nature allowed only a limited range of possibilities, then similar structures (relationships) within nature must occur repeatedly. Finding analogies in nature was unavoidable:

> The question of the reality of analogies in nature derives most of its interest from its application to the opinion, that all the phenomena of nature, being varieties of motion, can only differ in complexity—and therefore the only way of studying nature is to master the fundamental laws of motion first, and then examine what kinds of complication of these laws must be studied in order to obtain true views of the universe.[71]

There *must* be repetitions of fundamental laws throughout the sciences, including laws of thought. Maxwell was aware that this might push analogical resemblance beyond mere similarity, with *analogy* verging on becoming *identity.* He stated: 'The resemblances between the laws of different classes of phenomena should hardly be called analogies, as they are only *transformed identities*.'[72] Whether this meant that all forms found in nature sprang from one common root, or whether it resurrected the possibility that the mind imposed *its* patterns onto nature was a question unresolved.

---

69 Ibid..

70 Maxwell considered the variety of *forces* and *influences* which acted on human beings and affected voluntary actions, (C&G1, 240). The notion that *mind* could mystically or mesmerically bend nature to its will (distinct from the notion that free will could intervene in nature) was in vogue, and Maxwell was aware of a populist notion that the human mind was the only true cause of *all* phenomena, and that physical causes were only apparent. He rejected that idea.

71 JCM, 'Analogies in Nature', (C&G1, 241; Harman, I, 380).

72 Ibid. (Italics ours).

Marischal College, Aberdeen

(1856–1860)

Chapter 12
Professor in Aberdeen

In an era of few Universities it may have surprised some of Maxwell's Cambridge colleagues to learn that Aberdeen had two. There was the original King's College, founded in 1495 when Bishop William Elphinstone, Chancellor of Scotland, petitioned Pope Alexander VI on behalf of King James IV. There was also the post-Reformation Marischal College, founded in 1593 as a second and more Protestant institution. The modern University of Aberdeen was created in 1860 by the union of King's and Marischal, a merger which affected Maxwell's career.

In February 1856, Maxwell was encouraged by James Forbes of Edinburgh to apply for the chair of natural philosophy at Marischal. His father was also keen for him to return to Scotland. The appointment lay with the Crown through the Lord Advocate and Home Secretary, and Maxwell found it amusing that, in theory at least, the successful applicant had to be theologically orthodox. He wrote tongue-in-cheek to his father: 'As to theology, I am not aware that the mathematicians, as a body, are guilty of any heresies.'[1] The necessary certificates and testimonials were gathered, presented, and proved effective. The appointment was made in late April 1856, with Maxwell starting duties in November of the same year.

John Clerk Maxwell never knew that his son's candidacy had been successful. He died at Glenlair on the third of April 1856 at noon. James was with him. Later that day James wrote to Aunt Jane, telling her that his father had been sleepless and confused during the night, though able to rise and dress for breakfast. After speaking to Sandy, the Glenlair gardener, he sat in the drawing-room, gave a short cry, and died soon afterwards.[2] The next day, as the new laird, James led the daily prayers

1 JCM, *Letter to John Clerk Maxwell,* 15 Feb 1856, (C&G1, 251; Harman, I, 393).

2 JCM, *Letter to Jane Cay,* 3 Apr 1856, (C&G1, 253). JCM, *Letter to Mrs Blackburn,* 3 Apr 1856, (C&G1, 253).

which all staff attended. His uncle, Robert Dundas Cay, came to help, and, in a letter to his sister Jane, described how James had slipped smoothly into his new role, introducing small innovations which were intended to enhance the spiritual life of the household:

> It is beautiful to see the feeling of all the people towards him, all thinking for him, and trying to assist him in every way, and he trying to carry on everything as before: or, when he wants to make a change, his anxiety, lest people should think he disapproves of the former customs. For instance, he wished to have the servants in for prayers every evening, instead of our reading by ourselves and reading to them separately; he was quite afraid they should think his doing so would look as if he thought it was wrong, it not having been done before.[3]

Maxwell stayed at Glenlair for most of the spring and summer of 1856, settling affairs and familiarising himself with his new responsibilities.[4] There were people to direct. There were estate houses to be repaired. There was a visit to Belfast to see his cousin and close friend, William Dyce Cay (1838–1925), who was starting studies in engineering. Cay had won the prestigious Straiton Gold Medal in mathematics at Edinburgh, and became a highly successful civil engineer.

Maxwell also made a brief return to Cambridge, tidying up his remaining work at Trinity, as well as teaching at the Working Men's Institute.[5] In August there was the British Association gathering at Cheltenham, at which he presented a paper 'On Compound Colours and the Effect of mixing Blue and Yellow Light'.[6] Amidst this activity, visitors came and went at Glenlair, both family and Cambridge friends. All of this meant that it was late autumn before he moved to Aberdeen, lodging with Mrs Buyers at 129 Union Street, a short distance from Marischal.

The pain of bereavement went deep.[7] Cousin Jemima refers to it in her *Memoirs* for September 1857, when James visited her and Hugh

---

3 Robert Dundas Cay, *Letter to Jane Cay,* 8 Apr 1856, (C&G1, 254).

4 JCM, *Letter to Litchfield,* 4 July 1856, (Harman, I, 410).

5 J.J. Thomson, 'James Clerk Maxwell', *A Commemoration Volume,* 12.

6 Cf. Harman, I, 412–413.

7 McCartney, 'I remember Years and Labours as a Tale that I have Read', 292, interprets Maxwell's 1856 poem 'Recollections of Dreamland', as addressing the death of his father.

Blackburn at their remote Highland retreat of Roshven.[8] She wrote that James found the loss of his father incalculable, and commented that, although he had thrown himself into his new appointment at Marischal, he was prone to fits of melancholy. During his stay, he expressed his sorrow in verse:

> Alone on a hillside of heather,
> I lay with dark thoughts in my mind.
> In the midst of the beautiful weather,
> I was deaf, I was dumb, I was blind.

The Marischal Teacher

Classes started in November. Maxwell walked to the college each morning, always taking the same side of the street and always crossing at the same point. Repetitive habits were a feature of his personality, first noticed in Edinburgh. He was described by one of his Aberdeen students as: elastic in gait as if on springs; easy or even careless in dress; his expression a mixture of the serious and mirthful; and his personality marked by brightness and freshness.[9]

His classroom in college was a corner-room in the upper tier of a wing.[10] Following custom, he opened the morning lecture with one of the Church of England collects or with another short prayer, both of which came to him naturally. Academically, he had his hands full. He lectured fifteen hours each week. He pursued his own areas of research. He started to revise the syllabus. Very soon he established a rhythm for the year: six months in Aberdeen, and the other six at Glenlair where, as well as running the estate, he worked on new scientific ideas in the peace and quiet of his rural retreat.

In addition to college work, there were extra-mural classes. On Monday evenings, at the Aberdeen School of Science, starting at ten minutes past eight o'clock, he gave free weekly lectures on natural philosophy to artisans, focussing on the principles of mechanics.[11] He

---

8 Fairley, *Jemima,* 47.

9 George Walker, 'James Clerk Maxwell', *Galloway News & Kirkcudbrightshire Advertiser,* 18 May 1883.

10 Robert Walker, 'James Clerk Maxwell', 202.

11 *Aberdeen Press & Journal,* 27 Oct 1858. Garnett, *Heroes of Science: Physicists,* 286.

proudly told C.J. Monro that the Monday evening attendees did better than the university men on false balances and their quantity of work![12] The artisan course comprised twenty-four lectures, one evening a week, with the students' fees set at eight shillings per annum.[13] These classes meant that, although F.D. Maurice's network of Working Men's Colleges did not extend to Aberdeen, Maxwell was able to continue his support for the concept of further education for all, which he saw as an important part of his Christian responsibilities. The Aberdeen School of Science had been established in concert with the Department of Science and Art, which was under the Committee of the Privy Council on Education. Though not independent of the State as were Maurice's Working Men's Colleges, it had similar aspirations, and aimed to give:

> Young mechanics, tradesmen, etc., instruction in those branches of science which are of the greatest practical utility in the chief Arts, Trades, or Professions followed by workmen in this locality; and also to afford instruction in Practical Science to others who may be prevented, by their means or avocations, from attendance on such Classes during the day.[14]

Many of its lectures were held in Marischal's classrooms, and the library of the Mechanics' Institution was also made available to students. The school was open to ladies, with teachers, including pupil-teachers, 'particularly invited to the opportunities afforded by this school of extending their knowledge of a class of subjects extremely useful to them in their profession.'[15]

One of Maxwell's unusual duties was supervising attendance at Professor Pirie's lectures on Practical Religion.[16] Earlier in the nineteenth century, the Murtle Foundation endowed various charitable trusts in and around Aberdeen, and established lectureships with the intention of deepening the piety of university students. During the autumn of 1856—Maxwell's first term at Marischal—Professor

12 JCM, *Letter to Monro,* 26 Nov 1857, (C&G1, 292; Harman, I, 573f).

13 Reid, 'Maxwell at Aberdeen', in Flood (ed.), *JCM: Perspectives on his Life and Work*, 27.

14 *Aberdeen Press & Journal,* 27 Oct 1858.

15 *Aberdeen Herald and General Advertiser*, 24 Oct 1857.

16 JCM, *Letter to Tait,* 3 Dec 1856, (Harman, I, 481).

William Robinson Pirie (1804–1885) gave the Murtle Lectures on 'Practical Christianity', whilst the college principal, Professor Daniel Dewar (c.1783–1867), gave the Murtle Lectures on 'Evidences of Christianity.'[17] All science students had to attend some theology lectures, and all theology students had to attend some classes in natural philosophy. Maxwell's attendance was mainly to ensure good behaviour among the students!

The Murtle lectures are mentioned by Dr Robert Walker (1842–1920), one of Maxwell's former students, who became Librarian and Registrar at Aberdeen University plus Secretary to the University Court. He was twin brother to the Revd George Walker (1842–1919) of Castle Douglas, and when Robert visited George at his Manse, only seven miles from Glenlair, he occasionally visited Maxwell.[18] The brothers had attended Maxwell's last session in Marischal over the winter of 1859/60.

Some years later, Robert published an account of the irreverent student attitude to the Murtle lectures.[19] Writing in the *Aberdeen University Review* for 1915/16, he recollected that the compulsory lectures were held every Thursday afternoon, but were 'somewhat inadequately appreciated'. Because of the unruly student behaviour, staff-members—including the professors of Greek, natural history, natural philosophy, and moral philosophy—took turns in keeping order! Walker added: 'All that I can now recall of these courses of lectures is the deep, raucous voice of the principal, Maxwell's father-in-law, dilating on the instability of human greatness as illustrated in the career of Saladin, whom our Richard of the lion-heart overthrew.'

Concerning the other Aberdeen University, Maxwell's affability resulted in good personal relationships with its lecturers, though inter-collegiate camaraderie was unusual enough to be worthy of comment in a letter to Aunt Jane: 'I have been keeping up friendly relations with the King's College men, and they seem to be very friendly too. I have not received any rebukes yet from our men for so doing, but I find that the families of some of our professors have no dealings, and never had, with those of the King's people. Theoretically we profess

17 *Banffshire Journal & General Advertiser,* 30 Sep 1856.

18 Robert Walker, 'James Clerk Maxwell', 193.

19 Ibid., 204.

charity!'[20] He also commented on a stiffness in Aberdeen society: 'No jokes of any kind are understood here. I have not made one for two months, and if I feel one coming I shall bite my tongue!'[21]

Tutoring Tribulations

Maxwell was a superb scientist and brilliant mathematician, but a poor lecturer. This mattered little at Cambridge since Cambridge students relied heavily on their personal tutors. At Aberdeen the lecture was the main instrument of education. It was not that Maxwell disliked teaching undergraduates. Nor was it that he wanted the chore of lecturing out of the way in order to do research work. He loved teaching. He took trouble in preparation. He thought up interesting demonstrations—enjoyed even more by the students when they didn't work. He was enthusiastic about sharing ideas. But he struggled to express himself clearly in an oral presentation. His mind was apt to jump ahead too often. Concepts accumulated too quickly for his listeners to follow easily.[22]

For those with the ability to keep up with him, there were riches to be discovered. And he was eager to help. For example, although professors had unlimited access to the library, the students were only allowed two volumes at a time. To circumvent this rule, Maxwell took out books for his students, explaining, when challenged, that all the students were his friends![23]

Kindness could not hide deficiency. The Revd George Reith—father of Lord John Reith, first Director General of the BBC—who studied under Maxwell, wrote in 1914 that Maxwell was a 'noble-souled Christian gentleman with a refined delicacy of character that bound his class to him *in a devotion which his relatively meagre qualities as a teacher* could not undo.'[24] That verdict was universal, even among Maxwell admirers.

Another student was a local Episcopalian, David Gill (1843–1914), who went on to become Her Majesty's Astronomer at the

---

20 JCM, *Letter to Jane Cay,* 27 Feb 1857, (C&G1, 264).

21 JCM, *Letter to Campbell,* 22 Dec 1857, (C&G1, 297).

22 J.J. Thomson, 'James Clerk Maxwell', *A Commemoration Volume,* 15.

23 C&G1, 260.

24 Jones, 'The Complete Physicist: James Clerk Maxwell', 20. (Our italics).

Cape of Good Hope Royal Observatory. Gill adored Maxwell, but he was not blind to his hero's deficiencies. In a speech to the 1908 London dinner of the Aberdeen University Club, Gill opined that Maxwell was one of two or three of the greatest geniuses who had lived since the days of Newton, 'and yet they did not understand him in Aberdeen somehow!'[25] The audience laughed when Gill admitted that Maxwell was no teacher; that his lectures could be terrible; and that many of his experiments failed! But, Gill added, the experiments were far more interesting in the failure than if they had worked! And those who chose to stay behind after the class 'used to get a most delightful hour or two, and learn an immense deal that we never forgot; a great deal that we did not understand at the time but that came back to us afterwards.' In these sessions Maxwell expanded on what he had tried to teach in the lecture; showed experiments he was currently working on; and shared problems that awaited solution. He could be so engrossed in these informal sessions that he forgot time, resulting in Mrs Maxwell arriving, wondering why he had not come home to his dinner. She then escorted him home, whether he wanted it or not.

Gill recalled Maxwell's humour and fondness of paradox. In Gill's opinion, the humour could backfire, with some of Maxwell's statements not being taken seriously.[26] He cited Maxwell's proposal to his 1858 class, that a truly scientific standard of length should be measured in wavelengths of light instead of in yards or metres. This was regarded by some as a huge joke. Fittingly, it was Gill's task, near the end of his life, to urge the International Bureau of Weights and Measures to define the metre in terms of wavelengths of light.[27]

Gill had Maxwell for only two years (1858–1860) but was a life-long devotee. On one occasion, when visiting the shop of Glasgow optics specialist James White, and standing at the counter looking at

---

25 Gill, 'Address to London Dinner of the Aberdeen University Club', Nov 1908 (*Aberdeen Press & Journal, Thursday* 19 Nov 1908). Cf. Forbes, *David Gill: Man and Astronomer,* 17.

26 Cf. Walker, 'James Clerk Maxwell', 209: 'Many persons . . . often asked themselves whether he was in earnest or joking—a habit which certainly rendered it all the more difficult to feel sure that one understood aright so entertaining a conversationalist.'

27 Forbes, *David Gill: Man and Astronomer,* 13f.

some instruments, he felt a hearty slap on the back.[28] Looking round he saw the beaming face of his old professor. Maxwell introduced Gill to his companion who was none other than William Thomson (Lord Kelvin).[29] The three discussed apparatus before Maxwell and Thomson invited Gill to join them for breakfast. The star-struck Gill told his family that it was one of the proudest moments of his life.[30] Maxwell was a referee when Gill applied for the post at the Cape of Good Hope, describing him as one of his ablest students.[31]

Gill was not the only enthusiast who confessed that his hero was a poor lecturer. George Walker of Castle Douglas said the same: 'It has been frequently asserted that Maxwell was not successful as a teacher: on this point I would testify *that I know*.'[32] Walker added ruefully: 'He might have taken his own advice and given it to us a little thinner!' At the same time: 'There was such simplicity and directness (usually due to geometrical methods) in his proofs of even difficult theorems, that no student who set himself to profit by his instructions could fail to do so.'

George's brother, Robert, attended lectures during Maxwell's later professorship at Cambridge. He commented: 'I cannot remember any laboratory, other than the lecture-table round which a party of us, after the lecture, used to crowd, listening with delight to what often developed into a further protracted elucidation of the subject of the lecture.'[33] He added that Maxwell was always most lucid when he fell on geometrical methods. However, 'whether he can be said to have been successful as a teacher is a question not

28 In 1869 the Glasgow firm of 'James White, Optician and Philosophical Instrument Maker to the University', moved from 95 Buchanan Street to 78 Union Street. Maxwell and Thomson met often at White's. Cf. JCM, *Letter to W. Thomson,* 1 Oct 1869, (Harman, II/2, 498); & JCM, *Letter to W. Thomson,* 5 Oct 1869, (Harman, II/2, 501).

29 William Thomson became Lord Kelvin in 1892, after Maxwell's death in 1879, hence we refer to Thomson rather than Kelvin.

30 Forbes, *David Gill: Man and Astronomer,* 22f.

31 JCM, *Letter of Reference for David Gill,* 16 May 1878, (Harman, III/2, 650).

32 George Walker, 'James Clerk Maxwell', *Galloway News & Kirkcudbrightshire Advertiser,* 18 May 1883 (Italics original).

33 Robert Walker, 'James Clerk Maxwell', 193, 202.

easily answered, owing to the difficulty of discovering, and starting-from, some common measure between the ignorance and incapacity of his pupils and his own vast knowledge and ready acumen. He certainly presumed and reckoned on the willing mind; and I am certain that no pupil who brought that could fail to profit. Many profited immensely.'

In 1860 it was his lecture-room failings which cost Maxwell the coveted chair of natural philosophy at Edinburgh. James Forbes had left to become Principal at St Andrews, and Maxwell asked Forbes to support his application.[34] As things turned out, it was Maxwell's friend, Peter Tait, who was successful. When the *Edinburgh Courant* reported on the process, it noted that though Maxwell was one of the most remarkable men known to the scientific world there was another quality which was desirable in a professor, namely the power of oral exposition and the ability to speak clearly to ordinary students. The *Courant* concluded: 'We have little doubt that it was a deficiency in this power in Professor Maxwell which made the Curators prefer the claims of Mr Tait.'[35] Tait was a powerful lecturer, with 'clearness of vision, great simplicity and frankness of speech, a dash of humour of a very delicate and subtle nature' contributing to his rare and exceptional talent.[36] The two school-friends remained close both socially and academically. They critiqued each other's proposed publications, with Maxwell occasionally giving his views in humorous verse.

In 1931, the physicist J.J. Thomson contrasted Maxwell's reputation in the lecture-room with the clear papers he presented at scientific meetings.[37] Thomson suggested that may have been because conference lectures were read from a prepared text, whereas in class lectures he found it difficult to keep to the 'severely pedestrian tramp

34 . JCM, *Letter to Forbes,* 8 Nov 1859, (Harman, I, 623); JCM, *Letter to Faraday,* 30 Nov 1859, (Harman, I, 627).

35 Cited in: Knott, *Life and Scientific Work of P.G. Tait,* 16. Cf. J.J. Thomson, 'James Clerk Maxwell', *A Commemoration Volume,* 15.

36 Fergusson, *Chronicles of the Cumming Club,* 203.

37 Cf. G.T. Stokes, *Testimonial for J.C. Maxwell,* 25 Feb 1856, (Harman, I, 402), recommending Maxwell to Marischal: 'Having once been present when he was giving an account of some of his geometrical researches to the Cambridge Philosophical Society . . . I was struck with the singularly lucid manner of his exposition.'

along one road necessary for such teaching, and may have soared off into higher regions beyond their ken.'

Sir Ambrose Fleming (1849–1945), who was taught by Maxwell at Cambridge from 1877 to 1879, acknowledged that though Maxwell had a wide knowledge of the history of science, and could illustrate the simplest facts by a telling anecdote or phrase, he 'had too much learning and too much originality to be at his best in elementary teaching.'[38] Fleming's verdict was that Maxwell's teaching was a delight for those who could follow him. Garnett was of the same mind, as he wrote in *Nature*, days after Maxwell's death:

> He had not the power of making himself clearly understood by those who listened but casually to his pithy sentences, and consequently he was not a so-called popular lecturer. Nor was he a most successful teacher of careless students. But when he had those about him who could enter into his mind, and, receiving the golden truths from his lips, could alloy them in such a way as to make them acceptable to the ordinary student, no better teacher could be desired, even for the most elementary instruction. His wonderful imagination was of great value, not only in supplying illustrations for didactic purposes, but in suggesting analogies and opening up new fields for research.[39]

Karl Pearson (less of a Maxwellophile) recollected that though Maxwell's books were splendid he simply could not lecture and seemed to have no planned course: 'After three weeks of his course on heat I beat a retreat.'[40] Another former student, Sir Richard Glazebrook (1854–1935), stuck up for his teacher:

> Teaching is not all done by lecturing. [Maxwell's] books and papers are vast storehouses of suggestions and ideas which the ablest minds of the past twenty years have been since developing. To talk with him for an hour was to gain inspiration for a year's work . . . The sure judgment of posterity will confirm the verdict that Clerk Maxwell was a great *teacher*, though lecturing to a crowd of untrained undergraduates was a task for which others were better fitted than he.[41]

38 Fleming, 'Memories of Clerk Maxwell', *James Clerk Maxwell: A Commemoration Volume, 1831–1931*, 121.

39 Garnett, 'James Clerk Maxwell', *Nature*, Vol. 21/524 (13 Nov 1879), 45.

40 Pearson, 'Old Tripos Days at Cambridge, as seen from Another Viewpoint', 32.

41 Glazebrook, *James Clerk Maxwell and Modern Physics*, 53.

Successful Seminars

In contrast with his college classes, Maxwell's extra-mural talks could be vibrant and full of interest. A few months into his Marischal professorship he lectured to a public audience in the Albion Street Chapel on: 'The Eye, its Construction and Uses'. The *Aberdeen Press & Journal* reported that this well-received popular lecture was illustrated by optical glasses, diagrams, and models of the human eye, as well as involving dissections of eyes from a bullock, a sheep, and a fish. The newspaper noted that the talk was remarkable for its clarity and simplicity, and for the enthusiastic attachment which the lecturer showed for his subject.[42]

Maxwell's main problem appears to have been in teaching the youngest students, many of whom simply could not grasp what he was trying to put across. For more advanced students he had voluntary classes, conducting them on a seminar basis. These were far more successful, as his strength lay in his practical demonstrations. One student, an Aberdeen farmer, recalled an experiment with electricity, during which Maxwell mounted a 'volunteer' on an insulated stand while (as expressed in the Aberdeen brogue of the farmer) he 'pumpit him fu' o' electricity, so that his hair stood oot on end!'[43]

A favoured technique was to examine his students orally, to see what they really knew, before giving written tests each Tuesday. This imitated Sir William Hamilton's system. In November 1857 he wrote to William Thomson: 'My class is small this year. I have *viva voce'd* them all, and will begin written questioning on Tuesday. The second year (my old students) have been grinding in summer and have done well in examination. A good many of them come to me as an advanced class.'[44] A few weeks later he informed C.J. Monro that, though he had a small class with a 'name for stupidity', there was real

---

42 *Aberdeen Press & Journal,* 18 Mar 1857.

43 Jones, 'James Clerk Maxwell at Aberdeen', 59.

44 JCM, *Letter to W. Thomson,* 14 Nov 1857, (Harman, I, 556). See also: JCM, *Letter to Henry Richmond Droop,* 14 Nov 1857, (Harman, I, 557): 'I have a large advanced class for Newton, physical astronomy, the electric sciences, and high optics. … I have also a mechanics' class in the evening, once a week, on mechanical principles, such as doctrine of lever work done by machines etc. So I have 15 hours a week, which is a deal of talking straight forward.'

scope for exciting them into activity![45] His method, he explained to Monro, was to get into regular ways; to *viva voce* each student once a week; to examine the whole class in writing on Tuesdays; and to give them severe critiques on Wednesday morning. That regime, plus loads of exercises, seemed to work.

He told Monro that his class for physical astronomy, electricity and undulations, was very good, and he wanted to do them justice. This led him to introduce students' teas where he met and talked science with the students informally. The students' teas were his own idea.

In November 1857 he wrote to Aunt Jane: 'I have had two students' teas; at which I am becoming expert. I have also indulged in long walks and have seen more of the country. The evenings are beautiful at this season. There have been some very fine waves on the cliffs south of the Dee.'[46] These seminar/tutorial groups for able students continued through the winter, and he wrote to William Thomson at the end of January 1858: 'I have been lecturing on statical electricity to the second year, and next week I shall have half a dozen to study 'electrical images' over a cup of tea. I begin current electricity on Tuesday.'[47]

He was content. He commented to Litchfield: 'My *lines* are so pleasant to me that I think that everybody ought to come to me to catch the infection of happiness. This college work is what I and my father looked forward to for long, and I find we were both quite right that it was the thing for me to do.'[48] Approaching marriage added to his well-being.

---

45 JCM, *Letter to Monro,* 26 Nov 1857, (C&G1, 291; Harman, I, 573f).

46 JCM, *Letter to Jane Cay,* 28 Nov 1857, (C&G1, 293; Harman, I, 575). On JCM's advanced class, see: Gill, *A History and Description of the Royal Observatory, Cape of Good Hope,* (London, 1913) xxxi.

47 JCM, *Letter to W. Thomson,* 30 Jan 1858, (Harman, I, 579).

48 JCM, *Letter to Litchfield,* 5 March 1858, (C&G1, 304; Harman, I, 587).

Chapter 13
The Dewar Family

Soon after Maxwell arrived in Aberdeen he became acquainted with the family of Marischal's Principal, Daniel Dewar, and was invited to their home at 13 Victoria Street. Reid speculates that the initial contact arose because Dewar was reviewing *Reul-Eòlas*, a Gaelic book on astronomy written by D.M. Connell (1823–1905).[1]

Whatever prompted the invitation, Maxwell met the Dewar's second daughter, Katherine Mary (1824–1886). Matters moved apace, especially after September 1857 when he was invited to share the Dewar family holiday at Ardhallow, near Dunoon, which was the home of the Dewars' elder daughter Ann Gordon McCunn (1822–1883).[2] He stayed a week, before travelling to Roshven, further up the west coast, to visit Cousin Jemima and her husband Hugh Blackburn. In February 1858, he wrote to Aunt Jane about his engagement:

> This comes to tell you that I am going to have a wife. I am not going to write out a catalogue of qualities, as I am not fit; but I can tell you that we are quite necessary to one another, and understand each other better than most couples I have seen. Don't be afraid; she is not mathematical, but there are other things besides that, and she certainly won't stop the mathematics. The only one that can speak as an eyewitness is Johnnie, and he only saw her when we were both trying to act the indifferent. We have been trying it since, but it would not do, and it was not good for either. So now you know who it is, even Katherine Mary Dewar . . . So there is the state of the case. I settled the matter with her, and the rest of them are all conformable. I hope someday to make you better acquainted.[3]

In two further letters, one to C.J. Monro and one to Richard Litchfield, he wrote at length of how he and Katherine were getting to know each

---

1 Reid, 'Maxwell in Aberdeen', 36.

2 JCM, *Letter to W. Thomson*, 24 Aug 1857, (Harman, I, 537). In 1848 Ann married ship-owner/merchant John McCunn (d. 1866, aged 74).

3 JCM, *Letter to Jane Cay*, 18 Feb 1858, (C&G1, 303).

other, and of his joy at the coming wedding.[4] He said that though the engagement was based on only a year's experience, he and Katherine—without ever having uttered grand sentiments—were 'united in all things, high and deep.' Above all, he was filled with an 'infection of happiness.' He greatly respected Katherine's parents. The Dewars viewed him likewise as an eminently suitable match for their daughter.

Dewar was a churchman of strong evangelical persuasion who had risen from humble origins to the prestigious post of Principal of Marischal.[5] As an evangelical Christian, he expected his daughters to marry husbands of a similar outlook, and Maxwell being so readily welcomed into the family indicates Dewar's approval of the young professor's spiritual leanings.

On the second of June 1858, James and Katherine were married in Old St Machar's, Aberdeen, by the Revd John Clerk, minister of Dunoon and Kilmun. William Dyce Cay was best man. Friends and family, including Aunt Jane and Lewis and Frances Campbell, travelled north for the occasion. In a light-hearted pre-wedding note to Aunt Jane, James told her that the church doors would open at 12.15; performance commence at 12.30; luncheon on table at 1.00; guests stop eating at 2.00; closing with a formal dinner later in the day, of which he professed to know little![6] The *Aberdeen Press and Journal* recorded the event:

> At Aberdeen, on the 2nd June, JAMES CLERK MAXWELL, M.A., Professor of Natural Philosophy in Marischal College, to KATHERINE MARY, youngest daughter of Principal Dewar of Marischal College.[7]

For their remaining years in Aberdeen, James and Katherine stayed with her parents at 13 Victoria Street. During vacations they went

4 JCM, *Letter to Monro*, 27 Feb 1858, (Harman, I, 585–586); JCM, *Letter to Litchfield*, 5 March 1858, (C&G1, 304–307; Harman, I, 587–590).

5 For Daniel Dewar, see: (1) McNaughton, *The Scottish Congregational Ministry: 1794–1993*; (2) *FES*, III, 475; IV, 175; VI, 9; VII, 361, 363; (3) Wareing, "Daniel Dewar, *DSCHT*, 240; (4) Ian R. MacDonald, *Aberdeen and the Highland Church (1795–1900)*; (5) Donald Sage, *Memorabilia Domestica: Parish Life in the North of Scotland.*

6 JCM, *Letter to Aunt Jane*, 24 May 1868: (Basel University Library).

7 *Aberdeen Press & Journal*, 9 Jun 1858.

to Glenlair. A formal Marriage Contract from 1858 indicates that Katherine was to have her own annuity of £250 per annum.[8]

Susan Place (c.1798–1876) of Yorkshire

Katherine's mother was Susan Place from Skelton Grange in Yorkshire. She married Daniel Dewar in October 1821, when he was minister of the Tron Church in Glasgow.[9] Her father was Edward Place (d. 1842), who, through marriage to Lady Ann Gordon (d. 1821), had acquired the Loch Dochart and Crianlarich estates in Perthshire. Lady Ann, who died eight months before Susan's wedding to Daniel Dewar, was a daughter of George Gordon (1722–1801), the Third Earl of Aberdeen.[10]

The Third Earl's son (Susan's uncle, Katherine's great-uncle) should have inherited the Earldom, but he was killed in a riding accident in 1791 resulting in the title transferring to George Hamilton-Gordon (1784–1860), the Third Earl's grandson. George Hamilton-Gordon is best known to history as Lord Aberdeen, who was British Prime Minister from 1852 to 1855, and during whose premiership Britain became embroiled in the Crimean War. Katherine's mother (Susan) was therefore cousin to the former Prime Minister. Lord Aberdeen was also the patron of the Church of Scotland parish of Ellon, though his personal religious feelings were for the Anglican Church. This connection resulted in Daniel and Susan Dewar's younger son Donald (1831–1862) being nominated as minister for Ellon, with his induction to the parish taking place in February 1860.

Daniel Dewar of Perthshire

Daniel Dewar was a Highlander, who was born in Glen Dochart, Perthshire, around the year 1783, though it is difficult to specify the date. According to family tradition his father was a blind piper who made a living by travelling through the country and playing at fairs and weddings. Daniel carried his father's fiddle case and acted

8 Dumfries and Galloway Archives, *Catalogue of the Clerk Maxwell of Middlebie Papers*, GGD56/49/15.

9 *Aberdeen Press & Journal*, 31 Oct 1821.

10 *Yorkshire Gazette*, 17 Feb 1821.

as guide on his journeys. Donald Sage recorded that an unknown wealthy individual took notice of the young lad, sending him to a school at his own expense.[11]

Spiritually, Daniel Dewar was influenced by the itinerant evangelist John Farquharson, and by the Haldane evangelistic movement which flourished in Scotland during the first quarter of the nineteenth century.[12] Years later, Dewar described Farquharson as the most wonderful man he ever knew, noting that Farquharson preached almost every day, and seldom without someone being awakened spiritually: 'I think I see him still on his black pony riding round Loch Tay from farm to farm, carrying the message of salvation to the people. Divine power accompanied his ministry. There was an awakening all-round the loch. Many were brought to Christ and continued steadfast and immovable in the gospel.'

In 1801 Dewar enrolled at the Robert Haldane Theological College. In 1804 he moved to Homerton Independent College in England, where one of the tutors was John Pye-Smith. After Homerton, he became a probationer for the Congregational Church's ministry in Perthshire, first at Breadalbane and then Aberfeldy. He then ministered to the Killin Congregational Church, one of many churches born out of the Haldane campaigns. Dewar soon left the Congregationalists to join the established Church of Scotland. After licensing by the Presbytery of Mull in 1812, his first Kirk post was as a missionary in Gaelic-speaking Strontian on the west coast of Scotland. Though barely a year there, he became known as an outstanding preacher.

In 1814 Aberdeen Town Council appointed Dewar as preacher in the Marischal College church of Greyfriars. In 1815 he gained his MA from Edinburgh University. In 1817 he took the chair of moral philosophy at King's College, Aberdeen, alongside his ministry at Greyfriars. In 1819, he moved to Glasgow as minister of the old Tron Church, made famous by Thomas Chalmers who had recently transferred to the new parish of St John's to put into practice his experiment of the Church taking full responsibility for the social as

11 Sage, *Memorabilia Domestica,* 306.

12 For John Farquharson and the Itinerant Evangelist movement, see: Donald Meek, 'The Independent and Baptist Churches of Highland Perthshire and Strathspey', *TGSI,* Vol. 56 (1988–1990), 269–343.

well as the spiritual needs of a parish. At the Tron, Dewar exercised a powerful ministry, filling Chalmers' shoes with distinction.

In 1814 and 1821 he was the invited speaker at meetings of the *Society in Scotland for Propagating Christian Knowledge.* And in 1820 he was guest speaker at the *London Missionary Society Annual General Meeting*: at the time, the *LMS* invitation to preach was one of the highest accolades in the British evangelical world.

Dewar was a distinguished Gaelic scholar. In 1831, along with Dr Norman MacLeod of St Columba's Church in Glasgow, he published *A Dictionary of the Gaelic Language in Two Parts.* The Dewar/MacLeod dictionary became the standard work in the field, highly respected well into the twentieth century. He also edited a Gaelic periodical. Much earlier, in 1806, he had published *A Collection of Gaelic Hymns for the Worship of God.*

In 1832, Dewar demitted as minister of the Tron, moving back to Aberdeen to take up the dual position of Principal of Marischal and its professor of church history, holding both posts until the union of Aberdeen's two universities in 1860, at which stage he retired. His elevation to the principalship of Marischal had not been unanimous in the college senate, though his return to Aberdeen was welcomed by the city's evangelical party.

Early in his Glasgow ministry, and alongside the noted Congregationalist minister, Greville Ewing (1767–1841), Dewar ministered to James Wilson (1760–1820) prior to Wilson's execution for high treason. Wilson was a weaver from Strathaven in Lanarkshire. When the Industrial Revolution undermined his livelihood, he, like others, campaigned for radical reform. He read Thomas Paine's *The Rights of Man.* He became sceptical of religion. And he distrusted Government. In time he became a member of the *Society of the Friends of the People,* and used to read portions of the satirical magazine *The Black Dwarf* to audiences at radical meetings.

In April 1820, the authorities pounced. *Agents provocateurs* enticed Wilson and others to march on Glasgow, thinking that French troops would assist them against the Government. Wilson fell into the trap and was arrested on a charge of high treason. He was found guilty and, though the jury submitted a plea for clemency, he was sentenced to death.

In Glasgow there was fury over Wilson's ensnarement and prosecution. On the day of his execution tens of thousands gathered in the city, shouting 'Murder! Murder!' and armed troops were needed to keep order. Prior to the execution the city magistrates, legal officials, and selected clergy met with Wilson in the city's Court Hall. Daniel Dewar prayed with the condemned man. The *Glasgow Herald* reported:

> Dr Dewar delivered a most energetic prayer in [Wilson's] behalf, from which it appeared that he was either in a state of impenitence or insensibility to the awfulness of his situation. After the prayer Wilson took a glass of wine, and Dr Dewar read various passages of Scripture, proving the love and the forgiveness of the Saviour to the chief of sinners, provided they confessed their guilt, and called on him for mercy. The Doctor then beseeched him in an impassioned and earnest manner to lay hold of the Saviour, and look only to him for forgiveness; and hoped that God of his infinite mercy would give him repentance of his sins.[13]

Part of the fifty-first psalm was sung, with Wilson appearing to join in the singing with great earnestness. The party then proceeded outside to the scaffold, with Ewing and Dewar still trying to comfort Wilson. After the execution, the angry crowd dispersed.

The Disruption of 1843 pushed Dewar into controversy. His sympathy with the principles of the newly constituted Free Church was well-known, as he had been one of the signatories summoning a Convocation in November 1842, at which much of the groundwork for a possible secession from the Kirk had been prepared.

In May 1843 he attended several sessions of the first Free Church General Assembly (though never signing any documents declaring personal adherence to the body), as well as attending sessions of the Kirk's own Assembly. During the months following the Disruption, he preached from Free Church pulpits as much as from Church of Scotland ones, declaring he was 'ready to save souls either in the one or the other.' In the heated religious climate of 1843, Dewar's oscillation between the Established Kirk and the Free Kirk was not viewed as Christian generosity or as the expression of an ecumenical

13 *Glasgow Herald,* 1 Sep 1820.

spirit. It was regarded as disloyalty and betrayal, and his ambivalence precipitated a crisis.

By October 1843, the question of the right of persons who had joined the Free Church to hold posts in the Universities was in the public eye. The Presbytery of St Andrews had summoned the physicist Sir David Brewster three times to appear before them, and when Brewster failed to attend it was argued that Brewster's clear commitment to the Free Church disqualified him from holding a professorship which traditionally required adherence to the established Church of Scotland. The case against Brewster fell on technical grounds.[14] But the mood was set. Dewar was viewed similarly and questions were raised. Did Dewar's informal alignment with seceders disqualify him as professor of church history at Marischal? Was his position as its Principal tenable? Yet the issue was not clear cut. The constitution of Marischal College did not in fact require its Principal to be a minister of the Kirk. But if Dewar's principalship was secure, what about his chair in church history? Did that require its occupant to be a minister of the established church? At that point Dewar stepped back from the brink, and he wrote a public letter clarifying his commitment to the established Kirk.

Dewar's decision was a great disappointment to Free Kirk enthusiasts. To their minds, he had betrayed his intentions. Why had he not nailed his colours to the mast as had Brewster? Why had he not sacrificed position and privilege for a greater cause? Dewar was ridiculed mercilessly by sections of the press. Following his letter of clarification, the *Fifeshire Journal* mocked: 'He has promised to be a good boy, and has declared that he has now cut all his former wicked companions.'[15] The controversy was remembered at Dewar's death a quarter of a century later, with one obituary stating: 'During the agitation which preceded the Disruption, he joined the non-intrusion party, among whom he was regarded as a leader, and was looked up to with great respect as a man of weight and influence; but his reputation was subsequently marred by the manner in which he shrunk and withdrew when the trying crisis came.'[16]

---

14 Gordon, *The Home Life of Sir David Brewster*, 99.

15 *Fifeshire Journal*, 12 Oct 1843.

16 *Greenock Telegraph & Clyde Shipping Gazette*, 1 June 1867.

Donald Sage accused Dewar of having a character defect which damaged his usefulness. Sage lamented that, having taken a leading part preceding the Disruption, Dewar turned back 'in the day of battle', and had remained in the Establishment fearful he would forfeit his positions.[17] In Sage's opinion, Dewar sat in a vehicle drawn by two horses: one was *ambition*, the other was *avarice.* Sage's cruel verdict was hurtful given that he and Dewar were friends. In defence of Dewar, it may be pointed out that Sage himself had left the Kirk, and that immoderate statements were commonplace at that time. Moreover, Dewar was not the only minister of evangelical persuasion who decided not to leave the Kirk. Calumnies aside, Dewar governed Marischal for another seventeen years after 1843.

Within the established Kirk, Dewar was seen as strongly evangelical, with his numerous publications, devotional and academic, revealing the tone of his theology: *Elements of Moral Philosophy and of Christian Ethics* (1826); *The Nature, Reality, and Efficacy of the Atonement* (1831); *The Work of the Holy Spirit in the Revival of Religion* (1839: republished 1859); *The Church: A Comprehensive view of the Doctrines, Constitution, Government, and Ordinances of the Church* (1845); *The Holy Spirit: His Personality, Divinity, Office, and Agency, Regeneration and Sanctification of Man* (1847); *A Manual of Family and Private Devotions* (1849); *The Evidences of Divine Revelation* (1854); *The Communion Services of the Church of Scotland* (1859); *The Believer's Treasury* (1862); and, *Elements of Systematic Divinity* (1867). This was Katherine's father. This was the man who approved of Maxwell as a son-in-law.

Katherine Dewar

Not all were happy with James' choice of a bride. Cousin Jemima inserted a testy and unsympathetic appraisal of Katherine in her *Journal.*[18] She wrote that James' marriage gave no satisfaction to his friends and relations, and that the lady was neither pretty, nor healthy, nor agreeable, though much enamoured of James. Jemima gossiped that Katherine's sister had brought about the match by telling James how much Katherine was in love with him, and that he—being of an

17 Sage, *Memorabilia Domestica*, 307.

18 Fairley, *Jemima*, 109.

affectionate and tender disposition—married her out of gratitude. Jemima also asserted that Katherine's mind afterwards became unsettled; that she alienated James from his friends; and that she was of a suspicious and jealous nature! In Jemima's opinion James was a saint in putting up with his wife.

John Arthur, in his book *Brilliant Lives,* discusses why the marriage was resented by some members of the Clerk and Maxwell families, concluding that the reason is difficult to establish. Perhaps James was expected to marry within closer circles. Perhaps the family felt he married on the rebound from his father's death. Perhaps the Dewars divided opinion. Was family reaction the reason why Campbell wrote that James suffered a crisis in 1857/58: a period coinciding with James' wooing and marrying Katherine?[19] We do not know.

Jemima's caustic comments apart, there is little evidence that James was other than happy in his marriage. In his Will, he spoke of the 'love, favour, and affection' he had for her.[20] Katherine's ailments did give him worrying times, and the available letters have frequent reference to Katherine's illnesses. On one occasion, James wrote to the American physicist Henry Augustus Rowland (1848–1901),[21] telling him that, though Mrs Maxwell had been exceedingly ill, she was improving and wished to have one of the American apples which he had brought on his visit to Glenlair.[22]

Rowland was by no means the only guest welcomed to Glenlair, London, or Cambridge. Campbell and Garnett's *Life*, and Harman's multi-volumes on Maxwell's *Scientific Letters and Papers*, cite numerous invitations for friends and colleagues to stay with the Maxwells or come for dinner.[23] Contrary to Jemima's claims, James

19 C&G1, 367.

20 JCM, *Will and Testament,* NRS/ SC16/41/35/120.

21 Rowland was a brilliant scientist, but slow to be appreciated in his own country. Maxwell recognised the value of Rowland's early scientific papers and had them printed in the *Philosophical Magazine*. In 1876, Rowland became the first occupant of the chair of physics at Johns Hopkins University, retaining the position until his death on April 16, 1901.

22 JCM, *Letter to Rowland,* 29 Nov 1875, (Harman, III/1, 253).

23 This is in tension with Ioan James, *Remarkable Physicists from Galileo to Yukawa,* 154: 'She became jealous of his scientific friends to such an extent that he could never invite them to his home.'

and Katherine were not reclusive, whatever controversies the marriage and Katherine's personality may or may not have sparked in sections of the family.

Married life was split between Glenlair and wherever James was employed: Aberdeen, London, or Cambridge. But Glenlair was the mainstay of their lives. R.V. Jones has suggested that Katherine's desire to live at Glenlair was why James relinquished his professorial position at Aberdeen in 1860. But if so, why did he immediately accept a position in London? James was wealthy, and there was no economic need to be employed. Katherine certainly enjoyed Glenlair, and was eager to assist in parish duties, though she felt a total failure after her attempts at cottage visiting.[24] When she shared this with James, he sat her down, told her not to worry, and to trust that God would bless her visits through just being herself. He quoted Milton:

> God doth not need either man's work or his own gifts;
> who best bear his mild yoke, they serve him best.
> His state is Kingly. Thousands at his bidding speed
> and post o'er land and ocean without rest:
> They also serve who only stand and wait.

In September 1860, James suffered from an attack of smallpox, perhaps contracted at a local fair where they bought a new pony for Katherine.[25] The pony was named Charlie—after the family relative Charles Hope Cay—and was a high-bred, spirited, light-bay Galloway, with arched neck and flowing tail.[26] James broke Charlie in, and the pony was frequently mentioned in later correspondence and was provided for in his Will. But the smallpox had been serious. During its course, Katherine was left alone with James, with the Glenlair staff only coming to the door of the sick-room. James stated afterwards that it was Katherine's tireless nursing which saved his life.

Five years later in September 1865, he was ill again. Whilst out riding he had scratched his head on the low-hanging bough of a tree. This brought a severe attack of erysipelas, a bacterial infection

---

24 C&G1, 430.

25 Ibid., 320.

26 Ibid., 319.

of the upper layer of the skin, characterised by a painful rash. Again Katherine nursed him. Each evening she quietly read to him a portion of Scripture, which at the time was the only reading or listening he could cope with.

The fair-haired Katherine was seven years older than James and neither physicist nor mathematician. Early in their relationship James assured her that he was not going to insist on her becoming a scientist, though he did give examples of spouses (both male and female) with little knowledge of their partners' interests before marriage, developing a similar passion:

> Mrs Sabine learnt mathematics of her husband after she was married, so she was not married for it. Murchison knew no geology when he was married, but his wife did a little; and there was a fall of a cliff in the morning early, and her maid told her of it, so she was for up; so Murchison got up too, and there were the great bones of an Ichthyosaurus in the broken cliff, and he was interested and took to geology. Before that he was an idle young officer.[27]

As things turned out, Katherine took an enthusiastic interest in her husband's activities. She helped in his laboratory and collaborated with him in experiments on light and viscosity. In several published papers James referred to observations made by (K) as well as by (J). The (K) was Katherine. The (J) was himself.[28] He wrote to Peter Tait in 1877: 'My better-half, who did all the real work of the kinetic theory is at present engaged in other researches. When she is done I will let you know her answer to your enquiry.'[29]

---

27 JCM, *Letter to K. Maxwell,* 16 Sep 1859, (C&G1, 313; Harman, III/2, 873). Maxwell's first reference is to Elizabeth Sabine (1807–1879) and her husband Edward (1788–1883). His second is to Charlotte Murchison (1788–1869) and her husband Roderick (1892–1871). Charlotte was a friend of Mary Anning (1799–1847), the self-taught palaeontologist and fossil collector, who discovered numerous spectacular fossils on the Dorset coast in southwest England.

28 Cf. C&G1, 318, 479 & 481. 'J' and 'K' mentioned in JCM, *Letter to Stokes,* May/June 1860, (Harman, I, 658). This was at Glenlair, before moving to King's College London. See also: JCM, 'On the Theory of Compound Colours, and the Relations of the Colours of the Spectrum', in which Maxwell records the observations of two individuals (J) and (K).

29 JCM, *Letter to Tait,* 29 Dec 1877, (Harman, III/2, 449).

James' confidence in Katherine's measurements is at odds with Pratt-Smith's interpretation of his poetry, which she has construed as James indicating that women could not measure accurately.[30] It is possible that the verses were written tongue-in-cheek. What is certain is that James knew, through Katherine's scrupulous work, that women were well able to give accurate data. At the same time, during most of his tenure at the Cavendish Laboratory in Cambridge, women were not allowed to use the laboratory facilities. In this he followed university policy, though, like many men of his era, who were accustomed to an exclusively male university environment, he was uncomfortable at the prospect of female students![31]

Katherine was frequently unwell in the 1870s, and the final years of James' life were clouded by her suffering.[32] Campbell noted how he was a natural sick-nurse to his wife, as he had been in the past to his friend Pomeroy, to his own father, and to others. At one point during Katherine's illness, he did not sleep in a bed for three weeks, yet still gave his lectures and carried on his work at the Cavendish Laboratory. Though there is little information about James and Katherine's relationship, what we do have points to a marriage characterised by sincere affection between husband and wife.

30 Pratt-Smith, 'Boundaries of Perception', 247/248.

31 Cf. Isobel Falconer, 'Cambridge and building the Cavendish Laboratory', [Flood (ed), *JCM: Perspectives on his Life and Work*], 86. Cf. Shuster, *A History of the Cavendish Laboratory, 1871–1910*, 35, quotes Garnett: 'For several years Maxwell would not open the laboratory to women. At last he gave his permission to admit women during the long vacation when he was in Scotland, and I had a class who were determined to go through a complete course of electrical measurements during the few weeks for which the Laboratory was open to them.'

32 C&G1, 372.

Chapter 14
A Shared Faith

Dating from just before their marriage until James' death, there was a regular exchange of letters between James and Katherine. The only letters extant are those *from* James to Katherine, which were quoted from by Lewis Campbell in his *Life of James Clerk Maxwell*. These yield valuable insights into their common faith, since they contain reflections on the Scripture passages which they both read, whether together or apart. At the same time, we note that what is available for us to read was chosen by Campbell, who did not have the same evangelical sympathies as his friend.

When away from Glenlair, James wrote almost daily, giving Katherine incidental details on people, fashions, and college customs; with those light-hearted observations complemented by comments on Scripture and deeper reflections on life. Their daily readings were based on a Bible Reading Plan, though which one is difficult to establish. The Bible they used was the Authorised Version.

In an 1858 letter sent prior to their wedding, James explained that his comments on Scripture were simply a help to understanding and were not to be seen as definitive. He was alerting Katherine that his style of reflection on Scripture might be different from what she was used to from her theologian father. As a lateral thinker, he could come up with the unusual:

> Now you must remember that all I say about texts and matters of that sort is only a sort of help to being together when we read, for I am not skilful to know what is the right meaning of anything so as to tell other people, only I have a right to try to make it out myself, and what I say to myself I may say to you. [1]

Personal devotions were important to James, both as a Christian and as a scientist. Each day he took time with God. Each day he read scripture. Each day he engaged in prayer. It was with this preparation that he set out on his scientific work of unveiling the works of the

1 JCM, *Letter to Miss K.M. Dewar*, 2 May 1858, (C&G1, 309).

Creator. In the letters published by Campbell, two major themes recur. One is the Christian life, with James reflecting on how the Spirit of God draws a person into God in union with Christ. The other is James' overall underlying piety, including awareness that sin was dealt with by Christ and that Christ's presence is at the heart of faith.

Thursday, 6 May 1858 [2]

James wrote to Katherine shortly before their marriage, discussing Isaiah 51 from the Old Testament, Galatians 5:19-23 from the New Testament, and musing on the nine-fold fruit of the Spirit—*Love, joy, peace, longsuffering, gentleness, goodness, faith, meekness, temperance: against such there is no law.* He compared Judaism and Christianity, asserting that Judaism made religion a thing of 'labour and wages', instead of an inward growth of faith working by love. He discussed the tension between '*desire of the spirit*' and '*desire of the flesh*', with the former 'tending towards God' and the latter 'towards the elements of the world.' Yet, in that tension he saw 'our training in this life', echoing what he had once discussed with Richard Litchfield about knowledge of both good and evil being a catalyst towards Christian growth.[3] Within this tension the Christian was challenged to stretch upwards towards God, rather than downwards:

> Like good trees, we stretch our best affections upwards till we see the sun, and breathe the air, and drink the rain, and receive all free gifts; instead of sending our branches after our roots, down among things that once had life, but are now decaying, and seeking there for nourishment that can only be had from above.

He brought his ordered, mathematical cast of mind to bear on the specific sequence of the fruit of the Spirit, detecting a progressive logic in the Apostle's thinking:

> See the order of ripening of the fruit. Now, *love* brings *joy* to ourselves, and this, *peace* with others, and this, *longsuffering* of their attacks, for why should we be angry? *Gentleness* is a higher degree of this, being active. *Goodness* is used in a less general sense than we use it. It seems something like 'good nature,' only better and more

2 Ibid., 6 May 1858, (C&G1, 310).

3 JCM, *Letter to Litchfield,* 15 Oct 1857, (C&G1, 282f; Harman, I, 540f).

> manly, and refers to the good disposition of a man among men. *Faith* is put in here as the result of good living; which is true, for it is nourished thereby. Then see what comes of adding faith to a good disposition, *meekness*, which we cannot afford to have without faith; and lastly, *temperance* or moderation, which is also founded on faith, and is a virtue that can never be perfect till all the rest are so.

He believed that the physical nature of human existence in this world was itself a means of knowing God better: 'Our flesh is God's making, who made us part of His world; but then He has given us the power of coming nearer to Himself, and so we ought to use the world and our bodies as means towards the knowledge of Him, and stretch always as far as our state will permit towards Him.' This spurned any notion that physical existence was unspiritual as such. James saw a positive aspect to physicality, not least because the physical sciences could enhance humanity's awareness of its Creator.

Sunday, 9 May 1858 [4]

In May 1858, James travelled south for Lewis Campbell's wedding and to be his best man. He wrote to Katherine describing a hectic Sunday in the parish. With Campbell only days away from marriage, he helped his busy friend:

> To-day we were called at seven; were down soon thereafter, and had everything leisurely and comfortably till ten, when we went to school. I had a class of youths just beginning to read, and some of them knew not what swine were, still less what a herd was.

Following the bible-class there was the church service, at which Campbell's friend the Revd Greville J. Chester (1830–1892) preached:

> At eleven to church. Lewis read prayers and lessons very well and distinctly, and Chester preached on James 1:15, '*Sin, when it is finished* [completed], *bringeth forth death*.' He showed up sin as the universal poison, and showed how it might be seen working death in several instances, and also in all, good and bad, in this life, and then turned to the next, and finally indicated the remedy, though not so clearly as Paul in Romans 8:2, which he should have read.

4 JCM, *Letter to Miss K.M. Dewar*, 9 May 1858, (C&G1, 311).

The comment, that Chester could have referred to Romans 8:2, is instructive. The Romans verse states, '*For the law of the Spirit of life in Christ Jesus hath made me free from the law of sin and death.*' James felt that Chester should have brought out more clearly that it is God's Spirit, working within the Christian, which produces change, rather than the believer himself or herself doing so. After lunch there was another service:

> In the afternoon Chester read prayers and Lewis preached on '*Ye must be born again*,' showing how respectable a man [Nicodemus] it was addressed to, and how much he, and all the Jews, and all the world, and ourselves, needed to be born from above (for that is the most correct version of the word translated again). Then he described the changes on a man new-born, and his state and privileges. I think he has got a good hold of the people, and will do them good and great good.

Monday, 10 May 1858 [5]

With the wedding due later that week, James wrote again to Katherine, reflecting on Ephesians 3:19: '*And to know the love of Christ, which passeth knowledge, that ye might be filled with all the fullness of God*'. This was the springboard for him expounding one of his favourite themes: aspiring to be drawn upwards to God and into Christ's love, rather than being pulled downwards:

> Paul can express no more: but read the last two verses and you will see this is not the crown, but only what can be asked or thought. What a field for ambition there is, for climbing up, or rather, being drawn up, into Christ's love, and receiving into our little selves all the fulness of God. Let us bless God even now for what He has made us capable of, and try not to shut out His spirit from working freely.

Gaal tried—wrongly in our view—to interpret this as James claiming that marriage was the highest point of the fruit of the Spirit, of being drawn into Christ's love, or of receiving all the fullness of God in this life.[6] James was referring more widely to the whole Christian life.

---

5 Ibid.

6 Gaal, *The Faith of a Scientist*, 50.

Thursday, 13th May 1858 [7]

Three days later, James reflected on how he and Katherine were reading the same passages of Scripture, though separated by many miles. The Scripture passage was three chapters on from the one discussed on the tenth of May, indicating that their reading plan covered one chapter per day:

> Here [Ephesians 6] is more about family relations. There are things which have meanings so deep that if we follow on to know them we shall be led into great mysteries of divinity. If we despise these relations of marriage, of parents and children, of master and servant, everything will go wrong, and there will be confusion as bad as in Lear's case. But if we reverence them, we shall even see beyond their first aspect a spiritual meaning, for God speaks to us more plainly in these bonds of our life than in anything that we can understand. So, we find a great deal of Divine Truth is spoken of in the Bible with reference to these three relations and others.

The latter sentence expresses James' trust in the Bible as setting the norms for human relationships.[8] With his own wedding only three weeks away, his reflections on family relationships are of interest, especially the comment: 'God speaks to us more plainly in these bonds of our life than in anything else that we can understand.' He was expressing an ideal in anticipation of his marriage to Katherine, founded on his own positive experience of family connections.

Sunday, 16 May 1858 [9]

James wrote to Katherine on Philippians 3, verses 13, 20 and 21. On verse 13, '*forgetting those things which are behind, and reaching forth unto those things which are before*', he repeated a favourite motif: the imperative to reach forward and upward into the knowledge and service of God. Similar thoughts were expressed on verses 20 and 21, '*For our conversation is in heaven, from whence also we look for the*

7 JCM, *Letter to Miss K.M. Dewar,* 13 May 1858, (C&G1, 311).

8 Gaal, *The Faith of a Scientist,* 51. Gaal (71f.) has an extensive section on Maxwell's view of Scripture within a Reformed context, but extrapolates beyond direct evidence.

9 JCM, *Letter to Miss K.M. Dewar,* 16 May 1858, (C&G1, 312).

Saviour, the Lord Jesus Christ, who shall change our vile body, so that it may be fashioned like unto his glorious body, according to the working whereby he is able even to subdue all things unto himself, where he wrote:

> 'Conversation' in v. 20 means going backwards and forwards, and refers to the 'walking' of the preceding verses. What a description of the power of Christ in the last verse, over 'all things,' and our vile bodies among the rest, and what a day it will be when He has done all His work and is satisfied. I think the more we enter together into Christ's work He will have the more room to work His work in us. For He always desires us to be one that He may be one with us. Our worship is social, and Christ will be wherever two or three are gathered together in His name.[10]

In the same letter he admitted frustration in not being able to share his faith more clearly with a friend, whose name Campbell elided from the published text:

> I have been vexed that I could not speak better to — —. I had a long walk with him, talking of what people have believed, and what was necessary to be believed. I hope we may come to understand each other, but more that he may come to the clear light. I wish I could speak to him wise words. He is so anxious to hear, and I to speak, and then the words are all wind after all.

---

10 Marston, 'Maxwell, Faith and Physics,' 290, cites one of Maxwell's Cambridge papers in which Philippians 4:8 is copied out in Maxwell's own hand with the note: 'Dr Abercrombie's book on this text has been the delight of her life.' The reference was to Dr John Abercrombie's devotional work *Think on these Things,* which was treasured by Katherine's sister, Anne McCunn. John Abercrombie (1780–1844), was an Edinburgh physician known to the Wedderburns (Fairley, *Jemima,* 100). His daughter, Susan, married John Bruce (former minister of St Andrew's congregation who left at the Disruption) in 1845 as the widowed Bruce's second wife.

Abercrombie had been an elder at Lady Yester's Chapel in Edinburgh, and after the Disruption became an elder in St. Andrew's Free Church under John Bruce, though he hesitated long and hard about leaving the Church of Scotland. Daniel Dewar's *Evidences of Revelation* was dedicated to Abercrombie. Maxwell's note was written on stationary from Ardhallow. For John Abercrombie's life see: Douglas Maclagan, *Sketch of the Life and Character of Dr Abercrombie,* (Edinburgh: Neill, 1854).

Friday, 13 April 1860 [11]

Campbell's selection skips two years, to a moment when James was in Edinburgh whilst Katherine was at Glenlair. It was when James' Aberdeen post was ending, and shortly after his unsuccessful bid for the Edinburgh chair. The passage for their daily reading was 1 Corinthians 12, in which the Apostle discusses spiritual gifts given to the Church, and teaches that the Church is made up of many and varying members, each of whom is indispensable:

> Now let us read chapter xii., about the organisation of the Church, and the different gifts of different Christians, and the reason of these differences that Christ's body may be more complete in all its parts. If we felt more distinctly our union to Christ, we would know our position as members of His body, and work more willingly and intelligently along with all the rest in promoting the health and growth of the body, by the use of every power which the spirit has distributed to us.

Gaal saw in these words a hope that different denominations might be one in Christ, and act together for the benefit of the whole body.[12] James would have concurred with such sentiments, but the extract is more about individuals working together for the health of the body of Christ rather than denominations.

What Torrance picked up was the concept of 'union to Christ'.[13] Torrance thought he detected, in James' use of the expression, not just a repetition of Pauline phraseology but an awareness that persons only truly become persons within relationships, with the crucial relationship being that of union with Christ. Torrance believed that James had *onto-relational* thinking in theological matters as he also thought onto-relationally in his physics. In Torrance's vocabulary, onto-relations occur when the *being* (Greek: *ontos*) of a thing is what it is only because of its relationships, with relations constitutive of the *being* of an entity. In other words, any entity—properly understood—exists and only exists, as part of a continuum in a state of dynamic interaction, and never in a state of static isolation. Existence-in-relationship makes 'being' what it is. As discussed later, this is important in Field Theory.

---

11 JCM, *Letter to Katherine,* 13 Apr 1860, (C&G1, 328). Campbell states, incorrectly, that the passage was from 2 Corinthians.

12 Gaal, *The Faith of a Scientist,* 53.

13 Torrance, 'The Significance of John Philoponos', 54.

What Torrance felt he detected in James' reference to 'union with Christ', was an awareness that he and Katherine had their existence only through being 'in Christ'. In other words, their relationship to Christ was not something *added* to who they were, but *constituted* who they were. They were nothing apart from who they were 'in him'. Similar wording was employed in an 1873 letter: 'There is One who is nearer to you and to me than we ever can be to each other; and it is only through Him and in Him that we can ever really get to know each other'.[14] Such sentiments persuaded Torrance to see 'union with Christ' as something which directed Maxwell beyond himself and his own subjectivity, to God. It directed him to God, in whom he found his true self, and it gave him freedom from human-centredness, granting 'the kind of objectivity he needed for critical scientific activity.'

Did Torrance read too much into what are a relatively small number of references to 'union with Christ' in Maxwell's letters?[15] Elsewhere, Torrance speculated that though Maxwell did not intrude theological ideas *directly* into his scientific theories, the Christian faith was so deeply entrenched in his thinking that it exercised a *regulative* role in the choice and formation of his leading scientific concepts.[16] In other words, theological concepts moulded his mind-set, consciously or unconsciously. On that basis Torrance was prepared to link 'union with Christ' (involving being-in-relation) to the way in which Maxwell viewed physical reality, with his faith giving him a conceptual framework which connected the two.

---

14 JCM, *Letter to Katherine,* Dec 1873, (C&G1, 387).

15 The following occur in Maxwell's letters. (1) 'Union with Christ': JCM, *Letter to Campbell,* 22 Dec 1857: 'You have to say that what men are and the nature of their actions depends on the state of their wills, and that by God's grace, through union with Christ, the contradictions and false action of those wills may be settled and solved, so that one way lies perfect freedom, and the other way bondage under the devil, the world, and the flesh, and therefore you entreat them to give heed to the things which they have heard.' (2) 'Union to Christ': JCM, *Letter to Katherine,* 13 Apr 1860. (3) 'In Him': JCM, *Letter to Katherine,* Dec 1873: 'It is only through Him and in Him that we can ever really get to know each other.' (4) 'Engrafted': JCM, *Letter to Katherine,* 14 Apr 1860. (5) 'The life of Jesus in you': JCM, *Letter to Katherine,* 22 June 1864.

16 Torrance, 'The Significance of John Philoponos', 54.

Saturday, 14 April 1860 [17]

The next day the reading was 1 Corinthians 13, with its celebrated description of Christian love. In the Authorised Version, the word *charity* is used instead of *love*. James wrote:

> Let us read about charity, that love which is so perfect that it remains when that which is in part shall be done away. May God purify our love, and make it fit for eternity, by grafting upon it the love of Himself, that so both the human root and the engrafted branches and the divine fruit may be holy to Him!

He alluded to John 15 and its imagery of Christ the vine and believers the branches (Gaal thought he had Romans 11:16–24 in mind).[18] With significant extrapolation, Gaal argued that James was discussing the nature of the Church, and God's mysterious work in the Church. In our view, James was simply sharing with Katherine the thought that their mutual love was strengthened, perfected, and given real substance through God grafting love of himself onto it.

Wednesday, 22 June 1864 [19]

There is a gap of four years to the next letter, when James was now working in London. Katherine was at Glenlair where he would join her for the summer. He again touched on issues which for him were at the heart of Christian faith, musing on the phrase: '*That the life of Jesus may be manifest in you*' (2 Corinthians 4:11).[20] In its original

17 JCM, *Letter to Katherine,* 14 Apr 1860, (C&G1, 329).

18 Gaal, *The Faith of a Scientist,* 54.

19 JCM, *Letter to Katherine,* 22 June 1864, (C&G1, 338).

20 2 Corinthians 4:7–18: 'We are troubled on every side, yet not distressed; *we are* perplexed, but not in despair; persecuted, but not forsaken; cast down, but not destroyed; always bearing about in the body the dying of the Lord Jesus, that the life also of Jesus might be made manifest in our body. For we which live are always delivered unto death for Jesus' sake, *that the life also of Jesus might be made manifest in our mortal flesh.* So then death worketh in us, but life in you. . . . For our light affliction, which is but for a moment, worketh for us a far more exceeding *and* eternal weight of glory; while we look not at the things which are seen, but at the things which are not seen: for the things which are seen *are* temporal; but the things which are not seen *are* eternal.'

context the text refers to Christians trusting that suffering can be an agent of positive change, enabling the life of Christ to become stronger and brighter within them, and leading to the love described in 1 Corinthians 13. The personal context was that Katherine had been ill:

> May the Lord preserve you from all evil, and cause all the evil that assaults you to work out His own purposes, that the life of Jesus may be made manifest in you, and may you see the eternal weight of glory behind the momentary lightness of affliction, and so get your eyes off things seen and temporal, and be refreshed with the things eternal!

He reassured Katherine that their love—even as imperfect human love—had a permanent quality, springing from Christ's own love for his people. It would not be lessened or cancelled by death, when they were transformed (in heaven) into the image of the Lord:

> Now love is an eternal thing, and love between father and son, or husband and wife, is not temporal if it be the right sort, for if the love of Christ and the Church be a reason for loving one another, and if the one be taken as an image of the other, then, if the mind of Christ be in us, it will produce this love as part of its complete nature, and it cannot be that the love which is first made holy, as being a reflection of part of the glory of Christ, can be any way lessened or taken away by a more complete transformation into the image of the Lord.

True human love was rooted in Christ's love. True human love was defined by, issued from, and existed within, Christ's love. To support his argument on the everlastingness of true human love, he returned to Paul's portrayal of perfect divine love.

> I have been back at 1 Cor. 13. I think the description of charity or divine love is another loadstone for our life to show us that this is one thing which is not in parts, but perfect in its own nature, *and so it shall never be done away*. It is nothing negative, but a well-defined, living, almost acting picture of goodness; that kind of it which is human, but also divine. Read along with it 1 John 4, from verse 7 to end; or, if you like, the whole epistle of John and Mark 12:28. (Italics ours)

Thursday, 23 June 1864 [21]

Maxwell was concerned about how to live the Christian life, become more Christlike, and be drawn upwards to God. In his letters there are few references to 'salvation'. That was taken for granted in communications between believers. However, in this letter he expressed his belief that it was the righteousness of God which put a believer right with God, and which brought no condemnation (Romans 8):

> Think what God has determined to do to all those who submit themselves to His righteousness and are willing to receive His gift. They are to be conformed to the image of His Son; and, when that is fulfilled, and God sees that they are conformed to the image of Christ, there can be no more condemnation, for this is the praise which God Himself gives, whose judgment is just.

Here he echoed verses such as Romans 8:1; 8:29, etc. Importantly, he emphasised that it was *God* who made the believer what he or she should be, through the believer submitting to the righteousness of God. Their part was to be willing to receive the gift of being conformed to the image of God's Son, i.e. the image of Christ. When that was the case, then '*there can be no more condemnation*' (cf. Romans 8:1). And, because these things are true, then the Christian life can be lived:

> So, we ought always to hope in Christ, for as sure as we receive Him now, so sure will we be made conformable to His image. Let us begin by taking no thought about worldly cares, and setting our minds on the righteousness of God and His kingdom, and then we shall have far clearer views about the worldly cares themselves, and we shall be continually enabled to fight them under Him who has overcome the world.

Why did he highlight conforming to the image of Christ? In a lengthy excursus Gaal has argued that James was consciously repeating central tenets of Reformed theology, namely: that being created in the image of God determines our humanity in an exclusive manner; that the original purpose of our creation was to be 'a reflected likeness, a human who mirrors the Creator's will and glory'; that this perfection no longer exists in us because of the Fall; that it has been realised afresh in Jesus Christ; and that the purpose of God in salvation is to

21 JCM, *Letter to Katherine,* 23 June 1864, (C&G1, 338–339).

unite us to Christ in whom we are created anew after his image.[22] Gaal concluded that this 'is how we can understand Maxwell's theology' and he saw it as Maxwell placing himself squarely in the Reformed tradition.

Though Maxwell may well have agreed with the doctrine Gaal ascribed to him, what he really had in mind was the *Shorter Catechism's* emphasis, which interpreted the *imago Dei* in an *ethical* rather than a *metaphysical* sense.[23] In his letter to Katherine, it was the *ethical* dimension which was uppermost in his thinking.

Sunday, 26 June 1864 [24]

A few days later, James expressed the thought that if sin had to be paid for, he would be helpless—recalling Psalm 130:3: '*If thou, Lord, shouldest mark iniquities, O Lord, who shall stand?*' He was thankful that Christ bore, abolished, and cancelled sin. These thoughts came to him as he reflected on 2 Corinthians 5:10: '*For we must all appear before the judgment seat of Christ; that everyone may receive the things done in his body, according to that he hath done, whether it be good or bad*':

> Note in ver. 10 that the judgment is according to what we have done, so that if we are to be counted righteous, we must really get righteousness and do it. Note also that we are to receive the things done in the body, not rewards or punishments merely, but the things themselves are to be brought back to us, and we must meet them in the spirit of Christ, who bore our sins and abolished them, or else we must be overwhelmed altogether.

In the same letter he told Katherine of attending Pastor 'Baptist' Noel's church in London, where he enjoyed the simplicity of the message and the worship. This was not the Maxwells' regular place of worship whilst in the Capital, which was Palace Gardens Chapel. But Pastor Noel's seriousness, and his evangelical preaching, appealed to him:

---

22 Gaal, *The Faith of a Scientist,* 60.

23 *Shorter Catechism,* 'Q&A. 35: What is sanctification? Sanctification is the work of God's free grace, whereby we are renewed in the whole man after the image of God, and are enabled more and more to die unto sin, and live unto righteousness.'

24 JCM, *Letter to Katherine,* 26 June 1864, (C&G1, 339).

> I have come from Mr Baptist Noel. The church was full to standing, and the whole service was as plain as large print. The exposition was the Parable of Talents, and the sermon was on John 3:16. The sermon was the text writ large, nothing ingenious or amusing, and hardly any attempt at instruction, but plain and very serious exhortation from a man who evidently believes neither more nor less than what he says.

Tuesday, 28 June 1864 [25]

This letter, with its reflections on Christ as human and divine, is discussed in Chapter 4, when considering Thomas Crawford's influence.

December 1873 [26]

Maxwell reflected on the analogy between human marriage and Christ's union with the Church (Ephesians 5:28–33). He repeated the importance of the presence of Christ, reminding Katherine of what must have been a frequent thought between them: that it was in their being 'in Christ' that they would know each other best:

> I am always with you in spirit, but there is One who is nearer to you and to me than we ever can be to each other, and it is only through Him and in Him that we can ever really get to know each other. Let us try to realise the great mystery in Ephesians 5, and then we shall be in our right position with respect to the world outside, the men and women whom Christ came to save from their sins.

The piety expressed in Maxwell's letters was Christ-centred in belief, and Christ-centred in Christian living. Crucially, James emphasised the importance of both he and Katherine being 'in Christ', and that it was in that union with their Lord that they became their true selves and were drawn ever closer to their Lord and God.

---

25 JCM, *Letter to Katherine,* 28 Jun 1864, (C&G1, 339).

26 Ibid., Dec 1873, (C&G1, 387).

Chapter 15
Science and God's Universe

Maxwell spent the early autumn of 1856 at Glenlair, preparing his Marischal College inaugural lecture, sitting beside a roaring fire and aided by liberal doses of coffee and anchovies.[1] His Aberdeen professorial career was to start at eleven o'clock on Monday the third of November, and he wanted to address wide-ranging questions. When the day came, an array of academic prominenti were present: Daniel Dewar, Principal of Marischal; Dr Piazzi Smyth, the Astronomer Royal of Scotland; a throng of Marischal's other professors, plus a large number of students. What they saw was a young man of twenty-five, who was junior by at least fifteen years to his Marischal colleagues. Two days later the *Aberdeen Press & Journal* gave an extensive account of the speech.[2]

Aberdeen Inaugural Lecture (1856)

Maxwell began by referring to the accelerating progress of science; its expanding boundaries; and the difficulties faced by students in keeping up with new discoveries. He made two introductory points.

First, the students of 1856 need not worry about the future of science. Despite the great advances under way there was still enormous scope for fresh explorations. Science had not yet discovered everything! Its progress had simply created a larger circle of knowledge with an expanding boundary. The task was increasing, not decreasing: 'We have not only to reduce to order the regions that have been conquered, but to keep up constant operations on the frontier, on a continually increasing scale.'

Second, the sheer mass of material which students had to study might result in them losing contact with the higher aim of the study of science at a university. University was not an academic conveyor-belt, but was about providing qualifications in other ways, such as serving God in Church and State. At the very start of his professorial life, Maxwell set the study of natural philosophy within a religious and

1 JCM, *Letter to C.J. Monro,* 14 Oct 1856, (C&G1, 263; Harman, I, 416).

2 *Aberdeen Press & Journal,* 5 Nov 1856. For full text: Harman, I, 419–431.

civic context: faith and duty were what gave purpose and meaning to life in general, and to education in particular.

He then moved to the main substance of his lecture, which had three main heads: the nature and limits of natural philosophy; the type of evidence underlying such sciences; and the relation of physical science to other departments of the university.

1 Nature and Limits of Natural Philosophy. The goal of university study was not to produce technocrats, but graduates with wide horizons: 'Are we to study many things rapidly, or a few things calmly? Shall we acquire what information we can, and leave deep thinking to professional men? Or shall we employ our limited time on one or two subjects, and be content to be behind the age in general information?' He saw the method of teaching as crucial, since the aim was not to cram a mass of disconnected data into a student's mind, but to:

> Learn something of the first principles of the works of Nature, the rudiments of the Creation, so that we may no longer be oppressed with the multitude of wonders, but rather confident that, as we have recognized the operation of general principles in some instances, so by induction we shall discover more and more that the whole system of Nature is disposed according to a wonderful order.

For Maxwell, higher education should identify fundamental principles and laws of nature. It should discover why these worked as they did, and, through recognising these general principles, grasp the overall system and order of nature. Because there were *principles* to discover, then:

> It is needless to bewilder our minds with an accumulation of detached facts; for in the light of the great laws of nature these facts are no longer detached and bewildering, but are necessary consequences of the principles which they illustrate.

In other words, once the logic and rationale of nature was grasped, then what previously had been viewed as random phenomena could now be seen to be linked, enabling the enquirer to comprehend both the whole and the parts more easily. Moreover, in becoming familiar with even just one or two great physical laws, we could begin to view the universe as expressing the highest principles of order and beauty. Once detected, this inner harmony of nature would lead us on. Nature was no longer 'a mere assemblage of wonders to excite our curiosity' but 'a systematic museum designed to introduce us step by step into the fundamental principles which are displayed in the works of Creation.'

For Maxwell, *deductive* logic enabled the mind to analyse and understand the *parts*, but it was *inductive* logic which enabled comprehension of the *whole*. Torrance viewed Maxwell's inductive approach as illustrating how the best scientific minds worked with an intuitive grasp of the whole, and were never content with a plodding analysis of individual parts: instead, the mind had to leap beyond its data, and it was possession of this instinct which constituted the difference between the great and the less great.

2 ***Type of Evidence in Natural Philosophy.*** Maxwell emphasised the vital relationship between the empirical and the theoretical. The theoretical relied on the empirical, and the empirical gave substance to the theoretical. Pure thought was never enough: 'I have no reason to believe that the human intellect is able to weave a system of physics out of its own resources *without experimental labour*' (Italics ours).

As far as Maxwell was concerned, pure theory, detached from roots grounded in empirical reality through physical experiments, resulted 'in an unnatural and self-contradictory mass of rubbish.' Strong language! The task of science was to uncover the intrinsic logic and order of the universe which was given to it by its Creator, and in this task experimentation was vital. Moreover, since the laws of nature were not 'arbitrary and unconnected decisions of Omnipotence' but were 'essential parts of one universal system' brought into being by 'infinite Power', then the universe has an inherent order, which—in Maxwell's mind—issued from its divine source.

Maxwell explained to his audience that when we see how one part of science regularly harmonises with other parts, and that there is a consistency with basic principles, then we move beyond acknowledging accidental *illustrative analogies* within nature to something more. We recognise that the created order has *laws* which make sense, and we recognise that the human mind can comprehend these laws by means of measurement, weight, calculation, and discovery. He found it wonderful 'that man's reason should be made a judge over God's works'. By this he meant that human reason had somehow been given the ability to understand and analyse what God had made, and to think in a way which echoed nature itself. The mind was correlated with nature. At this point he was drawing on ideas expressed in his

'Analogies in Nature' paper of a few months earlier, in which he had mused on how the human mind could think in harmony with the structures of nature. Coming from the Scottish Common Sense realist tradition, Maxwell also affirmed that scientific knowledge can be true (objective) knowledge, though always incomplete:

> Man has, indeed, but little knowledge of the simplest of God's creatures. The nature of a drop of water has in it mysteries within mysteries, utterly unknown to us at present. *But what we do know, we know distinctly.* We see before us distinct physical truths to be discovered; and we are confident that these mysteries are an inheritance of knowledge—not revealed at once, lest we should become proud in 'knowledge and despise patient inquiry—but so arranged that, as each new truth is unravelled, it becomes a clear, well-established addition to science, quite free from the mystery which must still remain, to show that every atom of creation is unfathomable in its perfection. (Italics ours)

Returning to the religious dimension, Maxwell saw the study of the world as a spiritual duty. Science helped to fulfil God's original command in Scripture, which was to subdue the earth and have dominion over the creatures (Genesis 1:28). He interpreted 'dominion' in terms of taming the mysteries of the universe through science, with the scientist peering with awe into unsearchable depths, and treasuring what his 'little line and plummet' could reach, admiring 'the wisdom of Him, who has so arranged these mysteries that we find [first] that which we can understand' making it possible for us to gain 'an ever-increasing stock of known truth concerning things whose nature is absolutely incomprehensible.'

The structure and order of nature stimulated the enquirer to worship the One who gave the universe its logic. It was the peculiar function of physical science to lead humanity 'to the confines of the incomprehensible' and 'to bid us behold and receive it in faith, till such time as the mystery will open.'

Maxwell saw his work as a spiritual calling, as much as a profession. And, precisely because all scientific knowledge is incomplete, the enquirer needed to retain confidence (faith) in the orderliness and logic of nature in its totality. Incompleteness of knowledge meant that consistency could not be proven. That required faith. For Maxwell, it was belief in God the Creator, imparting order to his creation, which gave the basis for that assurance.

3 Natural Philosophy and other Disciplines. Maxwell repeated his concern that in the modern world—full of amazing discoveries—there was too much to teach, and too little time in which to teach it. A traditional liberal education anticipated familiarity with all departments of existing knowledge, but that was now impossible. Education had to change with the times. Yet, even though in a modern university there was no room for formal disputations—once such a key component of university life—he was glad to see that language and grammar were still considered primary elements of a proper education. Applause broke out when he said: 'The right use of speech is the first thing to be taught. We must first speak grammar, and then speak the truth.' Knowledge was impossible without clarity of expression.

He emphasised the necessity of mathematics for the study of natural philosophy. At the same time—again stressing the importance of the empirical—he commented that 'many, who care not for algebra delight in geometry; and those who are unfamiliar with calculation will yet reason rightly about forces when they meet with them in actual experiment.' In saying this, Maxwell was not only referencing the notion that geometry trained the intellect, but was underscoring his own commitment to keep the empirical and the theoretical together. He also had in mind Michael Faraday who, with little ability in mathematics, had produced remarkable results. Ironically, Maxwell's own future work would ensure that no future physicist—however talented experimentally—could make a lasting contribution without advanced mathematics.

Addressing the students, he declared: 'My duty is to give you the requisite foundation, and to allow your thoughts to arrange themselves freely. It is best that every man should be settled in his own mind, and not be led into other men's ways of thinking under the pretence of studying science.' This brought more applause. He continued:

> By a careful and diligent study of natural laws, I trust that we shall at least escape the dangers of vague and desultory modes of thought, and acquire a habit of healthy and vigorous thinking, which will enable us to recognise error in all the popular forms in which it appears, and to seize and hold fast truth, whether it be old or new.

He was of the view that theology students would benefit by studying nature: 'They will learn that though the world we live in, being made by God, displays his power and goodness even to the casual observer, yet that it conceals far more than it displays, and yields its deepest meaning only to patient thought.' Continuing this theme, he stressed:

> Constant attention to the foundation of our beliefs will also be of service in showing us the limits of our knowledge and of our reasoning powers; and when either the one or the other is assumed to be greater than it is, we shall be able to detect the assumption which underlies all the logic and rhetoric which seems at first so formidable, and, by disowning the cunningly-devised and plausible wisdom of man, we shall enter into possession of that secret wisdom, of which we know only this, that it is not our own.

He closed as he began: human knowledge can be true and objective knowledge, and all human knowledge is ultimately rooted in God.

The *Aberdeen Press & Journal* commented that the lecture was favourably received. Harman has noted that Maxwell's view of science was strongly religious in tone, and that he was anxious to emphasise the importance of a unified science of physics in which physical phenomena can be explained in terms of the structure and laws of motion of a mechanical system.[3] At that stage, Maxwell still thought in terms of 'mechanical' arrangements.

Address to Third-Year Students (1857)

Maxwell made similar points in his opening address to third-year students for the 1857 session.[4] Looking wider than the course syllabus with its inevitable exams, he stressed that the purpose of university study was threefold: to win our livelihood; to help our fellow men to do their duty; and to worship God better. In science itself, students should not be satisfied until they were able—through understanding causes—to predict what would happen under circumstances which have not yet occurred. Only then had they penetrated the logic of the event.

---

3 Harman, I, 19. Harman adds: 'His philosophical remarks echo expressions in Whewell's *Philosophy of the Inductive Sciences*.'

4 JCM, *Introductory Lecture to Third-Year Students*, Nov 1857, (Harman, I, 542–547).

It was grasping that inner rationality of nature which enabled humanity to understand its role as 'the appointed lord over the works of Creation'. Comprehending fundamental principles through knowing, submitting to, and fulfilling 'the laws which the Author of the universe has appointed' was critical to this role of dominion.[5] Maxwell was again alluding to Genesis 1:26–28 which states that God gave humanity *dominion* over the earth, with Maxwell interpreting *dominion* as *curatorship*. The curator of an exhibition provides interpretation of the objects, giving sense to the whole. Similarly, humanity interprets the created order, giving sense to it by discovering the laws which its Author has embedded within it. This is close to what Barth and Torrance would later express: that men and women are the creatures set on the boundary between heaven and earth through whom the created order comes to know itself; and that, through science enabling the created order to know itself, humanity offers the worship of all creation to God.

For Maxwell, part of the glory of science—through searching out natural laws and applying them to human needs—was in being a driver of human progress, enabling the advance of civilisation. Humanity should not halt in its scientific understanding until 'the whole earth be subdued, and every man's reason enlightened with physical truth.' He did not equate *subduing* with *exploiting*. What he had in mind was the ability of scientific research to bring vast expanses of natural phenomena under intellectual understanding, and persuade nature through experiment and reason to reveal its secrets.

He then connected right thinking and the basic principles of human knowledge (laws of thought), with consideration of the rightness or wrongness of human action (laws of moral philosophy). Such a change of direction may appear idiosyncratic to a twenty-first century readership, but Maxwell reflected a nineteenth-century desire to find a basis for ethics in logic. That was what Sir William Hamilton had sought, and although Maxwell preferred to look to Scripture for the basis of his ethics, he was still interested in attempts to root morality in logic and metaphysics. In Maxwell's eyes, the study of humanity and of right and wrong was the ultimate goal of all academic disciplines, and true science should always be a step towards a true ethics as well as towards a true theology.

5 Ibid., (Harman, I, 543).

Concerning scientific method, he again stressed to his students the indispensable connection between the empirical and the theoretical. In the past, humanity had tried to understand the natural world through the power of the mind alone, but for Maxwell that was a dead end. Against reliance on pure thought: 'patient experiment effected what unaided reason could never do'. He told the students: 'Reason led by Experiment rose to a higher level than Reason disdaining Experiment could ever reach.'[6] He acknowledged the historic change in the method of enquiry: from one of pure reason to that of reason informed by, and based upon, experimental, empirical truth. (At Aberdeen he pioneered the practice of students conducting experiments rather than only observing, and introduced the same practice in London and Cambridge.)[7]

An enemy which had to be slain was *relativism*. Maxwell accepted that, at any given moment, our understanding of nature is incomplete, and every theory requires revision. But incompleteness and necessary revisability did not mean that truth was totally subjective. He was aware that some thinkers (such as continental positivists) used incompleteness and revisability to foster relativism: 'Truth, they said, is that which each man knoweth or believeth to be true. So that if a man sincerely believes a statement to be true, which to another man appears not to be true, then it is true for the first man and untrue to the second.'[8] Maxwell rejected that position.

He also rejected the notion that all that can be known about things is how they *compare* to other things, rather than *what they are in themselves*. Moving beyond Sir William Hamilton's dictum that things are only known in their relation to other things, he stated: 'The modern philosophy, confessing the impossibility of pronouncing anything absolutely great or small, confines itself entirely to expressing the relations between the things themselves.' Against this, Maxwell argued that the scientist was able to access absolute criteria, not just comparative criteria: 'Man is indeed the measurer of the universe, but he does not use *himself* as his standard of measure nor his own variable feelings as an indication of real existence.' Instead, as Maxwell pointed out, the scientist

6 JCM, *Introductory Lecture to Third-Year Students*, Nov 1857, (Harman, I, 545).

7 James, *Remarkable Physicists from Galileo to Yukawa*, 153.

8 JCM, *Introductory Lecture to Third-Year Students*, Nov 1857, (Harman, I, 546).

uses objective measurements which are independent of his or her feelings: e.g., a thermometer with its precise markings rather than just 'feelings' of hot or cold. He recognised that, in continental Europe, positivism—after Kant—held that the true nature of things could never be known. But Maxwell reminded his students that all material things are capable of exact measurement; that such measurements lead to laws of nature; and that the objective criteria of 'number, measure, and weight, are the elements of all knowledge of the material world.'[9] This was Maxwell the realist scientist. He was convinced that scientific method accessed objective knowledge. In this he stood full-square in the Scottish tradition.

Robert Walker recalled Maxwell, on other occasions, pointing out that in a university, 'We are bound to recognize not only the unity of Science itself, but the communion of the workers of Science.'[10] In other words, it is the *community* of scientists which provides the checks, balances, questioning, and mutual appraisal which are required for advance to take place—even if some of the greatest insights originate with individuals working in isolation.

The Rings of Saturn, and Kinetic Theory of Gases

Aberdeen was aware that it had an exceptional man in its midst. Soon after his arrival, Maxwell won the prestigious Adams' Prize, and the city took pride in its young professor.[11] The *Aberdeen Herald and General Advertiser* noted the ripples of fame extending beyond Scotland:

> The Adams Prize has just been awarded to Mr James Clerk Maxwell of Glenlair, Professor of Natural Philosophy, Marischal College, Aberdeen. The subject of the prize essay on this occasion was the theory of Saturn's Rings, and we understand that Mr Maxwell has treated the rings as fluid masses. . . . Though yet a young man (it being barely three years since he was an undergraduate of Trinity College, of which he is now a Fellow) he has given such proof of rare genius for scientific pursuits that the mathematicians not only of his own country, but of Europe, are interested in his further progress.[12]

9 Ibid.

10 Robert Walker, 'James Clerk Maxwell', 211.

11 *Aberdeen Press & Journal*, 28 Sep 1859.

12 *Aberdeen Herald and General Advertiser*, 6 June 1857.

The Adams' Prize had been founded in 1848 by members of St John's College, Cambridge, in honour of their colleague John Couch Adams (1819–1892). Adams and Urbain Jean Joseph Leverier (1811–1877) had discovered simultaneously, by calculation alone, the position, orbit, and magnitude of the planet Neptune, before it was observed physically. The Prize gave £130 (a year's stipend for a clergyman) and was awarded every second year to the writer of the best essay on a set subject in pure mathematics, astronomy, or other branch of natural philosophy. It was open to any who had ever been admitted to a degree at Cambridge.

Before going to Aberdeen, Maxwell told Litchfield that he was looking at the problem of the motions of Saturn's Rings, which he found to be a 'stiff but curious subject'.[13] Saturn's rings were a puzzle. How were they constituted? Were they rigid? Were they fluid? Were they composed of diverse matter?[14] How could they remain stable without breaking up, drifting away, or collapsing into Saturn itself? The French philosopher and astronomer, Pierre-Simon Laplace, had conjectured that a uniform solid ring would be unstable, but could be stable if its mass were unevenly distributed. In Maxwell's Adams' Prize essay, he proved mathematically that Laplace's intuition was correct.[15] In terms of original scientific work completed whilst at Marischal, the 1856 Adams' Prize essay on the Rings of Saturn was his major triumph.[16]

After intense work on the topic at Glenlair through the summer and autumn of 1856, which continued in Aberdeen, Maxwell met the submission deadline of the sixteenth of December. Though he was the only candidate that year, he was awarded the prize since his paper was a *tour de force* of mathematical analysis. His conclusion was: 'The only system of rings which can exist is one composed of an indefinite number of unconnected particles revolving round the planet with different velocities according to their respective distances.' He showed that there were two possible arrangements: the particles might be in a

13 JCM, *Letter to Litchfield,* 4 July 1856, (C&G1, 261; Harman, I, 410).

14 Harman, I, 438, footnote.

15 Flood, 'Introduction', in Flood (ed.), *JCM: Perspectives on his Life and Work,* 10.

16 JCM, 'On the Stability of the Motion of Saturn's Rings', (Harman, I, 438–479). The Prize was awarded in 1857.

series of narrow rings; or they might be moving through each other irregularly. In the first case, the destruction of the system would be very slow. In the second case, it would be more rapid but with a tendency towards a narrow ring arrangement which could retard the process.[17]

Maxwell's paper became the definitive word on Saturn's Rings, and observations by NASA's *Pioneer 11* in 1979, the *Voyager 1* and *Voyager 2* flybys in 1980/1981, and the *Cassini* probe of 2009, confirmed his prediction that the rings are composed of particles and are neither a solid body nor a continuous fluid.[18] More recent research has homologated his judgement that the ring particles are not stable, but are being pulled by gravity into Saturn and may vanish entirely over the next 300 million years.

Studying the movement of particles in Saturn's Rings prompted Maxwell to consider the kinetic theory of gases, and in 1859 he read a paper on 'Illustrations of the Dynamical Theory of Gases' to the British Association in Aberdeen, publishing it in 1860. His interest in this field made him aware that statistical models were required in such complex scenarios, since not all molecules in a given volume of gas had identical energy.[19] It had been assumed, by the likes of Rudolf Clausius (1822–1888), that, though particles in a gaseous state flew around in every direction, they all had the same uniform speed.[20] Maxwell thought differently. His insights were extended by

---

17 NVN1, xv.

18 Jones, 'The Complete Physicist: James Clerk Maxwell', 11.

19 Harman, I, 26: '[After some of Maxwell's work was criticised by Clausius in 1862, Maxwell was led] to a new formulation of the kinetic theory of gases, [and] to discussions of the essentially statistical nature of the Second Law of Thermodynamics, and to speculations about the limitations of statistical knowledge.' Coulson, 'Interatomic Forces', *Clerk Maxwell and Modern Science,* 44, similarly points out that Maxwell did not *initiate* the notion of a Kinetic Theory of Gases, but developed it considerably.

20 Coulson, 'Interatomic Forces', *Clerk Maxwell and Modern Science,* 45. McCartney, 'I Remember Years', 297, noted that later scientists, including J.J. Thomson, doubted Maxwell's argument but not the conclusion. Thomson, 'James Clerk Maxwell', *Commemoration Volume (1931),* 14, wrote: 'Though his proof … has been criticised no one doubts the truth of his result—like some other men of great physical insight, his instincts were better than his reasons.'

Ludwig Boltzmann (1844–1906), with the later Maxwell-Boltzmann Equations for the distribution of velocities in a gas laying the basis for the science of statistical dynamics which governs modern molecular physics.[21] Flood has claimed that Maxwell produced the first ever statistical laws, with the 'Maxwell Distribution' leading to a new approach in physics.[22]

On the debit side, Maxwell's skilled application of mathematics to physics in the Adams Prize paper left the likes of Michael Faraday at a disadvantage. Faraday sent his young friend a *cri de coeur*: 'There is one thing I would be glad to ask you. When a mathematician engaged in investigating physical actions and results has arrived at his conclusions, may they not be expressed in common language as fully, clearly, and definitely as in mathematical formulae?'[23] Faraday added that on a one-to-one personal level Maxwell conveyed his conclusions simply, and that he (Faraday) could think and work from these. Faraday pleaded: 'Would it not be a good thing if mathematicians, working on these subjects, were to give us the results in this popular, useful, working state, as well as in that which is their own and proper to them?' Alas. Even for a brilliant experimentalist such as Faraday, the future depended on advanced mathematics.

The 1859 British Association Meeting

The British Association came to Aberdeen in 1859, with the new Mechanics' Institution and the whole of Marischal College given over to its meetings. The opening took place in Aberdeen's sumptuous Music Hall with a packed audience of two-thousand. The Hall, with its imposing façade of six Ionic columns, had recently been completed and was magnificently decorated. Banners, bearing the Aberdeen arms of Bon-Accord, floated on either side of the 1,000 pipe Henry Willis Organ, and the hall's new acoustic properties proved a resounding success.

---

21 For a clear exposition see: *JCMFN* (Issue 3, Spring 2013).

22 Flood, 'Introduction', in Flood (ed.), *JCM: Perspectives on his Life and Work,* 12. Whittaker, 'Maxwell's Famous (or Infamous) Demon', (Flood, ed.) *Perspectives,* 168f, noted that though Maxwell introduced statistical methods, it was Boltzmann (under Maxwell's influence) who produced a genuine statistical mechanics.

23 Faraday, *Letter to JCM,* 13 Nov 1857, (C&G1, 290).

His Royal Highness, Albert the Prince Consort, was Honorary President of the British Association that year. The 1851 Great Exhibition in London had shown the Prince's interest in technology and science, and he was delighted to accept the role of the Association's president. He entered the hall wearing the riband and star of the Scottish Order of the Thistle, accompanied by Professor Richard Owen who was the retiring President. Attendees included: Charles Lyell, William Hamilton, David Brewster, Roderick Murchison, William Thomson, Michael Faraday, John Herschel, Admiral Fitzroy, T.H. Huxley, and Samuel Wilberforce. These leading lights in the world of science had made their way north to the Granite City for this historic meeting. Huxley and Wilberforce were to cross swords the next year in their memorable stand-off over Darwin's theories.

Prince Albert, in his fifty-minute speech, alluded to the egalitarian and outward-looking philosophy of the Association:

> This Association is a popular Association, not a secret confraternity of men jealously guarding the mysteries of their profession, but inviting the uninitiated, the public at large, to join them, having as one of its objects to break down those imaginary and hurtful barriers which exist between men of science and so-called men of practice.[24]

Reflecting on the benefits of such gatherings, he commented that meeting together produced 'new thought by the contact of many minds, as the spark is produced by the friction of flint and steel.' He emphasised that advancement in the sciences came, not through a process akin to 'the action of the monarchy of a paternal Government, but the republican activity of the Roman Forum', and that such gatherings:

> Drew forth the philosopher from the hidden recesses of his study [and called in the] wanderer over the field of science to meet his brethren, to lay before them the results of his labours, to set forth the deductions at which he has arrived, to ask for their examination, to maintain in the combat of debate the truth of his positions and the accuracy of his observations.

The Prince highlighted the cooperative nature of scientific work. He pointed out that science progressed as the community of scientists

24 HRH Prince Albert, 'Speech to the British Association', 14 Sep 1859: *Aberdeen Press & Journal*, 15 Sep 1859.

interacted, communicated, cross-examined, checked, encouraged, and inspired one another. His speech set the tone perfectly.

Leading members of the Association along with all the foreign guests were invited to Balmoral on the closing Thursday, where Queen Victoria was in residence. Archibald Geike remembered with special pleasure the stage-coach ride up Royal Deeside to her majesty's highland residence.[25] The Queen herself, in her daily *Journal,* wrote of the event, which she termed a Fête, and which around two hundred attended. She noted the arrival of 'huge omnibuses and other vehicles, laden with Philosophers.'[26] The Royal Family was arrayed in Royal Stewart tartan and there was a guard of honour of the 93rd Regiment. Victoria described it as a beautiful sight in spite of the frequent slight showers, with the 'wild notes of the Pipes', and the background of mountains, rendering the scene striking in the extreme.

Three hundred and sixty-one papers, short and long, were read or made available during the Aberdeen gathering, with Maxwell busily involved as one of the secretaries of the Mathematical and Physical Science section. His own paper, 'The Dynamical Theory of Gases', outlined his law for the distribution of velocities of molecules in a gas, the so-called Maxwell Distribution. He also spoke on 'The Mixture of the Colours of the Spectrum', and read a paper 'On an Instrument for exhibiting the motions of a Ring of Satellites' and demonstrated his device for showing the motions of a ring of satellites. Along with David Brewster and William Thomson, he was appointed to a committee researching statistics connected with colour-blindness.

25 Geike, *A Long Life's Work,* 70. Geike also recalled that at the conference, Brewster strayed over-confidently out of his own discipline, committing a geological *faux pas*: 'Sir David Brewster created some astonishment by communicating a paper on "A Horse-shoe nail found in the Old Red Sandstone of Kingoodie." He was too grave a man of science to be capable of a joke.' Geike added, 'The geologists kept their gravity with praiseworthy patience.'

26 Victoria, *Journal,* (http://www.queenvictoriasjournals.org), 22 Sep 1859.

Chapter 16

Revivalism, Romanism, Reflections

As the new proprietor of Glenlair estate, Maxwell became a heritor for Parton Kirk. One of his first responsibilities was helping to find a minister for the parish when the Revd Thomas Leckie died of pneumonia on the first of July, aged only forty-four. The patroness of the parish, Miss Xaveria Glendonwyn, looked to her factor and the heritors for advice. In a letter to Richard Litchfield, Maxwell told Litchfield he had to:

> Enquire into the merits of the younger clergy and the sentiments of the parish on the subject; for our minister died unexpectedly this week and there are no resident proprietors in the parish except the patron of the living who is a lady of the Romish persuasion who has been for a year in Edinburgh and denies herself to all her friends.'[1]

In 1857, aided by advice from Maxwell and other available heritors, the factor, Alexander Fergusson, counselled Miss Glendonwyn to present the Revd John Young as the new minister.

Young seemed to be a good appointment. He had been parish minister of Haggs, near Stirling, since 1844. More recently, he had been recommended by the Church of Scotland to the British Government for the prestigious post of minister at St Clements, Berbice, in British Guiana, South America, which was one of the most financially lucrative posts available anywhere to a Church of Scotland minister.[2] The position was attractive to Young, who previously had been declared bankrupt in 1853, when he was described not only as a Minister of the Gospel but a dealer in minerals and a merchant and trader.[3] Young accepted the Government post, but for some reason changed his mind almost immediately and chose Parton instead. It was a surprising turn of events. Yet all seemed well. However, in just over a

1 JCM, *Letter to Litchfield,* 4 July 1856, (Harman, I, 410). Campbell omits this section of the letter (C&G1, 261).

2 *Falkirk Herald,* 1 Jan 1857.

3 *Glasgow Free Press,* 19 Mar 1853. *FES,* II, 422.

year, Young was deposed by the Kirk's General Assembly, having been found guilty of conducting himself in a manner inappropriate for a minister of the Christian Gospel. He was accused of drunkenness, profane swearing, irreverent language, and public disorderly conduct, to the 'great scandal of religion and disgrace of his profession.'[4] And not just once! He was deemed to have acted such on many occasions: in streets, railway stations, and public houses, stretching back to his time as minister of Haggs!

Intriguingly, according to the available evidence, the complaint was not brought by the Parton people, nor by the Parton patroness, nor by the Parton heritors.[5] In fact, when the local Presbytery initiated proceeding against John Young—before the General Assembly became involved—a petition signed by over one hundred parishioners was submitted in his defence. After ejection from his parish, Young preached to open-air audiences before moving to Glasgow where he died in 1859. The question arises: if the charges against Young were well-founded, why were Maxwell and the other heritors unaware of his character before they advised Miss Glendonwyn to nominate him?

The heritors were back to square one, with Maxwell's first major involvement as a parish heritor having been a sobering experience. A new minister was still needed. The next choice was the Revd William Law of Auldfield Chapel, Pollockshaws, but he withdrew close to the date of appointment.[6] Finally, the Revd Henry Alexander Pattullo was inducted on the twenty-fifth of May 1859. By that time Miss Glendonwyn had died and her nephew Robert Glendonwyn Gordon (1824–1908) was Pattullo's presentee as the new patron.[7] A few years later, following complex legal cases involving the Gordons and the Glendonwyns, the right of patronage was lost by the families, with the erstwhile factor, Alexander Fergusson, assuming the responsibility. The patronage system in the Church of Scotland was phased out in 1874, with existing patrons offered compensation for the loss of their privileges. Fergusson declined the payment.

4 *Elgin Courier,* 4 June 1858.

5 *Falkirk Herald,* 16 Dec 1857.

6 *FES,* II, 422.

7 *Kirkcudbrightshire Advertiser,* 28 May 1886.

Revivalism

Ministerial confusion reigned in Galloway, but in Aberdeen Maxwell was in a city experiencing spiritual awakening. In late 1857, two months after the British Association meetings, and in the same Music Hall which had welcomed Prince Albert, a quite different event took place. It was connected to a wave of religious revivalism which had started in Ulster and was now impacting Scotland, especially Aberdeen and the north-east. The revivalist preacher Charles Finney (1792–1875) came from the United States to take part, with Aberdeen's Music Hall one of his venues, as announced by the *Aberdeen Press & Journal*:

> The Rev. C.G. Finney, a Professor in Oberlin College, Ohio, United States, a noted preacher, is at present on visit to Aberdeen. He preached in the Evangelical Union Church (Rev. F. Ferguson's), St. Paul Street, on Sabbath afternoon and evening, and again last night, and is again to preach to-day, and on Thursday and Friday evenings. Mr Finney's style is plain, familiar, and direct. He has had a large number of hearers. Mr Finney's meeting this evening . . . takes place in the Music Hall.[8]

Finney did not initiate revivalism in Aberdeenshire. That started a year earlier, with the inauguration of revival-seeking prayer-meetings. Various evangelistic campaigns had already began in the city, featuring invited preachers such as Grattan Guinness (1835–1910), Brownlow North (1810–1875), and Reginald Radcliffe (1825–1895), all of whom addressed huge meetings.

Kenneth Jeffrey's book, *When the Lord Walked the Land*, points out that the movement exhibited distinctive characteristics in different areas: the city reaction was different from that of the farming interior, which in turn was different from that of the seaports. Meetings in the fishing villages could be highly emotional, whereas congregations elsewhere were extremely orderly. Jeffrey suggests that may have been because the fishing village services were conducted apart from the established churches, whose ministers tended to try and ensure that emotionalism did not surface.

Maxwell was interested in the revival. We do not know if he went to Finney's event in the Music Hall, but it took place within term-time and he may well have attended, perhaps along with his father-in-law,

8 *Aberdeen Press & Journal*, 16 Nov 1859.

Daniel Dewar. What we do know is that several of his fellow academics were heavily involved in the movement.

One was William Martin (1816–1890), Marischal's professor of moral philosophy.[9] When Maxwell first arrived in Aberdeen, he and Martin were regular companions on daily walks in and around the city until Martin became unwell.[10] Martin saw his professorial position as an opportunity to influence students spiritually. He was also a preacher of some renown, and during vacations preached in far-flung districts, including Orkney and Shetland. The year before Maxwell came to Marischal, Martin published a book on Butler's *Analogy,* commending it to his Sabbath Class at Church. This may explain why Maxwell was reading and admiring Butler's *Analogy* soon after his arrival in Aberdeen.[11] In connection with the revival, Martin was instrumental in inviting Reginald Radcliffe to the city in November 1858, and tried to get the use of one of Aberdeen's large parish churches. When rebuffed he hired the small mission hall in Albion Street for Radcliffe's first meetings.[12]

Another academic who was a keen supporter of revival meetings was Dr William Pirrie (1807–1882), Regius Professor of Surgery at Marischal. Pirrie not only went to Reginald Radcliffe's meetings but attended services addressed by Brownlow North, James Smith of Greyfriars, and D.T.K. Drummond—the same D.T.K. Drummond who admired Dean Ramsay so much.[13]

---

9 *Aberdeen Press & Journal,* 21 Feb 1890.

10 JCM, *Letter to Aunt Jane,* 27 Feb 1857, (C&G1, 264).

11 JCM, *Letter to Campbell,* 22 Dec 1857, (C&G1, 294; Harman, I, 576).

12 Jeffrey, *When the Lord Walked the Land,* 57. Also, J. Brown, 'A Godly Heritage: Revival in North-East Scotland', Part 1, *The Believer's Magazine,* April 2016.

13 Cf. *Aberdeen Press & Journal,* 18 Jul 1860: 'Revival of Religion: Open-air meetings at Huntly in the Castle Park, Huntly (kindly granted by Her Grace the Duchess of Gordon), on the 25th and 26th July, to unite in Prayer for the outpouring of the Holy Spirit on the district and land, and for the Preaching of the Everlasting Gospel. [The following will take part in the services] . . . William Pirrie M.D, F.R.S.E, Regius Professor of Surgery, Marischal College, Aberdeen; . . . Reginald Radcliffe, Esq; Rev. D. T. K. Drummond, English Episcopal Church, Edinburgh; . . . Brownlow North, Esq; Rev. James Smith, Greyfriars' Established Church, Aberdeen . . . etc.' See also Pirrie's Obituary in the *Dundee Courier*, 22 Nov 1882, which stated that Professor William Pirrie,

Things changed after the Revd James Smith (1805–1871)—minister of Greyfriars, the collegiate church for Marischal College—was moved by the revival, prompting him to open the doors of his church to Radcliffe. This brought the revival into the heart of college life. Smith was then censured by his Presbytery and Synod on a technical breach of church law in allowing an unqualified person to preach in Greyfriars. When the matter was taken to the 1859 Church of Scotland General Assembly, the Assembly agreed with the lower church courts in terms of church law, though speeches supportive of Smith's intentions were made by senior Kirk figures.[14] Despite the Assembly's rebuke, Smith continued to allow Radcliffe to preach in Greyfriars, with William Martin presiding at several meetings and conducting the devotions.[15]

The most powerful support for revivalism within Marischal came from Maxwell's father-in-law, Principal Dewar. Given the controversy raised in Presbytery, Synod, and General Assembly concerning the use of Greyfriars church for revivalist preaching, the fact that Dewar threw his weight behind the movement so publicly was significant.

In December 1859—a year after Radcliffe arrived in Aberdeen, seven months after James Smith defended his position at the General Assembly, and a month after Finney's meetings in the Music Hall—Dewar republished one of his monographs, entitled *The Work of the Holy Spirit in the Revival of Religion, and the Duty of Christians in regard to It.* The monograph sold for fourpence (postage free for three copies) and was distributed widely.[16] Dewar had written it twenty years earlier, following a remarkable communion season in Kilsyth which affected him deeply. That was when Kilsyth experienced a season of revival during the ministry of the Revd William Burns (1779–1859). Remembrance of these dramatic days meant much to Dewar, and he was excited at the possibility of them being repeated. He recalled the huge gatherings of between ten and twelve thousand persons; of twelve

---

an elder of the Free West Church, Aberdeen, was a supporter of evangelical and philanthropic movements in the city, frequently making large donations to the Young Men's Christian Institute and the Young Women's Christian Institute. Professor William Pirrie should not be confused with Professor William Robinson Pirie, also at Marischal.

14 *Scotsman*, 24 May 1859.

15 The Greyfriars' elders were divided on the issue. See: Greyfriars Kirk Session Minutes: 1851–1868, 4 & 7 Jul 1859, (NRS/CH2/492/2/251–256).

16 *Aberdeen Press & Journal*, 14 Dec 1859.

hundred taking the sacrament on the communion Sunday; of over eighty new professions of faith on one day alone; and of baptisms in the local mill pond. He recollected William C. Burns (jnr) (1815–1868) from Dundee preaching to thousands by moonlight. He remembered the ministry of Dr Cesar Malan (1787–1864) who had come from Geneva. And he recalled himself preaching from a high exterior flight of stairs to thousands who were gathered in the fields before him.[17]

Dewar's openness to what happened at Kilsyth came from a deeply-engrained sympathy with evangelistic campaigns, rooted in the formative influence of the itinerant preacher John Farquharson. Though the Kilsyth revival had also seen various ecstatic phenomena such as swooning and convulsions and shouting—not all of which he was comfortable with—his dominant memory was of a mighty movement of the Spirit of God.

In May 1860, Dewar committed himself irrevocably to the revivalist cause when, at the Church of Scotland General Assembly, he moved that it should give thanks to God for the revivals. Only a year earlier the General Assembly had rebuked James Smith for allowing Greyfriars to be Radcliffe's preaching base. Now Dewar was showing which side of the fence he was on. His carefully worded motion, approved by the Assembly, proposed that:

> While by no means committing itself to an approval of all that may have accompanied recent religious movements, [this Assembly] agrees to recommend to all ministers and elders of this Church special watchfulness and prayer in regard to the progress of vital religion in their parishes—*watchfulness*, lest that which is holy may be abused, or that which is good discouraged; —and *prayer*, that God may be pleased still farther to revive his own work in the midst of the years, and to grant unto this Church and the world a yet fuller outpouring of his Holy Spirit.[18]

Revivalism would be discussed in the Dewar household, which now included James and Katherine after their marriage. Like his father-in-law, Maxwell welcomed the movement, though he was wary of over-excited religious fervour, with his remarks on some of the features

17 Cf. *Scotsman,* 2 Oct 1839 & *Fifeshire Journal,* 3 Oct 1839.

18 *Aberdeen Press & Journal,* 30 May 1860.

of the 1843 Disruption testifying to his concerns on that front.[19] Though his biographer, Lewis Campbell, was not so evangelical as Maxwell, Campbell conceded that Maxwell was deeply interested in what was taking place.[20] Campbell commented that Maxwell's contact with evangelical friends in England (possibly a reference to the Taylers among others) had prepared him to sympathise with such experiences, and that his Calvinistic reading had familiarized him with 'the language used'. However, Campbell added that Maxwell 'was in no wise distracted from his professional duties by this or anything else; and although he referred to it in conversation, it has left no trace in any of his remaining letters which I have seen.' Campbell knew that Maxwell was interested in the revival, but—as with the Otley experience—he was anxious to portray his friend as broad-church.[21]

Romanism

At the opposite pole to evangelical revivalism was Roman Catholicism. John W. Arthur, in his *Brilliant Lives,* has analysed Maxwell's complex Presbyterian and Episcopal ancestry, showing that on both parents' sides he came from a lineage of intermarriages with Catholic families, which had taken place even during times of strong anti-Catholic sentiment. In Maxwell's lifetime these links were more historical than active.

19 Cf. C&G, 410. Marston, 'Maxwell, Faith and Physics', 267, refers to a series of letters, penned by 'Omicron' in March 1859 to an anonymous recipient, titled *Five Letters on the Religious Movement in Aberdeen: with an Appendix on the Nature, Probability and Necessity of a Religious Revival.* Marston suggests the writer may have been Maxwell since the writer: (a) indicates he will meet the recipient at the British Association in Aberdeen in September; (b) has a high regard for Scripture; (c) displays familiarity with the established church, the Church of Scotland; (d) refers to theologians such as Jonathan Edwards (whom Maxwell read); and (e) discusses the Smith and Radcliffe situation. If the author were Maxwell, then the letters shed great light on his faith. However, we are hesitant about the attribution since others, such as William Martin, could equally fit the bill. The title of the publication has the phrase: '*An Appendix on the Nature, Probability and Necessity of a Religious Revival*'. From what we know of Maxwell from elsewhere it is unlikely he would have written such a title: it was not his style, and he had hesitations about aspects of revivalism.

20 C&G1, 260.

21 Ibid.

By the first half of the nineteenth century, various Relief Acts permitted Roman Catholics more participation in public life, and the profile of the denomination was increasing. Many landed families had covertly retained their Roman Catholic faith when it had been suppressed. Miss Glendonwyn belonged to such a one, and in 1787 had been baptised as an infant in the Munches domestic chapel, Dalbeattie, according to Roman Catholic rites.[22] Through property ownership she was patron of Parton parish, and when Parton Kirk became vacant in 1856 it was she whom Maxwell referenced in his letter to Litchfield, telling him that the living was in the hands of a lady of the Romish persuasion, 'who has been a year in Edinburgh and denies herself to all her friends.'[23] As we have seen, though she had all the legal rights of patronage, she did not interfere directly in the appointment of the new minister, delegating that to her estate factor and the heritors.

On the wider stage, the 1840s saw the defection of John Henry Newman from the Church of England to the Church of Rome; and the 1850s saw Pope Pius IX stamping his authority on his Church. Pius IX's papacy (1846–1878) was marked by an emphasis on submission to Rome in all matters of thought and faith. In 1854 he personally declared the doctrine of Mary's Immaculate Conception. In 1864 he issued *The Syllabus of Errors*, an unequivocal rejection of modernist trends in society. And in 1870 the Vatican Council declared the doctrine of Papal

22 The Maxwells of Munches, and the Glendonwyns of Parton, were both mentioned in an early eighteenth-century report to a Church of Scotland Commission of the General Assembly, which listed all the known Roman Catholic families in Scotland. Cf. *Lists of Popish Parents and their Children in various Districts of Scotland as given in to the Lords of the Privy Council and to the Commission of the General Assembly,* MDCCI–MDCCV: '*Parish of Buittle*: Barbara Maxuel relict of the deceased George Maxwell of Munches (appostat) . . . George Maxuel of Munches; Kathrine Maxuel, his spouse; William Maxuel, his son about 15 years of age now in France; George Maxuel his son about 14 years of age now in France; James Maxuel his son about 8 years of age; Barbara Maxuel his daughter about 12 years of age; Kathrine Maxuel his daughter about 10 years of age . . . *Parish of Parton*: Robert Glendoning Laird of Partoun; Agnas Herris his Lady; George Glendoning their son aged about 18 years of age; James Murray of Conheath son-in-law to the said Laird of Partoun; Agnas Glendoning his spouse, they alleging themselves to be married and if true by a popish priest as is thought.'

23 JCM, *Letter to Litchfield,* 4 July 1856, (Harman, I, 410).

Infallibility.[24] Though Papal Infallibility was only stated formally in 1870, the stance behind it had characterised Pius IX's papacy from the start. Writing to Litchfield, and aware of these trends, Maxwell reflected on the desire of some to submit to authority rather than thinking for themselves:

> As to the Roman Catholic question, it is another piece of the doctrine of liberty. People get tired of being able to do as they like, and having to choose their own steps; and so they put themselves under holy men who, no doubt, are really wiser than themselves. But it is not only wrong but impossible to transfer either will or responsibility to another; and, after the formulae have been gone through, the patient has just as much responsibility as before, and feels it too.[25]

Robert Walker commented on Maxwell's statement, asking rhetorically: 'Could the Protestant standpoint be stated more concisely, convincingly, and, I would add, charitably? Not only *wrong*, but *impossible*.'[26] For Maxwell, men and women had to think for themselves. They had to take responsibility for themselves.

The tendency of Roman Catholicism, under Pius IX's influence, to limit intellectual freedom and questioning, was contrary to Maxwell's instincts. Only a few years earlier he had conveyed to Campbell his conviction that *precisely because he was a Christian* he had to be open to every question which might arise, with no off-limits, and with no untouchable holy ground.[27] In his letter on Roman Catholicism to Litchfield, he touched on the danger of human beings attempting to pass responsibility for their thoughts and decisions in matters of faith and life to others. In the final analysis it was impossible to do so.

Reflections

Other facets of Maxwell's religious thinking during his Aberdeen years come in his regular correspondence with Lewis Campbell and Richard Litchfield.

In an 1857 letter to Campbell, Maxwell reflected—not for the first time—on the partial and contingent nature of all human knowledge.[28] He was aware of an error in his theory about Saturn's Rings, and had

24 *Ineffabilis Deus*, 1854; *Pastor Aeternus* 1870.

25 JCM, *Letter to Litchfield*, 5 March 1858, (C&G1, 304; Harman, I, 588).

26 Robert Walker, 'James Clerk Maxwell', 207.

27 JCM, *Letter to Campbell*, 7 March 1852, (C&G1, 178).

28 Ibid., 7 Aug 1857, (C&G1, 270–273; Harman, I, 532).

not yet tracked it down. This prompted him to reflect on the search for true and accurate knowledge in any area of life, not just the physical sciences. He likened humanity's search for solid knowledge to the experience of a college student. Humanity was still going through its *Lehrjahr* (apprenticeship year) in the knowledge of perfection, but still required a *Wanderjahr* (a year typically spent travelling abroad for wider experiences) even after gaining its first diploma. Maxwell applied this analogy to the spiritual life, where the first diploma was simply a certificate of 'having *Eyes* to see the *Work*, a *Conscience* to feel after *Right*, and *Faith* to believe in the *Word*; and to reach a station thereby where both those eyes and that conscience may be satisfied, or at least appeased.'[29] *Eyes to see the Work*, was the need for an awareness of purpose in life. *Conscience to feel after Right*, was the need for a (God-directed) moral compass. *Faith to believe in the Word*, was the need (for a Christian) to base his or her life in the teaching of Scripture.

In the same note he told Campbell that he had finished the first part of Jules Simon's book *Religion Naturelle.*[30] But he had reservations. Simon's natural religion appeared to evolve from a *subjective* concept of what 'perfection' must be, which was then exalted into Simon's personal notion of the divine. In Maxwell's opinion, Simon had no *objective* reference point. In his Aberdeen Inaugural Lecture Maxwell had argued strongly against basing science in purely mental ideas, stressing that objective, empirical data was essential to give any theory a solid grounding. In his appraisal of Simon, he took the same stance with theology. Ideas alone, however lofty, had little value without an objective basis. Against subjectivism Maxwell favoured another approach: 'Reason, taste, and conscience' had to be the judges of all 'knowledge, pleasure, and action.'[31] *Reason* had to determine our knowledge. *Taste* (a sense of the fittingness of things) had to determine our pleasures. *Conscience* had to determine our actions.

29 Ibid., (C&G1, 272; Harman, I, 532).

30 Jules Simon, *Essai Critique sur la Religion Naturelle*, (1851).

31 JCM, *Letter to Campbell,* 7 Aug 1857: 'My notion is, that reason, taste, and conscience are the judges of all knowledge, pleasure, and action, and that they are the exponents not of a code, but of the unwritten law, which they reveal as they judge by it in presence of the facts. The facts must be witnessed to 'by the senses, and cross-examined by the intellect, and not unless everything is properly put on record and proved as fact, will any question of law be resolved at headquarters.'

But how were these criteria to be evaluated? Here Maxwell appealed to an 'unwritten law' which is 'not a human code'. This was a reference to the biblical texts Jeremiah 31:33 and Romans 2:15 which speak of God's law being written on the hearts of men and women. Because of what God has written deep into the human heart, men and women are able to make right judgements. But for Maxwell, the actual facts, which that unwritten law assesses and evaluates, were 'witnessed to by the senses' and 'cross-examined by the intellect.'

Maxwell also argued that though facts (data) are important, they are but 'rough dead facts' from which 'living truth' has still to be drawn. Moreover, understanding 'living truth' depends on 'the utterer of facts' or the 'giver of reason'. In other words, knowledge is only possible because God (the utterer of facts and giver of reason) has given a rationality to reality, and has enabled the mind to know it. Maxwell later expressed his conviction—again against continental positivism—that the mind is able to go beyond merely gathering data, and is able to grasp the logic and structure within which such data rests (whether data be physical or spiritual) because the facts (data) have come from the One who gives reason, logic, and structure to reality.[32]

Whewell's emphasis on inductive logic lay behind much of what Maxwell wrote in this letter: whereas positivism limited science to no more than observation of empirical data, Whewell and Maxwell accepted that the mind could intuit and discern general laws beyond the immediately experienced phenomena.

Later the same year, he wrote to Campbell, who recently had been elevated by the Anglican Church from deacon to priest.[33] Maxwell referred to some of his current reading: Butler's *Analogy* (possibly

---

32 Cf. JCM, *Letter to M. Pattison,* 13 Apr 1868, (Harman, II/1, 362): 'You must see that my acquaintance with positivism whether as stated by Comte, Littre, or Mill is but slight ... [but] when [Comte] prescribes rules for the study of sciences already formed, such as astronomy . . . [then] all this part of the philosophic spirit is proscribed, and the astronomer is even forbidden to extend his views beyond the solar system. . . . The statements of certain positivists therefore about human knowledge and science in general are apt to make one think that they are convinced that new truths will fall into the old moulds, even in sciences yet unborn.' Cf. Marston, 'Maxwell, Faith and Physics', 271: 'By the 1870s, there was growing interest in the exclusive constraints on knowledge imposed by Comte's Positivism.'

33 JCM, *Letter to Campbell,* 22 Dec 1857, (C&G1, 293–297; Harman, I, 576f).

encouraged by William Martin), Herschel's *Essays,* Froude's *History,* and Buckle's *History of Civilisation.*[34] Concerning Campbell becoming a priest, the letter was both playful and serious:

> I take for granted that sometime on Sunday last you entered the second of the ecclesiastical transformations! May your life and doctrine set forth God's glory, and be the means of setting forward the salvation of all men!

Maxwell contrasted the work of a scientist and that of a clergyman. Scientists, such as himself, dealt with laws which were invariable, constant, and dependable, whereas clergy had to deal with human beings for whom 'the nature of their actions depends on the state of their wills.' Campbell's task, as Maxwell saw it, was to bring men and women to a place where: 'by God's grace, through union with Christ, the contradictions and false action of those wills may be settled and solved' always aware that in one way lies perfect freedom, and in the other way lies 'bondage under the devil, the world, and the flesh!'

Maxwell saw his scientific world as one of regular laws whereas Campbell's world involved human will, human decision, contingency, and uncertain behaviour. Whereas Maxwell could impact his world by working according to set rules, Campbell could only influence his by entreating his people 'to give heed to the things which they have heard.'[35]

Significant—especially given his use of the same concept in correspondence with Katherine—was Maxwell's use of the biblical notion of 'union with Christ'. He stressed to Campbell that it was through 'union with Christ' that men and women were transformed to be what God intends them to be. Gaal has argued that, for Maxwell, true human existence was thought of only in Christ and together with Christ, with the ability to live life, as it should be lived, entirely dependent on union with him.[36] Gaal asserted that Maxwell's emphasis on union with God, union with Christ, and union with one another, sprang from, and was directly related to his Field Theory notions.[37] But in this letter Maxwell was not thinking about ontology or metaphysics: he was expounding given scriptural concepts from a moral and ethical standpoint.

34 Harman, III/1, 119, footnote 5, stresses Maxwell's admiration of Butler, referring to Vol. I, 280, 576.

35 JCM, *Letter to Campbell,* 22 Dec 1857, (C&G1, 293ff).

36 Gaal, *The Faith of a Scientist,* 36.

37 Ibid., 45.

Towards the end of January 1858, Maxwell wrote again to Campbell about a priest's calling.[38] He began by teasing his old school-friend:

> Don't suppose that I intend to make you my confessor. It would not be just to you, for I do not like being confessed to myself; and I think everyone should bear his own burden, though willing to lighten that of others. Besides, confession brings into set words and distinct outlines, doubtful suspicions and half-formed thoughts, which would fade at once if they were not stirred up.

Then, an insightful section:

> The consciousness of the presence of God is the only guarantee for true self-knowledge. Everything else is mere fiction, fancy portraiture, done to please one's friends or self, or to exhibit one's moral discrimination at the expense of character.

Gaal interpreted Maxwell's statement, 'The consciousness of the presence of God is the only guarantee for true self-knowledge' as intentionally echoing the opening sentence of Calvin's *Institutes*: 'Our wisdom, in so far as it ought to be deemed true and solid Wisdom, consists almost entirely of two parts: the knowledge of God and of ourselves.'[39] In our opinion, this over-interprets Maxwell. He was simply emphasising that for him—and, by extension, for everyone—it was awareness of the presence of God which kept him honest with himself. As so often, Maxwell was concerned with moral and ethical action. Nothing could be hidden from God. Nothing could be pretended with God. Nothing could deceive God. The knowledge of the presence of God created humility and honesty. That was his experience.

In one of his communications with Richard Litchfield, Maxwell mused on the thought that, despite his love of science, friendship was far more important:

> It is in personal union with my friends that I hope to escape the despair which belongs to the contemplation of the outward aspect of things with human eyes. Either be a machine and see nothing but 'phenomena', or else try to be a man, feeling your life interwoven, as it is, with many others, and strengthened by them whether in life or in death.[40]

38 JCM, *Letter to Campbell*, 31 Jan 1858, (C&G1, 300–301).

39 Gaal, *The Faith of a Scientist*, 37.

40 JCM, *Letter to Litchfield*, 23 Sep 1857, (C&G1, 280–282).

A few weeks later, he discussed with Litchfield the linked concepts of 'The Good', 'Knowing Good and Evil', and 'Trying to be a Good Man.'[41] He wrote his letter after he had learned of the murders of his aunt's cousin John Wedderburn (1825–1857), his wife Alice (1833–1857), and their infant child John James (1856–1857) during the Indian revolt, an event to which he referred in his opening lines. Throughout his life Maxwell kept returning to moral and theological themes, and in this letter he pondered the relative merits of persons who had gained good character *after a bad history*, over against persons who had *never been in that position*:

> Some men redeem their characters by their deeds, and we praise them. Those that merely show their character by their deeds should be remembered, not praised; and a complete true man will live longest in the memory, and, I cannot but think, will be less changed in reality, than one who has doubtfully struggled with duplicity in his constitution, and has walked with hesitation, though along a good path.

Awareness of Good *and* Evil affected a person's rise to 'the highest state of all':

> I know that both [types of person] do deny and renounce themselves in favour of duty and truth, as they come to see them; and, as they come to see how goodness—having the knowledge of evil—has passed through sorrow to the highest state of all, they accept it as a token that they have found their true head and leader, and so, with their eyes on him, they complete the process called the knowledge of Good and Evil, which they commenced so early and so ignorantly.

Was Maxwell arguing that humanity could only know goodness through having had the knowledge of evil? He certainly stated that humanity, through becoming aware of pure goodness through its 'true head and leader' (God/Christ) was in a virtuous cycle in which knowledge of God/Christ enabled humanity to complete the process of the knowledge of Good and Evil. Maxwell's logic was:

- Awareness of the 'good' brings awareness of God.
- Awareness of God leads to higher knowledge of the 'good'.

41 JCM, *Letter to Litchfield,* 15 Oct 1857, (C&G1, 282f; Harman, I, 540f).

This spiralling reasoning appealed to Maxwell as a mathematician. It may also be that—in arguing that true knowledge of the 'good' resulted from a true knowledge of God/Christ—Maxwell had in mind Jesus' statement *'there is none good except one, that is God'* (Mark 10:18). But what might be the endpoint of this iterative process?

> What the completion of the process is I cannot conceive, but I can feel the difference of Good and Evil in some degree; and I can conceive the perception of that difference to grow by contemplating the Good till the confusion of the two becomes an impossibility.

Maxwell's point was that an ever-deepening knowledge of God stimulated knowledge of the 'good' to such a degree that confusing 'good' and 'evil' would eventually become impossible. He related this to personal identity, and to the fact that, for each person, their identity included their history, with that history including both good and evil:

> Then comes the mystery. I have memory and a history, or I am nothing at all. That memory and history contain evil, which I renounce, and must still maintain that I was evil. But it contains the image of absolute good, and the fight for it, and the consciousness that all this is right.

As a Christian believer, he was aware that the doctrines of redemption and salvation implied that part of his history and identity was evil and was to be renounced. At the same time, there was no need to renounce all his past, since that past also included the reflection or image of absolute good, plus the striving for it.

Was Maxwell rejecting the classical Calvinist doctrine of humanity's total depravity? Or was he celebrating the notion of common grace? Whichever was the case, his thinking was more about the positive side. What was important to him was that, as a Christian, he found his identity in God who is the 'good'.

Maxwell touched further on the issue of evil, a topic which he had discussed with Campbell in 1853 when he had alluded to the essential irrationality of evil: 'Men fall into absurdity as soon as they have settled for themselves the question of the origin of evil.'[42]

In that earlier letter to Campbell he had considered what was 'evil'? Was it simply 'unripe good' or a 'lack of goodness' rather than an objective entity in itself? He differentiated between *objective evil,* (the

42 JCM, *Letter to Campbell,* 20 Feb 1853, (C&G1, 182f).

abstract notion of evil) and *subjective evil,* (the human experience of pain and suffering). (a) Concerning *objective evil*: 'I say nothing with respect to objective evil, except that it is a part of the universe which it may be the business of immortal man to search out for ever, and still see more beyond. We cannot understand it because it is relative, and relative to more than we know.' What Maxwell meant was that *evil,* if it were to be properly defined, needed to be defined relative to the pure and absolute goodness of God, but that was impossible for us. (b) Concerning *subjective evil*: 'But subjective evil is absolute; we are conscious of it as independent of external circumstances; its physical power is bounded by our finitude, bodily and mental, but within these its intensity is without measure.'

Maxwell's statement that subjective evil is absolute was not, in this context, linked to the Calvinist doctrine of total depravity, but referenced how evil is perceived and experienced by us. His point was that, through actual pain and real suffering, sin and evil affect us far more than any abstract or theoretical notion. Gaal concluded that Maxwell's solution was to point to Christ, and reconstructed the following as Maxwell's probable line of thinking: 'If I see first Christ's love which is the only real absolute reality, then He can draw me nearer to Himself out of the depths of my own being, and in this way I will not look for my absolute reference point in myself.'[43]

During his Aberdeen years, Maxwell's reputation grew internationally, as well as throughout Britain. Despite that recognition, his post was not protected at the fusion of King's and Marischal in 1860, after which there could only be one professor of natural philosophy. In most departments the younger professor was retained, but not in natural philosophy. This was why Maxwell applied for the Edinburgh professorial chair vacated by his old supporter James Forbes. Edinburgh turned him down. Edinburgh's loss was London's gain. He was snapped up by King's College, London, becoming its professor of natural philosophy and astronomy.

43 Gaal, *The Faith of a Scientist,* 35. Maxwell expressed similar thoughts in JCM, *Letter to Katherine,* 10 May 1858, where he wrote of, 'being drawn up into Christ's love and receiving into our little selves the fullness of God.'

King's College London

(1860–1865)

Chapter 17
The Maxwells in London

The Maxwells arrived in London in October 1860, first staying at 7 Palace Gardens Villas, before setting up home at 8 Palace Gardens Terrace which has since been renumbered as No. 16 with a blue plaque commemorating Maxwell. The house was in fashionable Kensington, near Kensington Palace, and in an area now crowded with important Embassies. The recently built property had a large garret, ideally suited for James' experiments. Katherine was his willing helper. The 1861 census listed two staff, including a Scottish housemaid.

Among their neighbours was a Mrs Cecilia Chambers who lived with her son, George Frederick Chambers (1841–1915), at No. 2. A short time before the Maxwells arrived, and aged only nineteen, Chambers had published his first book, *A Review of Descriptive and Practical Astronomy*, which was a massive volume of six hundred pages. Later editions ran to three volumes. Though Chambers is not mentioned in the extant Maxwell letters, James would have been aware of the young man living only yards away. Other neighbours included stockbrokers, retired colonels, businessmen, lawyers, and assorted professionals.

Maxwell was soon part of the capital's science community. In March 1861, the mathematician Thomas Archer Hirst (1830–1892) recorded meeting him at an evening party in the home of the zoologist Dr William B. Carpenter (1813–1885), at which Hirst was introduced not only to Maxwell but to Hermann Helmholtz (1821–1891). Hirst described Maxwell as talkative with a broad Scotch brogue, who took great interest in his theory of ripples and currents. The topic fed directly into Maxwell's own interest in the flow of gases.[1]

From 1861 until 1862 the Palace Gardens household included Katherine's younger brother, Donald Dewar, who had come to London for medical treatment. Two of Katherine's brothers became ministers: Donald who was minister of Ellon in Aberdeenshire; and

1 Hirst, *Journal*, 24 Mar 1861, (Harman, II/1, 2).

an older brother Edward (1827–1855), who had been minister of Auchtergaven in Perthshire until his early death. When Donald arrived, much of the ground floor was given over to him and his nurse. He had been inducted as minister of Ellon only in February 1860, and his father, Principal Dewar, helped to cover parochial duties during his illness. Katherine was at Glenlair for part of the time, so James ate in a small backroom, usually breakfasting on porridge. Each day, when he returned from college, he helped to nurse Donald who lit up with pleasure whenever he saw him coming. If James and Katherine both needed to leave London, Mrs Dewar came and looked after her son.

Donald responded well to treatment, at least initially. This induced Maxwell to write with optimism to C.J. Monro in February 1862: 'When I wrote you last, we were expecting my brother-in-law whose leg was ailing and we wished him to come here for advice. The advice was as we feared—to have it off—so it was done just three months ago by Fergusson, and we think the disease is arrested, for the recovery has been very steady.'[2] At that point Donald was mobile on crutches, much of the pain had gone, and he was being measured for a new prosthetic limb. The mood was cheerful. When a tea-party was held, he was able to come upstairs and join the fun. A complete recovery was anticipated, with Donald becoming an active man 'with an American leg!' Sadly, this positive trajectory was not sustained. He died in June 1862 aged thirty, and after only two brief years in his parish.[3] His successor in Ellon was the spirited James Smith, formerly minister of Greyfriars and doughty champion of Reginald Ratcliffe and the revivalist movement.

Palace Gardens Chapel

Along with home and college there was church to attend. In Galloway, the Revd George Sturrock arrived in 1861, and he was to be friend and parish minister for the rest of Maxwell's life. Maxwell had been instrumental in bringing him from Aberdeen to Corsock.[4]

2 JCM, *Letter to C.J. Monro,* 18 Feb 1862, (Harman, I, 707). JCM, *Letter to H.R. Droop,* 28 Jan 1862, (C&G1, 332; Harman, I, 706).

3 *Aberdeen Press & Journal,* 18 June 1862.

4 Sturrock, 'Letter to Editor', *Glasgow Herald,* 29 May 1885.

But if things were now fine in Galloway, Maxwell was dissatisfied with much of the preaching he heard in London. He was a serious man when it came to faith. He never admired ingenious sermons, nor did he approve of preaching which played to the gallery. He disliked preachers with an after-dinner style of delivery. And he had no time for those who performed intellectual gymnastics in coming to the opposite conclusion of the Scripture verse they started with! His preference was for solid, biblically based, intelligent sermons. For him, the primary task of the preacher was to explain, affirm, and apply Scripture.

The London church which the Maxwells attended on a regular basis was John Offord's Palace Gardens Chapel, situated close to their home. Offord was an open-communion Baptist, meaning that a person did not need to be a Baptist to take the Lord's Supper in his church. Offord came from the south-west of England, and had ministered in Plymouth and Exeter before being called to London. Described as a zealous, hard-working pastor, and respected across denominations, Offord not only threw himself into the work at the Palace Gardens Chapel but was heavily involved with organisations such as the Evangelical Alliance.[5] He was a tutor at Spurgeon's Metropolitan Tabernacle Pastor's College (the training seminary set up by Spurgeon in 1856) and was in high demand as a speaker, with newspapers reporting his addresses at special meetings.

In September 1864, Offord spoke on 'Preaching' to the Open-Air Mission Society in the Adelphi Theatre, basing his remarks on 2 Corinthians chapter 5.[6] Reflecting on the methods used by some evangelists, he argued that there should be no frivolity, no lightness, and no unguarded expressions in preaching. Though preaching was about *Good News*, it was also about *Serious News*. The preacher was required to reflect the *gravity* of his message in how he spoke. The utter *seriousness* of sinful humanity's plight before God was central in Offord's presentation of Christian faith. He addressed the same Society in 1865, this time expounding core Christian doctrine,

5 Cf. *The Western Morning News*, 6 July 1869.

6 *The Atlas*, 1 Oct 1864. Offord's address had three headings: 'The Constraining Love of Christ'; 'The Account to be rendered to Christ'; 'The Ministry of Reconciliation committed to each Preacher.'

especially the difference between Justification and Sanctification.[7] This distinction is crucial in evangelical Protestantism and was stressed by several of Maxwell's ministers, including Thomas Crawford and Charles Tayler. For Crawford, Tayler, and Offord, the crux of evangelical doctrine was that men and women are justified (put right with God) purely by God's grace received through faith once and for all. Justification creates a new and irrevocable status before God for a Christian, which nothing can cancel. In contrast, sanctification is about the gradual process in a believer's life in which, through union with Christ, he or she matures spiritually and morally. This was what Crawford, Tayler, and Offord all taught, dovetailing with the Calvinist theology of Maxwell's Presbyterian roots.

In a letter to the Revd Charles Tayler, Maxwell confided that what he valued most in Offord's preaching was fidelity to Scripture, contrasting it with the fanciful preaching he endured elsewhere:

> At Cambridge I heard several sermons from excellent texts, but all either on other subjects or else right against the text. There is a Mr. Offord in this street, a Baptist who knows his Bible, and preaches as near it as he can, and does what he can to let the statements in the Bible be understood by his hearers. We generally go to him when in London, though we believe ourselves baptized already.[8]

Offord's method was to stick close to the original intended meaning of a text. This, for example, was the approach he took in his sermon on Revelation 3:20, '*Behold I stand at the door and knock.*' Offord knew that the text was a favourite for evangelistic sermons, and was widely used when addressing the unconverted. Though Offord did not rule out that application, he felt it was against the original context, which was an appeal to Christians who were lukewarm in their faith, rather than an appeal to outsiders. For Offord, context should determine interpretation:

---

7 Offord, *Justification and Sanctification: Their Relations and Distinctions.* (An address to Members of the Open Air Mission), 30 Oct 1865. Cf. *The Morning Advertiser,* 1 Nov 1865.

8 JCM, *Letter to Rev. C.B. Tayler,* 2 Feb 1866, (C&G1, 344).

> [Christ] addressed these words to a church in a special condition, and under special circumstances; and he has, doubtless, left them on the inspired record, that they may meet all like conditions, and be applied under all like circumstances, of the Church of God, until the end of time.'[9]

Offord's approach to exposition was based on accurate exegesis. This delighted Maxwell. He wanted preachers to dig into the Word of God in Scripture with the same honesty and respect for its inner logic and purpose as he dug into the Work of God in Nature, allowing it to speak for itself.

Several of Offord's sermons were reconstructed from notes taken by a member of his congregation, Madame Camilla Soares. These reveal a preacher who was clear and logical in his presentation. Unlike Charles Spurgeon (1834–1892) at the Metropolitan Tabernacle, Offord seems to have avoided illustrations, though he may have added lighter touches in actual delivery.[10] The hallmark of his pulpit-style was to treat Scripture with utmost seriousness, approaching it from within an evangelical mind-set. Maxwell approved.

Offord was not a rich man. When he died suddenly in July 1869, aged only fifty-nine, no adequate provision existed for his invalid daughters, and the Palace Gardens Chapel had to assist. The Chapel was then purchased for £5500 by the English Presbyterian Church, with the Revd Dr Augustus Schwartz its minister.[11] Schwartz was a converted Jew, who had been missionary to the Jews of Amsterdam for over twenty years, latterly working for the Free Church of Scotland. Schwartz's ministry in London proved brief. He died in the late summer of 1870, only months after his induction. The building was later bought by a Swedenborgian Group. All these developments occurred after Maxwell had resigned from King's College in 1865, though he and Katherine often returned to the metropolis.

---

9 Offord, 'Christ's Appeal to Lukewarm Saints', *Faithful Words: or, The Inner Life*, 2.

10 Cf. Offord, *Faithful Words: Or, The Inner Life: Its Nature, Fruits, and Conflict*, (London: 1867); Offord, *Notes of Sermons by the late J. Offord, taken by a Member of the Congregation*, (London: 1870).

11 *Dundee Courier*, 2 Sep 1870.

John Street Baptist Church

As well as attending Offord's church, Maxwell heard other preachers, such as Mr 'Baptist' Noel.[12] The Honourable Wriothesley Noel (1798–1873) came from minor aristocracy and had grown up in the Church of England. From 1827 until 1848 he was minister of the evangelically inclined Anglican congregation of St John's Chapel, Bedford Row, London. As a friend of Patrick Brontë—father of the three literary Brontë sisters, Charlotte, Emily, and Anne—Noel advised Charlotte and Emily when they were preparing to live on the Continent to further their French and their experience of teaching. He recommended an institution in Lille, though Brussels became their destination.[13] In 1849 Noel left the Church of England for the Baptists, becoming pastor of John Street Baptist Church in Bloomsbury. He served twice as President of the Baptist Union.

Maxwell approved both the style and content of Noel's preaching. He told Katherine that two addresses were given during the service. The first was an exposition on the parable of the talents. The second was a sermon based on John 3:16: '*For God so loved the world.*' The practice of two addresses: one a didactic explanation of a passage of Scripture (the Exposition), and the other a freer discourse (the Sermon), was common in non-Anglican churches, and widespread in Presbyterian churches in Scotland. In stating that the sermon was 'the text writ large', Maxwell was telling Katherine that Noel's method focused on revealing and explaining what was in the Scripture passage, rather than using the text as a launchpad for his own conjectures. Maxwell liked this. It was what he knew from Scotland. Moreover, as a scientist who sought to uncover and explain what was objectively *in* nature, he expected preachers to do the same with the Word of God. Just as scientists allowed nature to disclose itself as it is, so preachers should avoid forcing personal

---

12 JCM, *Letter to K. Maxwell*, 26 June 1864, (C&G1, 339). Gaal, *The Faith of a Scientist*, 58, 82, interpreted Maxwell as disappointed in Noel's treatment of the text. Gaal may have been unaware of the intended difference between the two addresses. Maxwell's phrase, 'hardly any attempt at instruction' defined the Sermon over against the Exposition. Maxwell was highlighting the difference of style and purpose between the two addresses.

13 Gaskell, *The Life of Charlotte Brontë*, Ch. 10.

notions onto Scripture. The task of both scientist and preacher was to reveal what lay within.

There is no record of the Maxwells attending Spurgeon's ministry though they may have, since Sturrock recorded that there were few of the non-conforming ministers of real note that Maxwell had not heard.[14] Cousin Jemima did hear the great preacher (possibly in the late 1850s) but she had less liking for evangelical preaching than had James. She wrote in her *Journal* that she and other members of the family walked several miles to Spurgeon's church one Sunday morning.[15] Admission was by ticket only as there was such a crowd, though the fashionable people—of whom Ruskin was one—had boxes. Spurgeon had no pulpit, but there was a platform on which—as Jemima described it—the preacher walked about, almost acted, suiting the action to the word. She recorded that Spurgeon had a fine powerful voice and could be heard all over the large building without exerting himself. The subject of the sermon was the Sabbath, and Jemima remembered him emphasising: 'Six days shalt thou work: hear this, ye lazy ones.' In her opinion, he was good, but not equal to many Scotch preachers.

In the afternoon, she went out to Blackheath and attended church with some friends. Her *Journal* documented a double scolding. One from her Anglican cousins for going to a dissenting place of worship. And one from her Presbyterian friends for travelling on a Sunday!

---

14 Sturrock, 'Letter to Editor', *Glasgow Herald*, 29 May 1885.

15 Cited, Fairley, *Jemima*, 124.

Chapter 18
King's College Scientist

The academic year was longer at King's College than at Marischal, stretching from early October to the end of June. This gave less time for Glenlair. Whilst classes were in session, Maxwell attended college for at least three mornings a week, delivering hour-long lectures to his regular students. There was a separate course on experimental physics which was spread over twenty lectures, plus an evening class which met once a week. He also helped at the Working Men's College. With his sympathy for Christian Socialism and its aim of improving the lot of ordinary workingmen, he enjoyed the extra-mural work. Even after his sudden retiral from King's in the spring of 1865, he taught the workingmen through the following winter.

King's College Inaugural Lecture

Maxwell delivered his London Inaugural Lecture in October 1860.[1] Unlike the Aberdeen equivalent, references to the divine were fewer and more oblique. Maxwell's aim was to spell out what natural philosophy was, and to detail its scope and methods. He was aware that King's College concentrated on applied science, so he talked of natural philosophy as it dealt with forces and matter. He speculated on whether chemistry could be reduced to mechanics, and—unknowingly anticipating sub-atomic physics—thought that might be possible if chemical elements were simply arrangements of a more basic kind of matter, but 'no one has hitherto been able to do this.'[2]

Concerning electricity and magnetism, he noted that an immense mass of facts had already been collected, reduced to order, and expressed within a number of experimental laws. However, the form under which these laws might ultimately appear—as deduced from fundamental principles—was as yet uncertain.[3] This throw-away

1 JCM, *Inaugural Lecture, London,* Oct 1860, (Harman, I, 662–674).

2 Ibid., (Harman, I, 664).

3 Ibid., (Harman, I, 667).

statement anticipated some of his best work whilst at King's, in which he consolidated masses of data into a series of comprehensive laws driven by recognised principles.

Was science running out of new findings? Had everything worth discovering been discovered? Maxwell reassured London, as he had reassured Aberdeen: 'The present generation has no right to complain of the great discoveries already made, as if they left no room for further enterprise. They have only given Science a wider boundary.'[4] He also stressed that care should be taken vis-à-vis metaphysics and physics, though insisting that a conversation between the two was necessary since the fundamental truths of physics demanded it:

> In this class we do not pretend to study metaphysics in a formal and direct manner; but if, by careful study of the laws of nature and their dependence on each other, we have been trained into watchfulness *over the processes of our own minds* and *clear habits of thought*, we shall come all the better prepared for the study of higher problems—whether they are presented to us in a metaphysical shape, or as they occur sooner or later to every thinking man.[5] (Italics ours)

In other words, no thoughtful person could avoid asking what lay behind natural laws. He cited Bacon, Descartes, and Leibniz as thinkers who kept physics and metaphysics together, albeit in different ways. Aware of how, in the past, his use of analogy had been misunderstood, he underlined the need to distinguish between 'experiments of illustration' and 'experiments of research.'[6] His lecture did have one clear reference to divinity when he remarked that students would see that:

> The laws of nature are not mere arbitrary and unconnected decisions of Supreme Power [but] form essential parts of one universal system, in which Infinite Power serves only to reveal unsearchable Wisdom and eternal Truth.[7]

Maxwell may have had Whewell's principle in mind: that a true natural philosopher believed that God had made a universe *capable*

4 JCM, *Inaugural Lecture, London,* Oct 1860, (Harman, I, 667).

5 Ibid., (Harman, I, 668f).

6 Ibid., (Harman, I, 669).

7- Ibid., (Harman, I, 670).

of examination, and that God had not arranged things in nature *arbitrarily*. As noted earlier, Darwin drew on that principle when he quoted Whewell on the title page of his *On the Origin of Species* (1859).[8] Whewell's point was that a theist, as a person believing in the orderly creative activity of God, should *expect* to see phenomena which were explicable through an extended series of natural causes and effects, rather than encountering phenomena which had no traceable roots. In other words, there were investigable dynamic processes lying behind the form of reality as we see it today: in the beginning, God had *not* made things exactly as they are now with no development since. In this vein, Whewell (and Maxwell following him) taught that the task of *science* was to understand how things are now, by unveiling the necessary chain of natural causes lying behind phenomena.

Maxwell believed that the trail of discoverable causes for phenomena in nature was inexhaustible, though the ultimate cause was divine. In this sense, he sympathised with materialists who sought material causes for all individual phenomena. But, unlike materialists, he believed that the 'whole' demanded a divine originator. Crucially, he emphasised that science also involved a commitment to understanding the internal logic of reality, and was never satisfied with simply cataloguing disconnected phenomena:

---

8 Whewell's fuller text, part of which was quoted approvingly by Darwin, is in Whewell, *Astronomy and General Physics considered with reference to Natural Theology*, (3rd ed., 1834), 356–357: 'We are not to expect that physical investigation can enable us to conceive the manner in which God acts upon the members of the universe. … Indeed, science shows us, far more clearly that the conceptions of everyday reason, at what an immeasurable distance we are from any faculty of conceiving *how* the universe, material and moral, is the work of the Deity. But with regard to the material world, we can at least go so far as this—we can perceive that events are brought about not by insulated interpositions of Divine power, exerted in each particular case, but by the establishment of general laws. … God is the author and governor of the universe through the laws which he has given to its parts … these laws and properties are … the instruments with which he works; [and] the institution of such laws, the selection of the quantities which they involve, their combination and application, are the modes in which he exerts and manifests his power, his wisdom, his goodness; through these attributes, thus exercised, the Creator of all, shapes, moves, sustains and guides the visible creation.'

> When we examine the truths of science and find that we can not only say 'this IS so' but 'this MUST BE SO for otherwise it would not be consistent with the first principles of truth', or even when we can only say 'this OUGHT TO BE so according to the analogy of nature', we should think what a great thing we are saying, when we pronounce sentence on the laws of creation, and say they are true, or right when judged by the principles of reason.[9]

Moving on, Maxwell touched on one of his lifelong principles: though knowledge is always *incomplete*, it can be *real and objective*: 'Man has indeed a very partial knowledge of the simplest real thing: the nature of a drop of water has in it mysteries within mysteries utterly unknown to us at present; *but what we do know, we know distinctly and scientifically*, and we may expect that more will be understood in due time' (Italics ours):

> [Though] every atom of creation is unfathomable in its perfection . . . it is the peculiar function of physical science to lead us, by the steps of rigid demonstration, to the confines of the incomprehensible, and to encourage us to apply our minds to that which we do not yet understand; since it is only to those who labour patiently and think steadily, that such mysteries are ever opened.[10]

Maxwell offered cautionary and prophetic advice: science should beware of 'assuming that the higher laws *which we do not yet know* are capable of being stated under the same forms as the lower ones which we do know.'[11] This was far-sighted, given how far physics would evolve in the wake of his own discoveries. Maxwell was open to science outside a Newtonian framework. Relativity theory and quantum theory proved him correct. New laws of physics broke apart the Newtonian system which, hitherto, was thought to be absolute truth, and within which all human understanding of the universe was thought to lie.

The London Triumphs

In 1860 Maxwell was awarded the Royal Society's Rumford Medal for his work on colour vision. In making the award, Major-General Edward

9 JCM, *Inaugural Lecture, London,* Oct 1860, (Harman, I, 670).

10 Ibid., (Harman, I, 670).

11 Ibid., (Harman, I, 671). (Italics ours)

Sabine (1788–1883), vice-president of the Society, commented that Maxwell had raised the composition of colour to the rank of a branch of mathematical physics.[12] In the same year Maxwell published his first papers on the kinetic theory of gases. His election to the Royal Society took place in 1861.

On the seventeenth of May 1861, he lectured at the Royal Institution, 'On the Theory of Three Primary Colours', identifying these as red, green, and blue over against Brewster's red, yellow, and blue. As part of his presentation, and assisted by the photographer Thomas Sutton (1819–1875), he gave the first demonstration of colour-image reproduction, projecting onto a screen the colour-image of a tartan ribbon.[13] Sutton had photographed the ribbon as three separate images: one through a red filter; another through a green filter; and the third through a blue filter. From the three negatives Sutton made three separate positives of black and white glass slides. Maxwell then set up three lanterns, one for each slide, and with the colour of light for each lantern corresponding to the filter used for the slide's original negative. The three images were then superimposed on the screen creating a full colour image. The result was good if not perfect. What it lacked was due mainly to limitations in the photographic materials. There was also luck. When the experiment was repeated a century later, it was realised that it had worked only because interference from ultra-violet light had produced a passable result for the red part of the spectrum.

In September 1861, the British Association met in Manchester, and set up a committee to determine the best way of defining and measuring electrical resistance. This was important as telegraph cables were being laid across the Atlantic to America, and precise knowledge of resistance was crucial, both for the physics and for the economics of the project. Though Maxwell was not on the original committee, he was co-opted as a key member the following year. The committee's final report, written by Maxwell and Fleeming Jenkin, was presented to the Association in 1863. This set norms for decades to come, with the new

12 Glazebrook, *James Clerk Maxwell and Modern Physics,* 106.

13 Dougal & Forfar, 'The Tartan Ribbon and Colour Photography': *NJCMF,* Issue 1 (Spring 2012). Cf. Flood, 'Introduction', and Longair, 'Maxwell and the Science of colour', in Flood (ed.), *JCM: Perspectives on his Life and Work,* 11 and 101–114 respectively.

trans-Atlantic cables of the mid-1860s benefiting from the science. The work appealed to Maxwell, who was uncompromising on the need for precision, and who set high standards at all his institutions, especially during his tenure as head of the Cavendish Laboratory in Cambridge.

It was in late 1861 and early 1862 that Maxwell published a key paper on physical lines of force, perhaps *the* key paper of his career so far. He followed it up in 1864 by offering further reflections on electricity, publishing these in 1865. He then presented two more papers on the theory of gases in 1866: one was his Bakerian Lecture on the viscosity of gases; the other was on the dynamical theory of gases. Niven suggests that the mental effort involved in all this work may have been one of the factors prompting Maxwell to step down from academia in 1865, though the exact reasons behind his sudden retiral are unknown.[14]

1861–1862: 'On Physical Lines of Force'

All of Maxwell's work at this time was inter-related, whether colour theory, gas theory, or electromagnetism. His ground-breaking ideas on light, electricity, and magnetism, came in stages, with his major contributions comprising his 1861/62 multi-part paper on Faraday's lines of force, and his 1864/65 monograph, 'A Dynamical Theory of the Electromagnetic Field.' The first led to the second, though with a massive leap of insight along the way. If we include his earlier and later publications, we can log his progress:

1855/56	*On Faraday's Lines of Force*
1861/62	*On Physical Lines of Force*
1864/65	*A Dynamical Theory of the Electromagnetic Field*
1873	*Treatise on Electricity and Magnetism*

It was his 1861/62 work which convinced Maxwell that light, electricity, and magnetism were interconnected. Within that pivotal multipart paper, he unpacked more of the implications of Faraday's notion of lines of force, pushing beyond what he had discussed in 1855/56. He was still exploring ideas picked up from Faraday and putting these in mathematical form, convinced that the mathematics which arose from Faraday's insights were objectively embedded in nature

14 NVN1, xvi,

itself. At this point he had not yet formulated the famed 'Maxwell Equations'. Moreover, as pointed out by several commentators, he was still thinking in terms of mechanical connections, even though more dynamic concepts were beginning to impinge on his thinking.[15]

His 1861/62 paper 'On Physical Lines of Force', was published in the *Philosophical Magazine* in four parts, split into five instalments. Harman suggests that Maxwell originally envisaged the paper being only in two parts, but, after further reflection at Glenlair during the summer of 1861, he realised he had to add more.[16] Part I was published in March 1861. Part II was published in two instalments in April and May 1861. After a long hiatus, Part III was published in January 1862, and Part IV was published in February 1862. The sections were:

> Part I: *The Theory of Molecular Vortices applied to Magnetic Phenomena.*
>
> Part II: *The Theory of Molecular Vortices applied to Electric Currents.*
>
> Part III: *The Theory of Molecular Vortices applied to Statical Electricity.*
>
> Part IV: *The Theory of Molecular Vortices applied to the Action of Magnetism on Polarized Light.*

Between the publication of Parts II and III, Maxwell not only reflected more deeply at his Glenlair retreat, but exchanged important correspondence with Michael Faraday in London and William Thomson in Glasgow. Maxwell's goal was a scheme which could explain, describe, and predict electromagnetic phenomena.

The model which he initially adopted was clumsily mechanical, involving 'molecular vortices' orientated along magnetic field lines, which he developed from suggestions made by Thomson in 1856. With his penchant for analogy, Maxwell's thinking was influenced by the pattern of swirls created in gases and fluids when motion took place, giving series of vortices adjacent to one another in the medium. His imagination may also have been stimulated by the swirls and eddies formed by running water in the streams on the Glenlair estate.

15 Cf. Torrance, *Divine and Contingent Order*, 152f; & Olson, *Scottish Philosophy and British Physics, 1850–1880*, 297.

16 Harman, I, 31.

The problem Maxwell was still trying to solve was: how could phenomena such as light, magnetism, and gravitation, exert a force on distant objects. He still thought in terms of a universal ether, and a complex model of swirls and rotations within the ether seemed to best explain how it might happen. However, though the vortex model was useful in some respects, what some readers of his paper failed to note was that Maxwell disallowed any *literal* interpretation of the analogy between magnetic action and fluid motion. Moreover, though his elaborate model was helpful, in the long run his mathematics would need neither vortices nor ether.[17]

After publishing Parts I and II, Maxwell wrote to Faraday.[18] As always, he expressed indebtedness to the older man, since it had been Faraday's amazingly original ideas about the nature of electricity and the nature of matter—expressed in Faraday's Friday evening lectures at the *Royal Institution* in 1844, 1845, and 1846—which had prompted Maxwell to move away from earlier theories about forces acting at a distance.[19] Now Maxwell had something revolutionary of his own to report to his hero. He told Faraday

---

17 Harman, II/1, 30f, comments: 'As he explained to Tait in December 1867, the ether model of his paper 'On Physical Lines of Force' had been "built up to show that the phenomena are such as can be explained by mechanism", but that the "nature of this mechanism is to the true mechanism what an orrery is to the Solar System."' Ref: JCM, *Letter to Tait,* 23 December 1867: Harman, II/1, 337. An 'orrery' is a mechanical model of the Solar System.

18 JCM, *Letter to Faraday,* 19 Oct 1861, (Harman, I, 683–688).

19 See Maxwell's earlier letter: JCM, *Letter to Faraday,* 9 Nov 1857, (Harman, I, 550–552): 'As far as I know you are the first person in whom the idea of bodies acting at a distance by throwing the surrounding medium into a state of constraint has arisen. . . . nothing is clearer than your description of all sources of force keeping up a state of energy in all that surrounds them. . . . you seem to see the lines of force curving around obstacles, and driving plump at conductors, and swerving towards certain directions in crystals, and carrying with them everywhere the same amount of attractive power spread wider or denser as the lines widen or contract.' JCM pondered the enigma of gravity: 'Does it require time? Is it polar to the "outside of the Universe" [i.e. does it also repel, somewhere?] or to anything? Has it any reference to electricity? Or does it stand on the very foundation of matter – mass or inertia? Then we feel the need of tests, whether they be comets or nebulae or laboratory experiments or bold questions as to the truth of received opinions.'

he had found the velocity of transverse vibrations to be 193,088 miles per second, which was remarkably close to the then accepted velocity of light of 193,118 miles per second. The implications were significant. In another letter, this time to William Thomson, Maxwell wrote excitedly: 'I have reason to believe that the magnetic and luminiferous media are identical.'[20] What had started as an attempt to devise a medium in space which could explain lines of force, had brought the realisation that the medium—whatever it was—had properties identical with light.

Following his correspondence with Faraday and Thomson, Parts III and IV appeared in early 1862.

In Part III Maxwell announced that he had established the nearness of the velocities of magnetic effects and light to such a degree that he could scarcely avoid the inference that light consisted in the 'transverse undulations of the *same medium* which is the cause of electric and magnetic phenomena.' In other words—as stated explicitly two years later—visible light was simply a particular manifestation of electromagnetic phenomena. Maxwell was beginning to see that light, electricity, and magnetism were not different things, but were part of a common electromagnetic field, travelling with the same finite speed. Faraday had always thought there might be a connection, whereas Thomson, who instinctively preferred 'mechanical' models, was sceptical.[21]

In Part IV Maxwell took his thinking further. For the next ten years this area of physics was at the heart of his work, culminating in his *Treatise on Electricity and Magnetism* (1873).

Maxwell's work confirmed Faraday's belief that the concept of *instantaneous* action at a distance was false, even for electricity and magnetism.[22] No object which was distant, or separated, from another object could exert a force on that other object *instantaneously*. Tait, in his reflective article on Maxwell for the *Encyclopaedia Britannica,* summarised the overall goal of his friend's work as being, as it had been for Faraday, to overturn the notion of instantaneous action

20 JCM, *Letter to W. Thomson,* 10 Dec 1861, (Harman, I, 695).

21 Reid, 'Maxwell at King's College, London', 63.

22 Cf. Harman, II/1, 20: 'In a lecture in early 1873 he stressed that the [theory of the electromagnetic field] had as its aim the resolution of the enigma in Newton's natural philosophy, the problem of action at a distance.

at a distance.[23] Forfar's opinion is that this was one of the most extraordinary breakthroughs in nineteenth-century physics, since many eminent physicists did not yet believe in electromagnetic waves travelling with finite speed, thinking that electric and magnetic effects were transmitted instantly.[24]

The possibility of instantaneous/simultaneous action at a distance has reopened through the notion of quantum entanglement, for which the 2022 Nobel Prize for Physics was awarded. Einstein—who was aware of how Maxwell's rejection of earlier notions of action at a distance had liberated physics to move forward—stalwartly resisted the implications of quantum entanglement all his life. Even as we write, basic questions are again being asked. What is time? What is space? What actually is the concept of distance in space-time?

23 Tait, 'James Clerk Maxwell', *Encyclopaedia Britannica* (9th ed.), Vol. 15, 647.

24 Forfar, 'Maxwell, Einstein, Newton, and Faraday', *NJCMF*, Issue 1 (Spring 2012).

Chapter 19
The Dynamical Theory

A start had been made. More work was needed. In December 1864, Maxwell presented another epoch-making paper to the Royal Society. This was his 'A Dynamical Theory of the Electromagnetic Field', which was published in the *Philosophical Transactions of the Royal Society* for 1865. Harman points out that the term 'dynamical' in the title represented a move beyond Maxwell's mode of thinking in his previous work ('On Physical Lines of Force') in which he thought predominantly of mechanical inter-connections within the ether. Maxwell was beginning to fulfil Herschel's 1845 prophecy: that a time was approaching when a merely mechanical view of nature would become impossible.

Maxwell's extraordinary work on electromagnetism looms so large that it can overshadow his multi-various other achievements such as colour theory and the development of statistical modelling. The James Clerk Maxwell Foundation's exhibition at 14 India Street in Edinburgh, attempts to redress this by having a whole room devoted to Maxwell's non-electromagnetic research. Nevertheless, his work on electromagnetism dominates the horizon, and it was his years at King's College, London, plus those immediately afterwards, which saw some of his most creative achievements in that field.

The pinnacle—though not universally recognised at the time—was his formulation of a series of mathematical equations which could account for all the known characteristics of light, magnetism, and electricity. His 1861/62 multi-paper laid the basis for these, but it was in this new paper, 'A Dynamical Theory', that he set out the developed, though incredibly cumbersome, versions of what have come to be known as the Maxwell Equations.

The significance of Maxwell's work was such that, in 2009, the Institute of Electrical and Electronics Engineers (IEEE) commissioned three brass plaques to commemorate his equations which now form the bedrock of modern physics. The plaques were fixed to three buildings, each with strong associations to Maxwell. One at King's College in London, another at Corsock

Kirk, and the third at 14 India Street. Each plaque has an inscription followed by a simpler form of Maxwell's original equations:

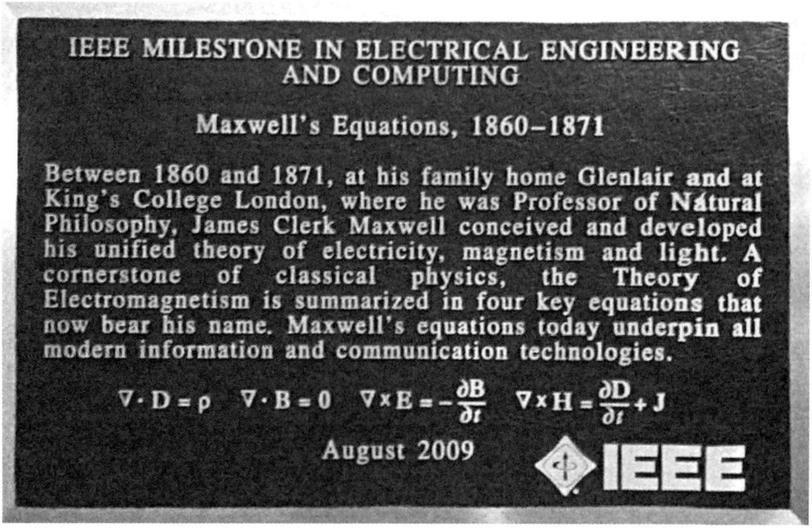

The equations have been described as among the most fundamental in the whole of physics since they unify the electric and magnetic forces.[1] It was the implications of these equations, including their profound impact on our understanding of the nature of physical reality, which was Einstein's focus in his 1931 paper, 'Maxwell and Physical Reality.'[2]

Like any insightful theory which touches on true fundamentals, Maxwell's Equations were packed with more than their discoverer was aware of, and demanded concepts yet to be developed. One related to the limit of the speed of light. In traditional physics, faster-than-light speeds were regarded as theoretically possible—though difficult to attain—since, in a Newtonian universe, velocities can simply be added or subtracted to determine the speed of an object in relation to an observer. Contra-intuitively, Maxwell's Equations allowed for an *identical* value for the velocity of any and every electromagnetic wave, *independent* of the observer. As Forfar has written:

> Einstein showed his debt to Maxwell by demonstrating that Maxwell's equations transformed correctly between two observers, provided the Lorentz transformation was used to change the viewpoint between the observers. It was such considerations that

1 Forfar, 'Maxwell, Einstein, Newton, & Faraday', *NJCMF*, Issue 1 (Spring 2012).

2 Einstein, 'Maxwell and Physical Reality', 66–73.

> enabled Einstein to state confidently that nature behaved according to the Lorentz's transformation and not according to the simpler Galileo/Newton transformation. Maxwell's equations, without any alteration, are compatible with Einstein's *'Theory of Special Relativity'*, whereas Newton's equations had to be changed. For example, the mass of a body now became dependent on its speed (as seen from Einstein's formula) whereas the mass of a body, under Newton, had been a constant, *m*, independent of the body's speed.[3]

Einstein also showed that Maxwell's Equations allow space and time to be described in a different way to Newton's system. Newton conceived of an absolute space and time which forms a container-box universe within which events happen but is itself unaffected by events. In contrast, the Maxwell Equations allow for a different universe: one in which space and time are seen as relational—with their qualities depending on the objects/events taking place—rather than being a neutral stage unaffected by them. Einstein's universe, which emerged from Maxwell's Equations, was relational, not absolute, with the *relationship* between entities part of what entities are.[4]

There were twenty Maxwell equations originally, far more than the elegant four of today. Some were reworked laws of electricity and magnetism which had been discovered by previous researchers such as Charles-Augustine de Coulomb (1736–1806), André-Marie Ampère (1775–1836), Hans Christian Oersted (1777–1851), Johann Carl Friedrich Gauss (1777–1855), and Michael Faraday. A few were formulated from scratch. In Maxwell's massive 1873 textbook *A Treatise on Electricity and Magnetism*, he reduced the number from twenty to twelve, but that was still too complex and confusing.

Since Maxwell, others have rewritten them in clearer fashion, using an assortment of notations. In 1884, Oliver Heaviside (1850–1925) did a major task of reworking and simplifying Maxwell's formulae, into four concise and elegant differential equations in two unknowns.[5] As such, the equations are sometimes known as the Maxwell-Heaviside Equations.

---

3 Cf. Forfar, 'Maxwell, Einstein, Newton and Faraday', *NJCMF,* Issue 1 (Spring 2012). Also, Arthur & Forfar, 'The Changing Notation of Maxwell's Equations', *NJCMF,* Issue 2 (Autumn 2012).

4 Cf. G.J. Whitrow, *The Nature of Time*, 84f.

5 Lindley, *Degrees Kelvin*, 271.

In his paper 'A Dynamical Theory of the Electromagnetic Field' Maxwell argued, more strongly than before, that the mathematics pointed to undulating but mutually supportive electric and magnetic fields. He repeated his conviction that the speed of such waves was—within experimental error—equal to the known speed of light. He restated his view that light was an electromagnetic disturbance in the form of waves propagated through the electromagnetic field, and that it behaved according to electromagnetic laws. Twenty years later his conclusions were confirmed experimentally by Hertz in Germany and by Lodge in England.

The significance of Maxwell's breakthrough was immense. Previously disparate phenomena were now viewed as a unified entity in the electromagnetic spectrum. He no longer depended on an intricate model of complex vortices, though he retained these in his mind as an aid to thought.[6] To avoid any doubt, he made it clear to his readers that, though mechanical phenomena might (by descriptive analogy) help them understand electrical ones, 'all such phrases in the present paper are to be understood as illustrative, not as explanatory.'[7]

With that caveat, it was this work which prepared the ground for the new physics of the twentieth century. Maxwell's journey was from a mathematical description of a mechanical world involving vortices, to a completely abstract model as set out in his 1873 *Treatise on Electricity and Magnetism.* As Flood expressed it: the physical imagery of the model was removed until all that remained was the mathematics; in fact, the mathematics was the model.[8]

---

6 Cf. Randall, 'Aspects of Maxwell's Life and Work', *Clerk Maxwell and Modern Science,* 9f. See Flood, 'Introduction', in Flood (ed.), *JCM: Perspectives on his Life and Work,* 15.

7 JCM, 'A Dynamical Theory of the Electromagnetic Field', Section 73, 487. Peierls, 'Field Theory since Maxwell', *Clerk Maxwell and Modern Science,* 28, points out that Maxwell struggled to move away from mechanical explanations. Peierls further stated (42), 'Maxwell's mechanical models have disappeared behind the equations he wrote down.' See also Siegel, 'Maxwell's Contributions to Electricity and Magnetism', in Flood (ed.), *JCM: Perspectives on his Life and Work,* 191, 203, where Spiegel emphasises that Maxwell moved to the abstract from the concrete.

8 Flood, 'Introduction', in Flood (ed.), *JCM: Perspectives on his Life and Work,* 16.

A Conceptual Revolution

One of the most accessible accounts of the importance of Maxwell's work is Einstein's short paper, 'Maxwell's Influence on the Development of the Conception of Physical Reality' (1931).[9] Einstein describes how in Newton's system physical reality was characterised by concepts of space, time, the material point, and force, with physical events seen as movements of material points in space. But Newton's system was essentially atomistic and mechanistic. Though it answered some questions, its adequacy had long been questioned, especially because of the properties of light, with its undulatory nature not fitting into Newton's system.

But it wasn't just Newton's *equations* which were questioned. His conception of *physical reality* was also under scrutiny. Newton's core idea (that physical reality was essentially atomistic) no longer sufficed. Nevertheless, a mechanistic approach remained, and physicists still tried to interpret events as the motion of inert bodies. It was inconceivable to imagine things in any other way.[10]

Then came the dramatic change which, in Einstein's view, should be linked with Faraday, Maxwell, and Hertz, for all time. In this revolution it was Maxwell who showed that everything known at that time about light and electromagnetic phenomena could be represented by a double system of partial differential equations, in which the electric and magnetic fields were themselves dependent variables. Though Maxwell had tried to find a way of justifying these equations through mechanical (analogical) models, in his final work only the equations themselves appeared as essential entities, independent of any visual model in the mind of the enquirer.[11]

Moreover—and this was of the utmost importance to Einstein—the 'field forces' to which the equations related, were now understood to be irreducible to anything else.[12] Such fields were in fact fundamental

9 Einstein, 'Maxwell's Influence on the Development of the Conception of Physical Reality', in Torrance (ed.), *James Clerk Maxwell: The Dynamical Theory of the Electromagnetic Field*, 29–32.

10 Einstein, 'Maxwell's Influence on the Development of the Conception of Physical Reality', 31.

11 Cf. Torrance, 'Christian Faith and Physical Science in the Thought of JCM', 233ff. Einstein realised that this formed a pattern for a new type of law.

12 Einstein, 'Maxwell's Influence on the Development of the Conception of Physical Reality', 31.

building blocks of reality. In Maxwell's 1873 Friday evening lecture to the Royal Institution, he stated: 'Gravitation, therefore, has as much right to be considered an essential property of matter as extension, mobility, or impenetrability.'[13] As with gravitation in particular, so with *fields* in general. In other words, a *field* was an elemental quality of reality. It was not to be viewed in terms of something else.[14]

Einstein claimed that by the turn of the century the concept of the electromagnetic field as an irreducible entity had been so firmly established that most theorists abandoned the possibility of there being some mechanical process lying behind it. The electromagnetic field was seen, not as a way of expressing another form of reality or another form of force, but as an essential building block of nature. To put it simply: in the same way as *matter* can only be described as *matter*, so the *field* can only be described as the *field*. As Rudolf Peierls (1907–1995) stated: the electromagnetic field is a part of physics which cannot be reduced to anything else.[15] This recognition altered how reality itself could be viewed, and lay behind Einstein's statement that before Maxwell, physical reality was thought of as material points, whereas after Maxwell they thought of physical reality as represented by continuous fields, not mechanically explicable (which Maxwell himself had initially sought to do).[16]

Thus, the *field* is an attribute of an object itself, and not an attribute of the supposed 'ether' surrounding the object. An object should no longer be defined as simply the three-dimensional item which is visible to our eyes: instead, the whole field-force-continuum associated with the object is implicated. The object does not *create* a field, as a cause produces an effect: instead, the field is *part* of the object. Nor is a field a 'field of objects' or a 'field of particles'. It is simply a *field*.

In the second volume of *A Treatise on Electricity and Magnetism,* Maxwell wrote of how Faraday had already sensed this: 'He never

---

13 JCM, 'On Action at a Distance' (1873), 48.

14 Harman, II/1, 26.

15 Peierls, 'Field Theory since Maxwell', *Clerk Maxwell and Modern Science,* 32.

16 Einstein, 'Maxwell's Influence on the Development of the Conception of Physical Reality', 31. In the new physics the ether was replaced by the electromagnetic field.

considers bodies as existing with nothing between them but their distance'; 'He conceives of all space as a field of force, the lines of force being in general curved'; 'He even speaks of lines of force belonging to a body as in some sense part of itself, so that in its action on distant bodies it cannot be said to act where it is not.'[17]

For Einstein, the change in the conception of reality initiated by Maxwell was the greatest change in physics since Newton. No longer should physical reality be thought of in terms of material particles. Instead, it was to be thought of in terms of continuous fields.[18] These cannot be visualised physically. They can only be represented by equations, and by the values which such equations generate.[19] In this new physics, all points in space have, or represent, a certain measurable force, with Maxwell finding that the formulae to be used were partial differential equations. The precocious Maxwell may have been moving towards this as early as 1856, when he wrote to his friend Henry Clerk:

> I think more of the distribution of lines of force in space, and less of any distinct centres to which they are supposed to tend. In fact, in many cases including all cases of electromagnets the lines of force return into themselves and do not run towards any centre whatever.[20]

Sean Carroll, research professor in the Walter Burke Institute for Theoretical Physics in the California Institute of Technology, put it this way: 'Our modern idea is that the world is not made of particles but is made of fields. A field, unlike a particle, is spread out everywhere throughout the universe. It is a *number* at every point in space and time. A particle has a location, but a field is spread out everywhere and is vibrating: it is changing its *value*.'[21] Quantum fields, vibrating at every location in space and throughout time, are for Carroll the

17 JCM, *A Treatise on Electricity and Magnetism,* Vol. 2, Section 529, 177.

18 Einstein, "Maxwell and Physical Reality", 71.

19 Gaal, *The Faith of a Scientist,* 28ff, pointed out that Maxwell's idea was to see the electric state of space as a different reality from the state of space without electrical effects. In other words, space was not a 'neutral' container. Faraday's insight was that in his mind's eye, he saw lines of force traversing all space where the mathematicians saw centres of force attracting at a distance. Faraday saw a medium whereas they only saw distance. Maxwell realised that these could be expressed as mathematically as the old mechanistic model.

20 JCM, *Letter to Henry Clerk,* 6 June 1856, (Harman, III/2, 869).

21 Carroll, *The Higgs Boson and Beyond,* 10. (Italics ours).

fundamental building blocks of nature; and, though we may still use the language of particle physics, we actually refer to a vibration in the field. The particle physicist thinks in terms of the value of an equation, rather than of an object.

Though Field Theory has evolved since Maxwell's day, what Einstein saw was that Maxwell's work changed everything.[22] Einstein believed that the theory of electrons as proposed by Hendrik Lorentz (1853–1928), plus his own special and general theories of relativity, flowed directly from Maxwell, and Einstein saw quantum physics in the same lineage though it differed in some respects from both Newton and Maxwell's programmes. Quantum theory was problematic for Einstein, since it did not claim to describe physical reality itself, but only indicated 'the probabilities for the occurrence of one of the physical realities to which attention is being directed.'[23] Einstein believed that science had the ability to know reality as it is in itself, closing his 1931 paper with words with which Maxwell would have heartily agreed:

> I am inclined to think that physicists will not be satisfied in the long run with this kind of indirect description of reality. . . . They must surely be brought back to the attempt to realise the programme which may suitably be designated as Maxwellian: a description of physical reality in terms of fields which satisfy partial differential equations in a way that is free from singularities.

---

22 Chen-Pang Yeang, 'The Maxwellians: The Reception and Further Development of Maxwell's Electromagnetic Theory', in Flood (ed.), *JCM: Perspectives on his Life and Work,* 222. Cf. Randall, 'Aspects of Maxwell's Life and Work', *Clerk Maxwell and Modern Science,* 14: 'Maxwell's description of reality is represented in his double system of partial differential equations in which the electric and magnetic fields appear as dependent variables. Since Maxwell's time, physical reality has been thought of as represented by continuous fields governed by partial differential equations and not capable of any mechanical interpretation. The advent of quantum mechanics, rather than the theory of relativity, has produced a situation in which we have to deal with amounts of energy concentrated in small spaces and this is in contradiction to Maxwell's equations and to the Newtonian system of mechanics. The quantities that appear in the laws of quantum mechanics make no claim to describe physical reality as such, but only the probabilities of the appearance of a particular physical reality.'

23 Einstein, 'Maxwell's Influence on the Development of the Conception of Physical Reality', 32.

The Nature of Reality

The common-sense view of matter is that it is 'stuff' which can be seen, touched, and detected by one or more of the five senses. Maxwell sensed that there was something more. This prompted his reaction to what his friends Tait and Thomson wrote in their 1867 textbook *Treatise on Natural Philosophy.* Tait and Thomson stated: 'We cannot, of course, give a definition of *matter* which will satisfy the metaphysician, but the naturalist may be content to know matter as *that which can be perceived by the senses.*'[24] Maxwell, who suspected that the issue might not be so simple, wrote in response: 'Matter is *never* perceived by the senses.'[25]

To be fair, Maxwell did not quote the entirety of Tait and Thomson's definition. What they actually wrote—both in the 1867 first edition of their *Treatise,* and in the new edition published a decade later—defined matter as: 'That which can be perceived by the senses, or, *as that which can be acted upon by, or can exert force.*'[26] The latter part of their definition did allow for a less materialist conception of matter and may well have been included by them because of discussions with Maxwell. But when Maxwell reviewed the new edition of their book for *Nature,* he wanted to dispel any idea that matter could be defined as something 'which can be perceived by the senses.'[27]

Maxwell's thinking had been changing for some time. In 1868, in a comment on Tait and Thomson's first edition, he revealed that he had chipped away so much at the idea of matter as 'stuff' that he could even entertain Bishop Berkeley's statement about matter as 'nothing but an enchanted vase of Circe fitted to receive Impulse and Energy.' It was not that Maxwell had become a Berkeleyan, but he saw Berkeley's rejection of matter as something and solid, as having merit to it. In this he echoed Faraday's April 1846 lecture.

What Maxwell was actually looking for was a *dynamical* rather than a *static* concept of matter. This was why he eventually argued,

---

24 Tait & Thomson, *Treatise on Natural Philosophy,* (1st ed.), 161.

25 JCM, 'Comments on Thomson and Tait's *Natural Philosophy,* Chapter II', July 1868, (Harman, II/1, 395).

26 Tait & Thomson, *Treatise on Natural Philosophy,* (New ed.), 219.

27 JCM, 'Review of Thomson & Tait, *Treatise on Natural Philosophy*', *Nature,* Vol. 20/505 (3 July 1879), 213–216. (*NVN2,* 776–785).

following Faraday, that matter might best be understood as something which was 'capable by its motion of becoming the recipient of momentum and energy'.[28] Moreover, as Harman noted, while Maxwell was endeavouring to establish the conceptual basis of his dynamical theory of the electromagnetic field, he was also anxious to refute any notion that the doctrines of materialism were sanctioned by the science of physics.[29] What he was moving towards was the idea that 'solid matter' is simply converging points of force, rather than discrete bodies in motion in space and time.

Whispers of this new thinking extended beyond academia. Maxwell's friend and former student, the Revd George Walker of Castle Douglas, later wrote in his local newspaper that Maxwell's ideas had led to a 'revolution in our conception of the nature of matter.'[30] Unlike William Thomson—whose mind-set was described by Sir John Randall as essentially material, in contrast to that of Maxwell's which was spiritual—Maxwell had an elasticity of mind which allowed him to entertain the thought that the division between matter and energy may not be as traditionally thought.[31] Einstein's iconic equation, $E=MC^2$ would demonstrate that matter and energy are two aspects of one reality.

Maxwell's work gave substance to the notion of fields of force as key to understanding electromagnetic phenomena. Faraday's ideas could no longer be dismissed as pretty pictures; they had to be respected as solid insights. All of this involved progressing from a strictly mechanical way of understanding how forces act. Hitherto, forces were thought to act by 'push' or 'pull' methods. In the preface of his *Treatise on Electricity and Magnetism,* Maxwell remarked on this change, paying tribute to Faraday:

> Before I began the study of electricity, I resolved to read no mathematics on the subject till I had first read through Faraday's *Experimental Researches on Electricity*. I was aware that there was supposed to be a difference between Faraday's way of conceiving phenomena and that of the mathematicians, so that neither he

28 Harman, II/1, 19f.

29 Ibid.

30 *Galloway News & Kirkcudbrightshire Advertiser,* 10 Nov 1882.

31 Randall, 'Aspects of Maxwell's Life and Work', *Clerk Maxwell and Modern Science,* 25.

> nor they were satisfied with each other's language. I had also the conviction that this discrepancy did not arise from either party being wrong. I was first convinced of this by Sir William Thomson, to whose advice and assistance, as well as to his published papers, I owe most of what I have learned on the subject.
>
> As I proceeded with the study of Faraday, *I perceived that his method of conceiving the phenomena was also a mathematical one,* though not exhibited in the conventional form of mathematical symbols. I also found that these methods were capable of being expressed in the ordinary mathematical forms, and these compared with those of the professed mathematicians. For instance, *Faraday, in his mind's eye, saw lines of force traversing all space, where the mathematicians saw centres of force attracting at a distance; Faraday saw a medium, where they saw nothing but distance;* Faraday sought the seat of the phenomena in *real actions going on in the medium,* where they were satisfied that they had found it in *a power of action at a distance impressed on the electric fluids.*[32] (Italics ours)

Though Maxwell never entirely rejected the notion of an ether, his mathematical description of reality no longer relied upon it. Yet, when he died in 1879, his electromagnetic theory of light was far from being accepted universally. Randall argues that it was slow to be adopted—both in Britain and in Europe—partly due to obscure presentation, and partly because of the success of the traditional action-at-a-distance theory.[33] Even Maxwell's friend, William Thomson, stated in his 1884 Baltimore Lectures on 'Molecular Dynamics and the Wave Theory of Light', that the 'so-called electromagnetic theory of light' had not helped science much hitherto.[34] It was Hertz and Lodge's positive demonstrations of Maxwell's ideas which paved the way for widespread acceptance.

---

32 JCM, *Treatise on Electricity and Magnetism,* ix–xi.

33 Randall, 'Aspects of Maxwell's Life and Work', *Clerk Maxwell and Modern Science,* 11, 15.

34 Thomson, *Baltimore Lectures on Molecular Dynamics and the Wave Theory of Light,* Preface, vii. 'If I can make a mechanical model, I can understand it.' Cf. Scott Walter, 'How Maxwell made his mark', *Nature,* Vol. 409/6818 (18 Jan 2001), 283–284.

Chapter 20
Darwin and Evolution

The Maxwells came to London in the autumn of 1860, arriving a few months after one of the most famous 'debates' of the nineteenth century. On Saturday the thirtieth of June 1860, at the Oxford University Museum, and coincident with the British Association gathering in Oxford that year, Thomas Henry Huxley (Darwin's Bulldog) and the Bishop of Oxford, Samuel Wilberforce, squared up over Darwin's findings. At the previous year's gathering in Aberdeen, the geologist Charles Lyell gave notice of Darwin's forthcoming publication *On the Origin of Species*:

> A work will very shortly appear by Mr Charles Darwin. He has been led to the conclusion, that those powers of nature that give rise to races and permanent varieties in animals and plants, are the same as those which, in much longer periods, produce species, and, in a still longer series of ages, give rise to difference of generic rank.[1]

Even before Darwin's book rolled off the presses into bookshops in late November 1859, there was an awareness that it had enormous implications for what was believed about the origin of humanity.

The issue was sensitive. During the Aberdeen gathering, the acting President gave his personal opinion that nothing said or published by geologists could be held as destroying belief in revelation, and that geologists could not prove that man had lived thirteen thousand or even ten thousand years ago. His remark was prompted by speculation as to whether mammoths, whose remains had been found at Ilford, had been contemporaneous with man.[2] By 1860, Darwin's ideas were the hottest of hot issues, with the theory of evolution by natural selection impacting faith, science, and society.

---

1 Lyell, 'On the Occurrence of Works of Human Art in post-Pliocene Deposits': *Report of the 29th Meeting of the BAAS* (1859), 95. Lyell advised caution in arriving at conclusions concerning the antiquity of the human race because of prehistoric bones in caverns alongside human bones.

2 *Caledonian Mercury*, 22 Sep 1859.

The occasion for the celebrated 1860 exchange was not a formal set-piece debate between Huxley and Wilberforce, but a lecture given by New York's John William Draper (1811–1882) on the intellectual impact in Europe of Darwin's theory, including the notion that human development was determined by natural laws rather than by divine intervention.[3] It had not taken long for Darwinism to move beyond biology, to theories about the development of society. Some already interpreted Darwinism as indicating the perfectibility of man, the omnipotence of reason, and the certainty of progress, as had been advocated by the French Encyclopaedists of the eighteenth century.

Huxley and Wilberforce's altercation took place in the wake of Draper's wide-ranging discourse, though the dry summary of Draper's lecture, as recorded in the official *Report*, has no hint of the volcanic eruption it created. The large library at the University Museum was crowded out; and, though there is no cast-iron proof of Maxwell's attendance, we know he was in Oxford for the meetings, delivering a paper on Bernoulli's theory of gases.[4] It is possible he was at the Huxley/Wilberforce showdown.

Some details of the post-lecture discussion on Draper's paper have become legendary, such as Wilberforce's jibe: 'Was it through his grandfather or grandmother that Professor Huxley claimed descent from apes?' Huxley countered that he: 'Would rather have an ape for an ancestor than a Bishop who distorted the truth!' The wording used by each is uncertain since no verbatim account was kept: but it was common for protagonists to play to the gallery, mock their opponent's position, and elicit cheering from the audience. All of this was normal practice in university debates, with speakers entertaining the audience to get support. Wilberforce and Huxley were doing nothing unusual in ridiculing each other's position, with the audience responding first to one and then the other. It was knockabout stuff, and part of the

3 Draper, 'On the Intellectual Development of Europe, considered with reference to the views of Mr. Darwin and others, that the Progression of Organisms is determined by Law', *Report of the 30th Meeting of the BAAS (1860)*, 115–116. The *BAAS* summary of Draper's paper explained that its object was to show that human cultural advancement did not occur accidentally or in a fortuitous manner, but was 'determined by immutable law.' Draper extrapolated from Darwin's theory of biological evolution, positing that an environmental determinism shaped society through irresistible forces.

4 *The Oxford Chronicle and Reading Gazette*, 7 July 1860.

expected drama. The newspapers described Wilberforce's intervention as eloquent and argumentative; with Huxley's reply styled as calm, dispassionate, and also argumentative. The term 'argumentative' was applied to both contributions: in context it meant that each speaker had presented his case with a powerful logic, and had spoken with energy and passion.

Academic accounts tend to view Wilberforce's joke as a last gasp attempt to save his position, with traditional Christian beliefs having been sent reeling by science. An image has been created of a defensive, anti-intellectual churchman, resorting to crude abuse; with the man of reason and science holding his ground with calmness and dignity. In fact, in 1860 there were many serious scientists not convinced of Darwin's theories, and not necessarily because of religious grounds. Moreover, Wilberforce was a thinker of some note, and a regular attender of British Association meetings; and in his intervention he raised serious scientific issues far beyond his famous one-liner.

Professional scientific hesitation over Darwin's theories was strong enough for the ninth edition of the *Encyclopaedia Britannica*—whose scientific sections were edited by Huxley himself along with Maxwell—treating the concept as still controversial. Only from the eleventh edition was it regarded as science.

Theists and Evolutionists

If some scientists were reluctant to accept Darwin, there were some Christians sympathetic to him. Among Maxwell's friends and associates these included Farrar, Goodwin, Litchfield, and Whewell, though not all were convinced about natural selection. Farrar was an enthusiastic supporter who conversed frequently with Darwin on science and faith, and who applied evolutionary concepts to the development of language. Goodwin (Maxwell's Cambridge minister) preached a special sermon in Westminster Abbey on the Sunday after Darwin's funeral, in tribute to the naturalist. Litchfield married Darwin's daughter, Henrietta. William Whewell, Master of Trinity, had encouraged Darwin when he returned from the *Beagle* voyage, though later he had strong reservations about natural selection. Maxwell's life-long friend, Peter Tait, was holiday golfing partner to 'bulldog' Huxley, despite differing from him on many issues. Contrary to how the nineteenth-century dispute between theists and evolutionists is often represented, there was substantial amicable interaction, mutual visits, and shared dinner conversation, alongside undeniably robust interchanges.

Maxwell was familiar with early, pre-Darwinian, evolutionary theory. He was fascinated by the anonymous *Vestiges of the Natural History of Creation* (1844), which set out an early theory of the development of species. Though *Vestiges* had been read with interest by Queen Victoria and Prince Albert, it received a hostile reception from the general public.

Maxwell's interest in geology had also introduced him to the thinking of Richard Owen (1804–1892) who had similar ideas. On one excursion round Arthur's Seat, during his university years in Edinburgh, Maxwell had expounded the *Owen Hypothesis* to Lewis Campbell.[5] Owen was a friend of the Wedderburn family, and comically described by Cousin Jemima as a very tall man who wriggled when talking.[6] Professionally, Owen was first friend to, and then implacable enemy of, Charles Darwin. Darwin commented: 'I used to be ashamed of hating him so much, but now I will carefully cherish my hatred and contempt to the last days of my life!'[7] After Darwin returned from his voyage on the *Beagle,* he had been introduced to Owen by Charles Lyell. At that point Darwin and Owen were on good terms, and Owen worked on some of the fossils which Darwin had collected in South America. Owen's conclusion was that these were of extinct giant rodents and sloths related to living species in the same region, and not to creatures in Africa as Darwin had first thought. Owen's observations pushed Darwin in the direction of evolution through natural selection.

During the 1840s, Owen had himself conjectured that species arose through an evolutionary process, but puzzled over the mechanism driving it. How could variants emerge to make evolution possible? He considered various possibilities: parthenogenesis, variations due to premature birth, and variations due to congenital malformations; but he was reluctant to publicise his thinking, possibly because of the reaction to Chamber's *Vestiges.* The idea of a progressive (improving) mutation

5 C&G1, 107.

6 Fairley, *Jemima,* 126.

7 Darwin, *Letter to J.D. Hooker,* 4 Aug 1872, (University of Cambridge, Darwin Correspondence Project).

of species was widely resisted, partly because it was believed that if God had created all things good, then living species could not be improved upon. *Vestiges* had been so controversial that Chambers' authorship was not acknowledged until after his death.[8] Despite the public mood, Owen published his book, *Nature of the Limbs*, in 1849, and was straightaway criticised for his evolutionary leanings and for suggesting that humans evolved from fish instead of being created directly by God. All of this was well before Darwin's *On the Origin of Species* (1859).

The young Maxwell was interested in new ideas, and was familiar with Owen's evolutionary views. Campbell recalled that 'all scientific theories had an interest for him', and that when Maxwell explained the *Owen Hypothesis* of types of creation, he not only had 'complete command of Owen's terminology' but was aware of the 'questions to which the theory led.'[9] Critically, Maxwell knew from Owen's work, even as a teenager, of the difficulty in finding a mechanism which could alter species over generations. This is to be borne in mind when we consider Maxwell's later hesitation over Darwin. His problem centred on the mechanism of natural selection which Darwin suggested drove it.[10] Maxwell considered that natural selection necessitated an impossible starting-point, in which every future possibility (mutation) was already present within the very first organism. Maxwell's awareness of Owen's difficulty may have been what alerted him to problems with Darwin's solution. Maxwell's mind-set often focussed on the initial starting conditions for a theory—whether in mathematics, physics, or biology—subjecting these to rigorous scrutiny.[11]

In passing, we note that some passages in Maxwell's paper 'A Discourse on Molecules', which he read to the 1873 British Association meeting in Birmingham, have been used to argue that Maxwell was a *creationist* rather than an *evolutionist*. That conclusion is not strictly valid. Maxwell did state that fundamental molecules had not 'evolved', and had been the same from the beginning of time, but not in a context of discussing Darwinism. Nor did Maxwell reject the possibility of change in species over time.

---

8 Cf. Geike, *A Long Life's Work,* 17.

9 C&G1, 107.

10 Stella Pratt-Smith, 'Boundaries of Perception', 253, misses this point, assuming that Maxwell was totally anti-evolution.

11 Cf. Marston, 'Maxwell, Faith and Physics', 285.

The Problem of Mutation

Long before Darwin published *On the Origin of Species* in 1859, Maxwell had been aware—from Owen's writings—of the difficulty in finding a workable mechanism to drive species-development. This fed his disquiet with Darwin's model. His problem was not with species-development as such, but with natural selection driving the process. Its required initial conditions seemed extraordinary. That was problematic for Maxwell, and he was not alone. Peter J. Bowler, in his book, *The Eclipse of Darwinism: Anti-Darwinian Evolution Theories in the Decades around 1900,* has shown that the tendency to accept evolutionism whilst rejecting Darwin's central premise of natural selection was widespread in the years immediately following publication of *On the Origin of Species.*[12]

Evidence of Maxwell's difficulties with Darwin is revealed in his contribution to a discussion at the Cambridge Philosophical Society in 1872.[13] This followed an address given by the local politician, traveller, and ethnologist, Neville Goodman (1831–1890). The lecture carried the unwieldly title: 'On the appearance of an extra digit on the hind limbs, and then on both fore and hind limbs, in two successive generations; and its bearing on the theory of Pangenesis.'[14] The issue under consideration was: how can variations exist in fundamental living structures for evolution to occur?

In his lecture, Goodman discussed possible implications for the theory of evolution arising from the case of a cow—belonging to a Mr Daintree of Huntingdonshire—which had three well-developed toes on each hind limb, besides the two ordinary rudiments which hang behind the foot. The cow's parents were both normal though its remote pedigree was unknown, but her calves had peculiarities.[15] Goodman posed two main questions: What mechanism lay behind this mutation? And, did Darwin's theories account for the phenomena? At the time, one of the favoured theories for explaining inter-generational changes

12 Bowler, *The Eclipse of Darwinism: Anti-Darwinian Evolution Theories in the Decades around 1900,* (Baltimore: John Hopkins University Press, 1983).

13 *Cambridge Independent Press,* 30 Nov 1872. Cf. *Proceedings of the Cambridge. Philosophical Society,* Vol. 2 (1872), 289.

14 Neville Goodman (obit.), *Cambridge Independent Press,* 3 & 10 May 1890.

15 *Cambridge Independent Press,* 30 Nov 1872.

to living organisms was Darwin's theory of *pangenesis*. All of this was before modern understanding of genetic mutations.

Darwin proposed his *pangenesis* theory in the concluding chapter of his book *The Variations of Animals and Plants under Domestication* (1868). He wrote it nine years after publishing *On the Origin of Species*, seeing pangenesis as the solution to a recognised problem in his theory of natural selection. In his pangenesis theory, Darwin proposed that every part of the body of a living organism emitted tiny particles called gemmules which migrated to the gonads and were transferred to offspring. These gemmules were then thought to develop into the appropriate body part as an offspring matured. Critically, the theory maintained that changes to various parts of the host body during an organism's life could affect a gemmule, cause a mutation, and be inherited. Goodman, who had taken his degree in the Natural Science Tripos at Cambridge in 1865, explained this to his audience, adding that the number of gemmules may be so great that not all are used in the next generation and may lie latent and undeveloped, only appearing in much later generations. He then applied the theory to the problem of the cow with an extra toe, touching on the thorny issue of how circumstances might change a gemmule.

After Goodman's presentation, the chairman made a few remarks on pangenesis, reading an extract from Darwin, and pointing out some of the difficulties of the theory, especially the sheer number of gemmules required. There was an obvious physical problem: how could such an immense number of different gemmules be physically contained in such a minute 'seed'?[16] And how could they be transmitted unchanged through successive generations? It was at this point that Maxwell spoke. He restated the physical difficulty of chemical molecules in *sufficient quantity* being packed in these small gemmules.

A lively discussion ensued, in which several leading Cambridge minds—none of whom were innately hostile to Darwin—explored difficulties surrounding his theory. Had Darwin really found a way of explaining how circumstances could force a mutation? It seemed

16 Cf. Marston, 'Maxwell, Faith and Physics', 277. Marston discusses the perceived limitations imposed by the finite size of molecules and implications for *pangenesis*: 'Ultimately, the resolution required the abandoning of Darwin's pangenesis in favour of Mendelian inheritance and the molecular structure of DNA.'

not. Maxwell added to the discussion, emphasising the problem of conceiving how external (environmental) circumstances could affect a gemmule and cause it to change (mutate); and if mutations could not occur, then all possible future characteristics of an organism still *had to be present in the original version of the organism,* with all possible variants having to exist from the beginning. But how? If all possible future variants did *not* exist from the beginning, then mutations were necessary. But how could these take place?

Two years later, Maxwell explored this further in a personal letter to Lewis Campbell. The burden of his argument was that he rejected the possibility of spontaneous mutations:

> If atoms are finite in number, each of them being of a certain weight, then it becomes impossible that the germ from which a man is developed should contain (actually, of course, not potentially, for potentiality is nonsense in materialism unless it is expressed as configuration and motion) gemmules *of everything which the man is to inherit*, and by which he is differentiated from other animals and men: his father's temper, his mother's memory, his grandfather's way of blowing his nose, his arboreal ancestor's arrangement of hair on his arms, and his more remote littoral ancestor's devotion to the tide-swaying moon.[17]

Maxwell's letter continued, rejecting the idea of 'structureless germs' as proposed by Francis Galton (who was related to Darwin, and who corresponded with Maxwell's cousin, Isabella, on ornithological issues):

> Francis Galton, whose mission it seems to be to ride other men's hobbies to death, has invented the felicitous expression 'structureless germs.' Now, if a germ, or anything else, contains in itself a power of development into some distinct thing, and if this power is purely physical, arising from the configuration and motion of parts of the germ, it is nonsense to call it structureless [simply] because the microscope does not show the structure: *the germ of a rat must contain more separable parts and organs than there are drops in the sea.* But if we are sure that there are not more than a few million molecules in it, each molecule being composed of component molecules, identical with those of carbon, oxygen, nitrogen, hydrogen, etc., *there is no room left for the sort of structure which is required for pangenesis on purely physical principles.* (Italics ours)

17 JCM, *Letter to Campbell,* 26 Sep 1874, (C&G1, 390; Harman, III/1, 127f).

Maxwell thought as a mathematician, statistician, and geometrician: there was simply not enough physical space for so many gemmules. Darwin, in a later letter to his son George Howard Darwin (1845–1912), disagreed with Maxwell's calculations concerning the size of molecules, though he respected Maxwell immensely.[18]

Maxwell was not hostile to Darwin, but neither was he satisfied that Darwin had established a mechanism allowing for mutation between generations. Mutation was vital to evolutionary theory, because the only alternative—that of all future possibilities being present in the very first seed—was impossible. This was not an outright denial of species-development. Maxwell was simply noting what seemed to be a contradiction in Darwinism.

Pain, Pleasure, Prayer, and Processes

Peter Pesic has claimed that Maxwell was so disturbed by Darwin that he would not allow Darwin's theories even to be discussed in his presence.[19] Frustratingly, Pesic has supplied no source reference. The assertion is surprising, given the various contributions which Maxwell made on evolution and his declared intention to examine every subject leaving no areas untouched. At the same time, we know from Karl Pearson that Maxwell was sensitive to biblical truth being questioned.

Whether or not Pesic's anecdote is accurate, Darwinism discomfited many Christian believers. One issue was that the principle of natural selection (survival of the fittest) indicated that randomness, rather than a benevolent divine purpose, was at the heart of biology: an idea which disturbed Victorian religion. Pesic described it as Darwin's veiled but unmistakeable disproof of divine providence with no purpose or direction guiding natural selection,

18 Charles Darwin, *Letter to G. H. Darwin,* c. 4 Sep 1876 (DCP): 'Do you remember calculating for me about number of molecules in a square of 1/1000 of an inch, in relation to Pangenesis; well I hear from Erasmus that, in *Encyclopaedia. Britannica,* Clark-Maxwell has been writing on this subject in relation to Pangenesis, & opposed to my views, but Erasmus says he cannot understand it, & that you might on your return read it.'

19 Pesic, 'The Bell and the Buzzer: On the meaning of Science', *Daedalus: Journal of the American Academy of Arts & Sciences,* (Fall, 2003), 40.

'only a battle to survive and to prevail in reproduction'.[20] Instead of a kindly providence overseeing nature, there was merely 'a blind, implacable play of random variations in mindless competition.' Pesic concluded that Darwin did not only present an alternative to Genesis, but undermined any account not based on chance. This opposed Christian faith on several fronts: it rejected the idea of divine purpose; it rejected belief in the goodness and harmony of creation; and it rejected the notion of the original perfection of creation.

It may have been in reaction to this 'nature red in tooth and claw' implication of Darwinism that Maxwell floated the possibility of more benign forces at work in nature. He outlined this, as an alternative to Darwin's mechanism of natural selection, at a meeting of the Eranus Club at Cambridge in 1876. He began:

> According to the simplest form of the Evolution theory, those individuals, tribes, or species, which have habits unfavourable to their success in the battle of life, die out, and help to improve the average quality of the remainder.[21]

Maxwell saw a problem with this aspect of natural selection. The problem being, that though the principle of natural selection may suffice for beings of great fecundity, it 'would involve a great waste of life among higher [but] less prolific forms.' As an alternative, he proposed that factors other than strength, cruelty, and physical adaptation should be considered; and that qualities such as taste, feeling, pain, and pleasure might also give a compensating advantage in the struggle for life.[22] He surmised that influences such as susceptibility to pleasure and pain could support a more sophisticated model than Darwin's brutal natural selection, going beyond blunt factors such as strength and endurance:

---

20 Ibid., 40. See also John Bowker's discussion on the move from teleology to teleonomy: Bowker, Did God create this Universe?', in Peacocke (ed.), *The Sciences and Theology in the Twentieth Century,* 107–111.

21 JCM, *Essay for the Eranus Club on Modified Aspects of Pain,* 31 Oct 1876, (C&G1 444–452; Harman, III/1, 403–409).

22 As a logician, Maxwell had to consider all possibilities, however hypothetical: so he dealt with the theoretical possibility that, 'there would be an equal chance that the tastes and feelings, when first developed, would be in favour of courses *leading to destruction*; but the individuals possessed by these tastes and feelings would be all the quicker exterminated for the good—that is to say, the more vigorous continuation—of the species.'

> By pursuing what we feel to be pleasant, and avoiding what we feel to be painful, we are following a course which was probably followed by many of our ancestors, and of which we know at least this much—that it did not bring them to such speedy destruction as to prevent them from leaving us as their descendants; and this is more than we can say for any absolutely new rule of life struck out by ourselves.

Maxwell was considering mechanisms which might apply in species-development, apart from that of natural selection as described by Darwin. He did not rule out evolution of some sort but argued that its driving force required deeper reflection. Towards the close of his Eranus Club paper he suggested an analogy with Christian faith, in which the spiritual discipline of self-control could be a selecting factor leading to the maturing (evolving) of life:

> In the Collect for the first Sunday in Lent, we pray for 'grace to use such abstinence' that, our flesh being subdued to the spirit, we may ever obey 'the godly motions' of Christ. Abstinence is to be used only as a help towards the subjugation of our lower nature, and the opening up of our higher nature to divine influence.[23]

In the development of the spiritual life, self-control and abstinence were positive and enhancing mechanisms. What Maxwell proposed was that in any development of physical life, then factors other than power, cruelty, and survival of the fittest might be at work. He was unconvinced that the randomness of Darwinism rather than divine intent shaped the course of life on earth, and—as alluded to in his musings on the Prayer Book collect—if there had been a development of species over generations then affirmative factors could have been decisive.

Some years later, the Scottish religious author Henry Drummond (1851–1897), echoed Maxwell's thinking, arguing in his books *The Greatest Thing in the World* (1894) and *The Ascent of Man* (1891), that the positive and altruistic quality of 'love' rather than the negative quality of 'selfishness' was the critical lever in both nature and the spiritual life, raising all life to something higher. Drummond did

---

23 Collect: 'O Lord, who for our sake didst fast forty days and forty nights: Give us grace to use such abstinence, that, our flesh being subdued to the Spirit, we may ever obey thy godly motions in righteousness and true holiness, to thy honour and glory, who livest and reignest with the Father and the Holy Ghost, one God, world without end. Amen.'

not cite Maxwell directly in reference to this principle, but we know he had read Maxwell's ideas on 'manufactured articles'. He was also conversant with Balfour and Tait's *The Unseen Universe* (1875), which he had referred to several times in his earlier work, *Natural Law in the Spiritual World* (1883). Drummond's attempt to prove that *altruistic love* was the major evolutionary principle has been heavily criticised.

Another problem for Christians was the Darwinian assumption that *evolution* of species implied *improvement* of species. This seemed to contradict the Genesis account which stated that God saw everything as 'good' when first created. But did 'good' mean 'perfection'? And did 'perfection' mean 'incapable of improvement'? If species evolved, and by implication improved, did that contradict the Christian revelation? As far as Maxwell was concerned, and as indicated in his thoughts on the Prayer Book collect, he was prepared to allow for growth and development even in something which was already good. Moreover, with his roots in the Calvinist tradition, Maxwell's Presbyterian realism accepted that the Fall had disturbed the goodness of God's original creation, and that it now exhibited the cruelty and dysfunctionality of a broken world. Though some Victorian hymn writers, influenced by romanticism, saw nature only in terms of 'all things bright and beautiful', that image was no part of Maxwell's Reformed heritage, which factored in the ravaging effects of the Fall.

There was a further problem which theists had with Darwinism. Darwin's materialist supporters claimed—that in putting forward a theory of how humanity came to be, solely in terms of a series of natural causes—Darwin had removed one of the great proofs for God: namely, that human beings are so complex that only a supernatural divine act can account for their existence. Had Darwin not proved that purely natural processes could produce astonishingly complex wonders without needing miraculous intervention? Did this not cancel any need for theism?

Maxwell was again untouched by this because his faith was never based on a 'God of the gaps'. He had always accepted that, in examining nature, the scientist could always find an antecedent natural cause for any natural phenomenon; and that proper evidence of the supernatural came from a consideration of the 'whole' rather than of the 'parts'.

Maxwell did not live into the era where a 'missing link' hypothesis was posited in order to retain a 'gap' within nature within which God alone could act and bring humanity into existence. He would not have seen that line of argument necessary. Thus, although some non-theists assumed that Darwinism shattered the design argument—since the design argument is predicated on there being complex entities which cannot be produced by natural processes and which were created instantaneously rather than by a process—Maxwell would not have seen it that way. He never appealed to breaks in nature as a basis for his faith. In his mind, every natural phenomenon was the product of innumerable natural causes. Darwin had not closed a gap which was necessary for the 'God-hypothesis'. All that Darwin had done—even if his hypothesis were correct—was to shed possible light on the internal, God-given, processes of nature.

Maxwell's faith did not need a 'gap' in nature. Indeed, very few Christians were believers because of the design argument. For most, faith came from elsewhere, and 'proofs' from nature were buttresses to faith, not essential foundations. For a Maxwell, nature could—if God so willed—be a totally closed and complete system with no gaps (putting aside completeness issues as raised by Kurt Gödel). For Maxwell, the fact that natural explanations could be found for each phenomenon never impinged on the God question.

The Victoria Institute

Maxwell's position, which allowed him an openness to the possibility of species development, made it difficult for him to become a member of the Victoria Institute. The Institute, otherwise known as the *Philosophical Society of Great Britain*, had been established in 1865, with the aim of promoting traditional Christian beliefs over against perceived implications of new scientific theories. The Institute was not officially opposed to Darwinism, but it did attract those who were sceptical of it. Its position was set out in the article 'Scientia Scientiarum', published in the first issue of the *Journal of the Transactions of the Victoria Institute* (1866).[24] The article stated that

24 Mitchell, 'Scientia Scientiarum', *Journal of the Transactions of the Victoria Institute*, Vol. 1 (1867), 5–29. Cf. McNatt, 'James Clerk Maxwell's refusal to join the Victoria Institute', 204–215.

the Institute's aim was 'to defend the truth of Holy Scripture against oppositions arising, not from real science, but from pseudo-science.' The founders saw themselves as guarding faith on two fronts: from theological liberalism, and from atheistic pseudo-science.

The Institute's first president was the Revd Walter Mitchell (c.1816–1874), who was widely respected as one of the most learned of the London clergy. Mitchell was well versed in science. He had been lecturer in natural philosophy at St Bartholomew's Teaching Hospital in London, and was the author of several scientific works on crystallography and mineralogy.[25]

In his 1866 inaugural address, Mitchell set out the Institute's philosophy.[26] He lamented that many laymen did not question the pretension of scientific dogma to be accepted as truth. Over against this submissive culture, the aim of the Institute was to challenge the assertion that science and Scripture were irreconcilable. Mitchell argued that proper Christian scientists trusted the book of nature *and* the Bible, since both were authored by the same Creator. He pointed out that Christians had been at the forefront in discovering the laws of nature, and he rejected the allegation that committed believers had closed minds and could not engage in an unprejudiced discussion.

Mitchell identified two critical questions: Were science and faith opposed in method or principles? Were science and faith opposed in facts? He rejected David Hume's critique of miracle, and he disallowed the assertion made by David F. Strauss that because the 'chain of endless natural causation can never be broken' then any supernatural intervention in the world was an impossibility. Against Hume and Strauss, Mitchell defended the freedom of God to break into the natural order to enact what may commonly be termed a miracle. That was what Mitchell meant when he stated that it could never be proved that natural laws apply everywhere. He did not mean that laws of nature were not universal, but was reacting to Hume and Strauss imposing limitations on the freedom of God to act within the created order. Maxwell would agree.

25 Cf. *The Morning Post,* 29 May 1874; *North Wilts Herald,* 1 June 1874.

26 Mitchell, *Inaugural Address read at the First General Meeting of the Members and Associates of the Victoria Institute, or Philosophical Society of Great Britain,* 24 May 1866, (London: Robert Hardwick, 1866).

On Darwin, Mitchell argued that the problem with Darwin's *On the Origin of Species* was that it fostered the idea of 'self-evolving powers of nature'.[27] He insisted that nowhere had Darwin incontrovertibly proved the existence of his law of natural selection. He further argued that a weakness in Darwin's theory was that it could not explain where life came from, and that Darwin shirked questions on the ultimate origin of life. Mitchell pointed out that Huxley had scornfully rejected the notion that humans had one (monogenetic) origin through Adam, despite elsewhere admitting that the hypothesis of a single origin was not unscientific. Mitchell's verdict was that Huxley simply wanted to reject anything linked to a possible supernatural origin.

Though Mitchell's primary argument focused on the issue of ultimate origins, he also pointed out that, whereas Darwin argued that the eye could be formed without skill in optics (the Blind Watchmaker position), the great Newton had been convinced there had to be a lawgiver for nature.[28] Mitchell asked rhetorically: who should be listened to most? Darwin, or the greatest scientist of all time (Newton). For Mitchell, the whole idea of 'laws' of nature indicated that they were not founded on chance but came from the hand of God.[29]

Maxwell could agree with much of Mitchell's argument, and several respected figures who were close to Maxwell belonged to the Victoria Institute. One was Maxwell's teacher, the physicist and mathematician George Gabriel Stokes (1819–1903), who in 1886 was simultaneously the Institute's president and president of the Royal Society.[30] Another was Sir John Ambrose Fleming, a student of Maxwell's at Cambridge in the 1870s, and the Institute's president in 1927. Fleming was a theistic evolutionist, believing in a divinely directed development of life. Missing from the list of Institute members during Maxwell's lifetime were notable 'North British' believers such

27 Mitchell, *Inaugural Address*, 6.

28 Ibid., 14.

29 Mitchell conflated defending creationism with defending theism.

30 McNatt, 'James Clerk Maxwell's refusal to join the Victoria Institute', 214, argues that Stokes broadened the position of the Institute to a point where Maxwell, if he had lived longer, might have felt able to join it.

as William Thomson, Peter Tait, Fleeming Jenkin, Balfour Stewart, and Maxwell himself.[31]

The Institute was keen for Maxwell to join. He was an outstanding scientist. He was a man of strong religious faith and churchmanship. He would be a great coup for the organisation. But Maxwell resisted. He felt that the Institute's stance, though commendable in some respects, militated against open enquiry. In 1875 the Institute made a final effort. Its secretary wrote to Maxwell, reminding him that amongst its members were the Archbishop of Canterbury, various prelates and leading ministers, several professors of Oxford, Cambridge, and other universities, and many literary and scientific figures. The extant version of Maxwell's reply to the Institute is incomplete, but expresses his position:

> Sir, I do not think it my duty to become a candidate for admission into the Victoria Institute. Among the objects of the Society are some of which I think very highly. I think men of science as well as other men need to learn from Christ, and I think Christians whose minds are scientific are bound to study science that their view of the glory of God may be as extensive as their being is capable of. But I think that the results which each man arrives at in his attempts to harmonise his science with his Christianity ought not to be regarded as having any significance except to the man himself, and to him only for a time, and should not receive the stamp of a society. For it is of the nature of science, especially of those branches of science which are spreading into unknown regions to be continually [developing] . . .[32]

Maxwell explained that his scientific work enhanced his awareness of the glory of God precisely because that work was free. As a young man he had written that there should be no untouchable 'holy ground' left unexamined, and that no Christian had cause to fear scientific advance. It was that principle which shaped his response. Maxwell had his own reservations about Darwinism, but he was open to the concept of a development of species. From school days he had welcomed serious consideration of new ideas, but membership of

---

31 The 'North British Evangelical' label was coined by Crosbie Smith, *The Science of Energy*, (Chicago: UCP, 1998).

32 JCM, *Letter to F.W.H. Petrie, Secretary of the Victoria Institute*, March 1875, (C&G1, 404f; Harman, III/1, 194). The extant manuscript ends abruptly.

the Victoria Institute seemed to imply *a priori* rejection of a specific scientific theory. He could not do that.

Jerrold L. McNatt criticised Maxwell for being too tolerant, writing that Maxwell's 'failure to detect the perils of theological liberalism is lamentable.'[33] As well as censuring Maxwell for his reluctance to join the Institute, McNatt suggested that Maxwell should have publicly condemned two writers—Rowland Williams and Henry B. Wilson—who were tried in church courts for their views on inspiration, justification, and the future state of the dead. McNatt concluded: 'Maxwell's faith was basically too personal, and his hesitation about speaking out concerning matters outside his area of expertise severely limited his influence at a critical time in church history.' This critique is unjust. Even McNatt acknowledged passages in Maxwell's letters where, for example, he expressed substantial disagreement with F.D. Maurice's side-lining of some tenets of the faith, despite admiring him in other respects.

McNatt suggested three reasons why Maxwell declined the Victoria Institute's invitation. First, because it opposed specific individuals whom he viewed as Christian comrades. Second, because its theology was to the right of Maxwell's broad evangelicalism. Third, because Maxwell's naïve admiration of the growing professionalism of science and theology had led him not only to oppose scientific materialism/naturalism (which was good), but to abandon any attempt to re-establish the dominion of theology over science (which in McNatt's opinion was bad).

For McNatt, Maxwell trusted the progress of science too much. He should have been more questioning. He should have been at the forefront of establishing a 'biblical science' over against 'secular science'. He should have pioneered a movement in which acknowledged biblical presuppositions overtly moulded the course of scientific theory. But Maxwell would never think like that. McNatt wrote from within the combative context of a polarised modern American religious culture, which was not Maxwell's. Maxwell did not think that a Christian who was a scientist had to set out faith presuppositions. For Maxwell, all *true* discoveries were God-given.

---

33 McNatt, 'James Clerk Maxwell's refusal to join the Victoria Institute', 214.

The Glenlair Years

(1865–1871)

Chapter 21
The Laird at Home

James and Katherine loved Glenlair. When James left Marischal he hinted at the attractions of a life totally free from academic duties: 'I shall do natural philosophy, whether in college or out of it, and I may perhaps publish more if I teach less.'[1] As it turned out he went to London, and it was a surprise in 1865 when he suddenly resigned his professorship and retired to Glenlair with the intention of spending the rest of his life away from formal academia. The reasons behind that decision remain uncertain. Was he tired of academia? Was the ethos of the College too utilitarian? Was the ability of the students unsatisfactory? Were his old problems of teaching duller minds resurfacing? Were there tensions with the College Governors? We do not know.[2] But retire he did, and life was busier than ever, causing him to remark to C.B. Tayler: 'I have now my time fully occupied with experiments and speculations of a physical kind, which I could not undertake as long as I had public duties.'[3]

Retirement lasted six years, before he was persuaded by Cambridge to accept a professorship and supervise the development of a new research facility, which in time became the Cavendish Laboratory. But from 1865 to 1871, James and Katherine enjoyed life at their Galloway retreat. They retained 8 Palace Gardens Terrace in London, visiting each winter and spring.[4]

The journey from Galloway to the south had become much easier after 1859 when the railway network reached Castle Douglas.[5] From March 1861 the line extended to the west and connected

---

1 JCM, *Letter to C.J. Monro,* 24 Jan 1860, (Harman, I, 644).

2 The issues are discussed by Reid, 'Maxwell at King's College, London', 65f. Also by Randall, 'Aspects of Maxwell's Life and Work', *Clerk Maxwell and Modern Science,* 16–21.

3 JCM, *Letter to C.B. Tayler,* 2 Feb 1866, (C&G1, 344; Harman, II/1, 249).

4 JCM, *Letter to W. Thomson,* 20 February 1868, (Harman, II/1, 346): '[In winter] we intend always to be in London as headquarters.'

5 JCM, *Letter to Stokes,* 9 July 1875, (Harman, III/1, 234): 'We arrived here [Glenlair] on 25th June breaking the journey at Carlisle. Mrs. Maxwell stood the journey very well and is making steady progress.'

with ferries crossing to the north of Ireland. Glenlair visitors were advised to take the train from Dumfries to either Dalbeattie or Castle Douglas, where they could get transport or be picked up by a gig sent by the Maxwells.[6] When travelling by train, he tended to choose third-class, preferring a hard seat, even though a cushion softened his pew in Corsock Kirk.[7]

The inheritance from his father, income from the estate, plus his pension of £400 per annum for life from the Crown following the loss of his Aberdeen post, meant he was financially comfortable. There is little doubt that the Glenlair years, free from the grind of university teaching, gave him opportunity to think at leisure and further his ideas in a way which would have been difficult, if not impossible, as an active professor.

Science was his delight. It was never a drudge. He was as enthusiastic as ever about exploring theories. He wrote up ideas. He produced books and papers. Most of all, he thought. All of which generated substantial correspondence with his peers throughout Britain and Europe. A special postal pillar-box, for the sole use of Glenlair House, was set into the stone wall across the Urr river. Campbell recalled Maxwell carrying letters to and from this rustic post-office in all weathers, and at the same time giving the dogs a run.

In the spring and early summer of 1867, the Maxwells took a tour of Italy, with James learning Italian in order to converse with the Italian physicist and politician Professor Carlo Matteucci (1811–1868), who pioneered the study of bioelectricity. The journey involved the inconvenience of their ship being quarantined at Marseilles.[8] During the 1860s there were several outbreaks of cholera in countries round the Mediterranean, resulting in ships having to stand off-harbour for several days or sometimes weeks. The ship might suspect the disease status of the port, or the port might suspect that of the ship. During the enforced hiatus, James helped with passengers' needs, acting

6 JCM, *Letter to Henry Augustus Rowland,* 19 July 1875, (Harman, III/1, 239): 'The nearest station is Castle Douglas about seven miles from this where I would meet you if you let me know by what train you will arrive there.' Also, JCM, *Letter to HAR,* 21 July 1875, (Harman, III/1, 239): 'Telegram received. You reach Dalbeattie at 11 Friday night. Come by Corsock post gig to Glenlair on Saturday morning start at six I shall expect you to breakfast.'

7 *Galloway News & Kirkcudbrightshire Advertiser,* 10 Nov 1882.

8 C&G1, 324.

as water-carrier. Back in Scotland, in October 1868 the Maxwells attended the laying of the foundation stone by the Prince and Princess of Wales for the new college of the University of Glasgow.[9]

Life was busy and pleasant. The Glenlair estate was supervised. The house was enlarged in line with plans drawn up by his father. During the summer months, James and Katherine rode almost every day, morning and evening, enjoying the surrounding countryside.[10] After dinner there might be readings from Chaucer, Spenser, Milton, or Shakespeare.[11] There was also the Glenlair tradition of regulating clocks according to local time—fifteen minutes behind Greenwich— with the noon meridian sun throwing the shadow of a stick onto a notched stone outside the door.[12]

The house benefited from several modern devices, including a system of electric bells which was installed in 1875. These replaced the traditional mechanical arrangement, which had fascinated James as a child. The new system had its problems, and in the summer of 1879 James wrote to the firm of D & J Graham:

> The connections of the electric bells which you put up for me in 1875 have given way, so that no current will pass from the battery to any of the bells. There does not appear to be any leakage by bad insulation but the return wire is broken somewhere. If I connect the bells to the battery by ordinary copper wires they ring very well. Can you send anyone to find the break and repair the wires? Send a superior workman who can do it without injuring the house more than is necessary. The nearest station is Castle Douglas.[13]

It was in 1872, at a meeting of the Iron and Steel Institute at the Corporation Galleries in Glasgow, that D. & J. Graham demonstrated models of their 'Patent Apparatus for Electric Bells, especially adapted for Ships, Mansions, Hotels, etc., as used with Sir William Thomson's Battery and Signalling Studs'.[14] James jumped at the chance of using this device, based on one of his friend's inventions.

9 JCM, *Letter to W. Thomson,* 6 Oct 1868, (Harman, II/1, 446).

10 JCM, *Letter to H.R. Droop,* 19 July 1865, (C&G1, 343; Harman, II/1, 227).

11 C&G1, 321.

12 Ibid., 372.

13 JCM, *Letter to D & J Graham,* 26 July 1879, (Harman, III/2, 831).

14 *Scotsman,* 8 Aug 1872.

During the 1870s one of the cottages on the Glenlair estate was severely damaged in a lightning strike. The discharge passed down the damp soot and blew out several stones from the base of the chimney, making its way to water in a ditch a few yards away. The cottage was built on a granite rock. This made James reflect on the best way to protect buildings which were erected on granite or other non-conducting foundations. His solution was to place a strip of metal on the ground all around the building, carry another strip along the ridge-stay from which pointed rods projected upwards, and unite this strip with the one on the ground through copper strips passing down each corner of the building. In effect, he enclosed the building in a metal cage.[15]

Lightning absorbed his attention more than once. After the Revd George Walker moved to Castle Douglas he was a regular visitor to Glenlair, as was his brother Robert. During one of Robert's visits, James gave him an example of the conditions necessary for a 'disruptive discharge of atmospheric electricity' by pointing out the different effects of a recent thunderstorm on two trees nearby. One very high tree was sound and intact because its roots ran into deep, damp soil. The other, despite being much shorter, had been shattered and destroyed, because its roots rambled over a bare and exposed dry rocky knoll.[16]

In Galloway, as in London, the Maxwells were a hospitable couple, regularly inviting local families from the Vale of Urr for an evening meal. James termed this: 'dining the valley in appropriate batches.'[17] A glimpse of what became the daily routine comes in a letter to Charles Cay, written a few months before he relinquished his London post:

> We have had very rough weather this week, which, combined with the dining, has prevented our usual airings. The ordinary outing is to the Brig of Urr. Katherine on Charlie, and I on Darling. Charlie has got a fine band on his forehead, with his name in blue and white beads.[18]

Visitors were impressed by the daily household prayers. One commented: 'The prayer, which seemed extempore, was most

15 Garnett, *Heroes of Science: Physics,* 294.

16 Robert Walker, 'James Clerk Maxwell', 196.

17 JCM, *Letter to C.H. Cay,* 5 Jan 1865, (C&G1, 341).

18 Ibid..

impressive and full of meaning.'[19] Sundays were for church, and Lewis Campbell described how, each Sunday, after returning from church, James buried himself in reading the works of the older divines—such as John Owen and Jonathan Edwards—noting that in theology, as in literature, while engaging with all phases, James' sympathies went largely with the past.[20] Campbell further commented that, whilst James was open to free thought on religious subjects, he did not believe in progress by ignoring differences, or by a lazy blending of clearly thought out traditional systems into a modern haze of a vague common Christianity. Campbell's verdict was that James always focussed on the permanent essence of things, whether in science, philosophy or theology.

Campbell's observations were acute. He put his finger on a critical aspect of his friend's philosophy. James Clerk Maxwell believed that proper scientific investigation—in any discipline—could produce objective knowledge. Though he was never averse to using the imagination, he dismissed 'what appeared to him as crude and half-baked theories about the highest subjects.' At the same time, he understood that all scientific theories and philosophical systems fell short of certainty. Applying the same thinking to his faith, he 'clung to the belief that love remains while knowledge vanishes away.' Typical of a man with evangelical sympathies, he judged preaching to be empty if it focussed on mere morality. Doctrine was important, yet admitted: 'I have no nose for heresy.' What marked his faith was trust, sympathy, and openheartedness.

Maxwell was disciplined, and produced an extensive output between 1866 and 1871. In 1866 he gave the Royal Society's Bakerian Lecture, 'On the Viscosity or internal friction of Air and other Gases.'[21] This included measurements made by himself and Katherine in their London attic.[22] In 1868 he published, 'On a Method of making a direct

19 C&G1, 322.

20 Ibid., 321.

21 *Philosophical Transactions of the Royal Society,* Vol. 56. Cf. *NVN2,* 1–25.

22 JCM, *Postcard to Tait,* 29 Dec 1877, (Harman, III/2, 559): 'My better half, who did all the real work of the kinetic theory is at present engaged in other researches. When she is done I will let you know her answer to your enquiry.'

comparison of Electrostatic with Electromagnetic Force: with a note on the Electromagnetic Theory of Light', which discussed the definitions of electromagnetic and electrostatic units of electric charge, and their relationship to the speed of light.[23] The same year he presented a paper, 'On Governors', to the Royal Society, which involved a mathematical analysis presaging control theory and cybernetics.

In 1870, Cambridge awarded him the William Hopkins Prize for original research. The same year he spoke to the Royal Society of Edinburgh on, 'Reciprocal Figures, Frames, and Diagrams of Forces', and was awarded the RSE's Keith Medal. 1870 also saw the publication of, 'On Hills and Dales', which was an early input to the mathematics of topology.[24] Soon afterwards his book *Theory of Heat* (1871) rolled off the press.

His substantial work *A Treatise on Electricity and Magnetism* was mainly shaped at Glenlair, though notebooks examined at King's College by Sir John Randall in the early 1960s show that its genesis was during Maxwell's time in London, with the massive volume not fully published until 1873.[25] The same was true of *Matter and Motion,* published in 1876.[26]

In addition to these books and papers, the concept, later described as Maxwell's 'demon', was first formulated during the Glenlair years in correspondence with Peter Tait.[27] Other Glenlair speculations included thoughts on the nature of multi-dimensional space.[28] He continued to attend British Association meetings, as well as acting as examiner in the Mathematical Tripos at Cambridge, with his examination papers acting as catalysts in the modernization of the Cambridge syllabus.

---

23 *Philosophical Transactions of the Royal Society,* Vol 58.

24 JCM, 'On Hills and Dales', *Philosophical Magazine,* Vol 40/269, 421–427.

25 Randall, 'Aspects of Maxwell's Life and Work', *Clerk Maxwell and Modern Science,* 2, 11.

26 Maxwell's unfinished, *Elementary Treatise on Electricity,* in which he attempted to show that physics could be studied without recourse to advanced mathematics, was edited after Maxwell's death by William Garnett and published in 1881.

27 JCM, *Letter to Tait,* 11 Dec 1867, (Harman, II/1, 328–334).

28 JCM, *Letter to C.J. Monro,*15 Mar 1871, (C&G1, 379; Harman, II/2, 617).

Chapter 22
Elder of the Kirk

Much had taken place at Corsock Kirk since the Revd George Sturrock arrived in 1861. Sturrock trained in Aberdeen where he was a city missionary during his student years, and where Maxwell first knew him, as did his father-in-law, Daniel Dewar. It was on Maxwell's invitation and Dewar's recommendation that Sturrock preached in Corsock Kirk in July 1861 for the congregation's approval as their local missionary. He proved satisfactory in every respect.

When Sturrock arrived, Parton Kirk was still the official parish church for the area. That remained the case until 1863 when Corsock was elevated into a parish. Until then, Corsock Kirk was categorised as a preaching-station, with its status only changing when enough endowments were secured for the support of both a building and a minister.[1] Moreover, when it became a full-status parish it was not automatic that Sturrock the local missionary *in situ* would continue as Sturrock the parish minister. He had to be reintroduced by the patrons of the parish—in Corsock's case the trustees—and reappointed by the Presbytery. Following a petition to the patrons, signed by the entire congregation, Sturrock was ordained by the Presbytery of Kirkcudbright on the twentieth of August 1863 as the first incumbent of the new parish.[2]

Sturrock threw himself into the work, and the congregation grew significantly during his ministry. It increased from sixty-three known communicants in 1863 to one hundred and ninety-three in 1899. As well as fulfilling his regular calling of pastoring and preaching, there were other duties which a country minister was expected to attend to. For several years Sturrock chaired the Corsock Mutual Improvement Society; he was chaplain to the Corsock Curling Club; he was chairman of the Village Hall committee; and he was present at almost every significant event in the community.[3]

1 Sturrock, *Corsock Parish Church,* 10-12.

2 Ibid., 11.

3 Cf. *Galloway News & Kirkcudbrightshire Advertiser,* 18 Feb 1881.

The Sturrock Family

1863 was an *annus mirabilis* for George Sturrock. The parish of Corsock was created. The church moved from preaching station to parish kirk. He was ordained as a minister of word and sacrament. And, on the twentieth of October, he married Catherine Mackenzie Stewart (c.1836–1900) in Aberdeen. The couple had four children: Margaret Mary (1864–1947), Charles Stewart (1866–1885), George Steven (1871–1872), and a second George Steven (b.1875).

The first baby George died tragically of unknown causes aged only five months in March 1872. When Maxwell heard news of the child's death he was in Cambridge, teaching and supervising the construction of the Cavendish Laboratory. Another son, also given the name George Steven, arrived in 1875.

The Sturrock's daughter, Margaret Mary, remained unmarried. Like her mother, she was an excellent pianist, regularly taking part at concert socials in the village hall.[4]

The oldest son, Charles Stewart, emigrated to Melbourne, Australia, in 1883. The following year his father and sister sailed out to visit him. His father had been awarded extended leave of absence by the Presbytery due to illness, with the family doctor recommending a long sea voyage.[5] During the visit George Sturrock posted home accounts of his experience, which were published in the *Galloway News & Kirkcudbrightshire Advertiser.* Heartbreakingly, soon after father and daughter returned to Scotland, there came news of Charles' death on the fifth of June 1885. He had contracted tuberculosis, and he died at only nineteen years of age.[6]

The Revd George Sturrock was progressive. This is evidenced not only by his interest in science, but by the considerable achievement of having a harmonium installed in the church to help lead worship, rather than relying solely on the traditional precentor. This was at a time when any instrumental music in Presbyterian churches in Scotland was controversial. With the elders and congregation agreeable to the experiment, Sturrock persuaded Presbytery to permit the innovation, and a 'very fine and suitable harmonium' was presented to the

4 Ibid., 18 Jan 1889.

5 Ibid., 18 July 1884.

6 Ibid., 26 Jun 1885.

congregation in November 1880 by Andrew Wedderburn Maxwell of Glenlair, along with William Barbour of Barley and the minister himself.[7] Wedderburn Maxwell—son of Aunt Isabella—had inherited the estate on James' death the previous year. The harmonium was to be used solely for worship, with the lady of the Manse, Mrs Sturrock, acting as organist. In 1899 a better instrument was required. It was then that Sturrock published his monograph *Corsock Parish Church: Its Rise and Progress,* as a contribution to a Bazaar to raise funds for a brand-new version.

New Parish, New Kirk, New Manse

When the new parish was created in 1863, James Clerk Maxwell was a founding trustee, along with James Crosbie of Merkland, the Revd James Maitland of Kells, Robert McCartney of Garcrogo, and John Fergusson of Kilquhanity. Maxwell also contributed to the minister's stipend, supplying around one-fifth of the local endowment.[8] He was equally active in procuring finance from elsewhere. In 1861 he wrote to his friend H.R. Droop in London explaining: 'If we can raise £1000, there is a fund already raised which will contribute £2000, so as to give a salary of £120 to the minister permanently; and, as the people are too poor to support the minister themselves, we hope to make the chapel independent of chance contributions in this way.'[9] He told Droop that a great part of the funds for the original construction of the church had been subscribed in London by friends of an English gentleman who had property near Corsock, but that that channel was now closed: 'If you can put us in the way of diminishing the deficit we shall be grateful, and I will see that the money goes to the fund, and that the names are duly entered, however small the contributions.'

The fund-raising efforts were successful. Corsock preaching station became a full-status Kirk. And James told his uncle, Robert Dundas Cay, that the congregation filled the building.[10]

---

7 *Galloway News & Kirkcudbrightshire Advertiser,* 9 Apr 1880 & 19 Nov 1880.

8 *Ibid.,* 29 May 1885.

9 JCM, *Letter to H.R. Droop,* 28 Dec 1861, (C&G1, 329).

10 JCM, *Letter to Robert Dundas Cay,* 21 Aug 1863, (Harman, II/1, 103).

An endowed church and stipend were not enough. A manse was also needed for the minister. No manse existed when Sturrock came to the Corsock preaching station in 1861, and that was still the case in 1863, even though he was now the fully ordained minister of the new parish. Perhaps with this in mind, Katherine held a celebratory ordination dinner at Glenlair which was attended by members of the Presbytery, and by most of the parish heritors.[11] During the evening the question of a house for the minister was raised. It was an issue of some urgency since a lady of the manse would soon be arriving. One guest remarked that a suitable cottage would cost £300, prompting James to throw out a challenge: 'If you will go on with it at once, I will give a fourth of that sum, or £75.' The challenge was accepted. Collections were taken at the kirk. Money was raised at local markets. Donations were received. By January 1865, James was able to tell Charles Cay that the Manse of Corsock was now finished, and that it stood near the river, not far from the deep pool where they used to bathe as youngsters.[12]

The Kirk at Parton was not forgotten. When Corsock's Communion Season was changed from July to the beginning of May—making James' attendance impossible as he had to be in Cambridge—he took the sacrament in Parton Kirk where his father had been an elder, and within whose civil parish Glenlair still lay.

Until 1856 Corsock also had no burial ground of its own, with interments taking place elsewhere. That was why James' mother and father, and later himself and Katherine, were laid to rest at Parton. Another obligation attaching to James as a heritor was subscribing to the maintenance of the horse-pulled hearse, which was used by the local undertaker to take the deceased to the place of burial.

Ordination to the Eldership

James became an elder of the Kirk as had been his father before him, though that was not an automatic appointment. He had to be judged suitable. His ordination to the eldership in the Corsock

11 Ibid..

12 JCM, *Letter to Charles Cay,* 5 Jan 1865, (C&G1, 341; Harman, II/1, 202). Also, Sturrock, *Corsock Parish Church,* 12.

congregation took place on the sixth of September 1863.[13] This was soon after Corsock had become a parish in its own right, and when a fully functioning Kirk Session consisting only of members of Corsock congregation needed to be formed. To get things started, the Presbytery had appointed George Sturrock, Alexander Knox (an elder from Parton), and other persons from adjacent parishes to form an initial Kirk Session. A major task of that temporary Kirk Session was to find men (not women at this stage) from the congregation who could become elders. Three names were proposed by the congregation: James Clerk Maxwell, Adam Callander, and John McCall.[14] Nomination by the congregation remained the rule in Corsock Kirk, though this was not always the case in Church of Scotland parishes, where nomination by the existing elders was more common.

Any *bona fide* member of a congregation could object to the ordination of a person to the eldership if there were reasons to doubt the candidate's manner of life or doctrinal beliefs. Accordingly, on Sundays the twenty-third and thirtieth of August 1863, an edict concerning the proposed ordinations was read to the congregation. On the day itself 'no person appeared to object' and the Kirk Session instructed the minister to proceed.[15]

During the service Sturrock preached from Romans chapter twelve: *'I beseech you therefore, brethren, by the mercies of God, that ye present your bodies a living sacrifice, holy, acceptable unto God, which is your reasonable service. And be not conformed to this world: but be ye transformed by the renewing of your mind, that ye may prove what is that good, and acceptable, and perfect, will of God.'* Following the sermon, the appointed questions were put to the three men, after which the minister, through solemn prayer, set them apart and ordained them to the office of the eldership. The minister then exhorted the elders and congregation alike on their respective responsibilities, before the new elders received the right hand of fellowship from the existing members of session. They were then required to sign the 'Formula' which summarised what they had agreed to in becoming elders, including allegiance to the

13 Mackay & Callander, *James Clerk Maxwell: Pathfinder of Modern Science*, 18.

14 Sturrock, *Corsock Parish Church*, 11, 15.

15 Corsock Kirk Session Records, 6 Sep 1863, (NRS/CH2/446/1/25).

Westminster Confession of Faith which was the Church of Scotland's subordinate standard of faith.[16] The Formula stated:

> I do sincerely own and declare the above *Confession of Faith*, approven by former General Assemblies of this Church, and ratified by law in the year 1690, to be the Confession of my Faith; and that I own the doctrine contained therein to be the true doctrine, which I will constantly adhere to; as likewise, that I own and acknowledge the Presbyterian Church Government of this Church, nor settled by Law, by Kirk-Sessions, Presbyteries, Provincial Synods, and General Assemblies, to be the only Government of this Church; and that I will submit thereto, concur therewith, and never endeavour, directly nor indirectly, the prejudice or subversion thereof; and that I shall observe uniformity of worship, and of the administration of all public ordinances within this Church, as the same are at present performed and allowed.[17]

---

16 F.A.J. MacDonald, 'Liberal, Broad, and Reforming Influences in the late Nineteenth Century', *The History of Scottish Theology,* Vol. 2, 419–431, discusses versions of the Formula from 1711, 1889, and 1910. What MacDonald omits is that although candidates for the ministry of Word and Sacrament ascribed to the 1711 Formula until 1889, elders subscribed to an older version which derived from 1694 (GA Act XI. Sess. 13, April 13, 1694). During the eighteenth century there were various attempts at harmonising wording for ministers and elders, but without success.

17 Kirk Session Minute Books sometimes recorded the words of the Formula, often on the last page where new elders could add their signatures. See, for example, the records of Edinburgh Morningside Kirk Session: 1839–1888, (NRS/CH2/486/1/386), and Moffat Kirk Session:1843–1862, (NRS/CH2/1523/1). The Corsock Kirk Session Minute Book does not include the Formula for the period in question. (It may have been written in another document which has not been preserved.) What the Corsock Kirk Session Minute Book *does* have is a copy of the new 1889 Formula which remained in force until 1910. That new (1889) Formula has two signatures appended (Samuel Henderson of Corsock, and George Clark of Knockbain, Corsock) in an entry from 21st June 1903, (NRS/CH2/446/1/125). The new (1889) Formula was less strict than the Formula which Maxwell signed, stating: '*I hereby declare my approbation of the Confession of Faith, as approved by this Church and ratified by law in 1690; and I promise to submit myself to the discipline and Presbyterian government of the Church as established by law, and will never endeavour, directly or indirectly, the prejudice or subversion thereof.*' F. Kingsley Elder's paper 'The Religious Outlook of James Clerk Maxwell' (2000), confuses the two Formulae, assuming that the Formula found on the last page of the Corsock Kirk Session Minute Book was what applied in 1863.

Corsock now had a workable Kirk Session, and the temporary elders who had been drafted in from neighbouring parishes, stood down from that role. The new Kirk Session comprised: George Sturrock, Alexander Knox, James Clerk Maxwell, Adam Callander, and John McCall.

What was James Clerk Maxwell agreeing to in signing the Formula? It focussed on doctrine, worship, and church government. It required agreement from its signatories that the doctrine taught in the *Westminster Confession,* the form of worship prevalent in the Church of Scotland, and the Presbyterian form of Church government, were founded on, and agreeable to, the Bible. James had been taught the *Shorter Catechism* as a child, and he was familiar with the doctrines contained in the *Westminster Confession*. But he did not sign his name lightly. He re-read the *Confession* carefully before adding his signature at his ordination. This was recalled in George Walker's article for the *Galloway News & Kirkcudbrightshire Advertiser*:

> In 1863, on the endowment (greatly through his own exertions) of the Parish Church of Corsock, he was ordained an elder there; and it is worthy of note that a man of such vast powers of mind had no scruples whatever, after a renewed study, in signing the *Confession*.[18]

In 1863, the wording of the Formula did not allow personal liberty of opinion on any doctrine which the *Confession* expressed. Such a conscience clause only appeared decades later. Yet, though the wording was unequivocal, the *de facto* situation was that signing had become regarded by many as indicating general allegiance to the core doctrines expressed in the *Confession*, rather than total agreement with every detail; it was this situation which prompted a new Formula in 1889 which was framed accordingly.[19] Nevertheless, because Maxwell accepted eldership in the Kirk only after having had a fresh study of the *Confession,* he appears to have been in substantial agreement with its approach.

---

18 *Galloway News & Kirkcudbrightshire Advertiser,* 18 May 1883, by George Walker, first published in *Life and Work*.

19 A.C. Cheyne, 'The Confession through Three Centuries', *The Westminster Confession in the Church Today,* 17–27. The 1889 Formula (which was to apply to both ministers *and* elders) used the phrase *I hereby declare my approbation of the Confession of Faith*', with 'approbation' meaning 'approval, commendation, praise', which was regarded by many as a weakening of adherence.

The Questions put to the elders implied there was harmony between the Presbyterian form of worship and Church government and the Bible. Yet the words 'founded on' and 'agreeable to' could be interpreted in one of two ways. They could mean that Presbyterianism was *demanded* by the Bible and that all other churches had it wrong. Or they could mean that Presbyterianism did not *contradict* biblical principles but emerged from them. Given Maxwell's personal spiritual journey, which included experience of Episcopalianism and other denominations, he may well have understood it in terms of the latter position.[20]

Duties of the Eldership

The first duty for Maxwell and his fellow elders was to set a date for Holy Communion. This would be the first time it would take place in Corsock Kirk. The date agreed was the twentieth of December 1863. As then customary in Scottish Presbyterianism, it was decreed that the Thursday before communion should be observed by the congregation as a preparation day of fasting, humiliation, and prayer.[21]

On the Communion Sunday itself, sixty-three communicants sat at three consecutive tables. This followed John Knox's model, introduced during the Scottish Reformation in the sixteenth century, in which communicants came to the front in relays to sit around a plain table as a group and receive the bread and wine together. Knox wanted the communion table to be regarded simply as a table for a meal received by the people of God, and not as an altar on which the sacrifice of Christ was re-enacted.

Maxwell's name appears in the Kirk Session records from 1863 until 1870. Though he was still an elder between 1870 and his death in 1879, he was largely based in Cambridge, and unable to attend its formal meetings. In fact, from the extant records—preserved in unbroken sequence—it appears that few official minuted meetings ever took place, with many decisions made informally by the minister and elders in conversation after the Sunday service.

20 Kingsley Elder, 'The Religious Outlook of James Clerk Maxwell', over-strongly insists that Maxwell preferred Presbyterianism.

21 Corsock Kirk Session Records, 12 Dec 1863, (NRS/CH2/446/1/26).

During the period of Maxwell's eldership there were thirty-five legally minuted Kirk Session meetings, with James present at five of these. As well as making arrangements for Holy Communion (which included approving candidates who wished to become new communicants) an important element was distribution of funds to the needy, though much of that burden had been removed from Kirk Sessions by the Scottish Poor Laws of the mid-nineteenth century. By Maxwell's time, the elders had four main areas of responsibility: pastoral oversight, moral discipline, parish education, and trusteeship.

1 *Pastoral Oversight.* As an elder, it was common for Maxwell to call on sick persons in the village, read the scriptures, and pray with them. Fifty years after his death, older members of the congregation recalled him arriving at their cottages, sporting a wide felt hat, and mounted on 'Dizzy', his black horse. One lady lit up with delight at the memory, exclaiming: 'Eh! He was a right nice mon!'[22] One newspaper article noted that he had the reputation of possessing: 'all the faculty for scriptural exposition and extemporary prayer that would mark a minister of the Kirk.'[23] In 1870 he took his turn in representing the congregation at the local Presbytery of Kirkcudbright and at the regional Synod of Galloway.[24]

2 *Moral Discipline.* Eldership involved dealing with moral discipline among church members and parishioners, and a majority of meetings for which written records were actually kept—the *quasi-legal* nature of such proceedings meant that formal minutes were a necessity—were concerned with individuals compeared on issues of sexual impropriety. The higher church court of Presbytery was notified if the Kirk Session had to deal with a case of adultery involving married persons. Of the five occasions when Maxwell's presence at a Kirk Session meeting was minuted, three involved discipline cases, with the other two dealing with Holy Communion arrangements.[25]

22 Riddell, *Clerk Maxwell of Glenlair,* 38.

23 *Galloway News & Kirkcudbrightshire Advertiser,* 10 Nov 1882.

24 Corsock Kirk Session Records, 8 May 1870, (NRS/CH2/446/1/30).

25 Discipline cases: 16 Jul 1865, 6 Jun 1869, 13 Jun 1869; Holy Communion arrangements: 15 Jan 1865, 25 Jun 1867.

On the sixteenth of July 1865, Maxwell was present when Jane McNaught—recently transferred to Corsock from Kirkcowan and Kirkpatrick Durham—was restored to church privileges and could again take Holy Communion, after having been suspended. On the sixth of June 1869, he was present when Thomas Melville was compeared because of his guilt in 'the sin of fornication'. Having expressed penitence, Melville received a solemn admonition from the minister. And, on the thirteenth of June 1869, Maxwell was one of the elders who interviewed a female parishioner named Donnely, on account of her fornication. After expressing penitence, she was admonished.

The records show that the elders took pains to trace the male wherever possible. Not only that he too might be disciplined, but that he might meet his responsibilities to any child born of the liaison. Often the males could not be traced, especially if they were itinerant labourers who moved from parish to parish. In former times guilty parties were dealt with harshly, enduring public humiliation in the Kirk on a Sunday. By the latter part of the nineteenth century, the penalties were less severe, though they still included rebuke by the minister—but now within the relative privacy of the Kirk Session rather than during public worship—and suspension from the sacraments. Yet the whole parish knew what was happening!

3 *Parish Education.* When Maxwell became an elder in 1863, the local Kirk Session still had responsibility for the parish schools. This had been the case in Scotland ever since the Reformation of the sixteenth century, and was in line with Knox's vision of a kirk and a school in every parish. The system changed when the 1872 Education Act (Scotland) made the State responsible for children's education.

The possibility of the State taking over from the Church had been years in the making, though James' father, John Clerk Maxwell, had vigorously opposed removing schooling from ecclesiastical jurisdiction. In 1854, John Clerk Maxwell had co-signed a letter to the *Scotsman*, expressing the conviction, 'founded on practical experience, that the present system of Parochial Schools (subject to defects capable of easy remedy) has worked well; and that, practically, children of all religious denominations are in the constant habit of enjoying the full benefit of the instruction given in them.'[26] The signatories declared it to be their strong opinion that, except

26 *Scotsman*, 18 Feb 1854.

for correcting defects in its working, and increasing its efficiency, the present system of Parochial Schools ought not to be interfered with, and that connection with the Church of Scotland ought to be maintained. Similar sentiments were expressed nationwide, resulting in a delay until 1872 when the Act was finally passed.

The post-1872 situation saw James Clerk Maxwell on the Corsock Board of School Governors, acting as Chairman at their meetings in the local Temperance Hotel. He described developments in a letter to Lewis Campbell:

> We are busy electing School Boards here. The religious difficulty [of denominational schools] is unknown here. The chief party is that which insists on keeping down the rates; no other platform will do. All candidates must show the retrenchment [reduction in costs] ticket.[27]

One of his fears concerning the new system was that harsh economics—keeping down the rates—would result in the closure of some rural schools which the old parish system had kept open, and he strove to prevent that happening for at least one of the schools in the local area. This was Merkland School, near Kirkpatrick-Durham. A majority on the Board wanted it closed, but Maxwell ran a successful campaign for its continuance, at least in the short term.

In December 1873, Sir Alexander Grant, chair of the Scottish Board of Education, after visiting the area and hearing representations, recommended to the National Board that the school at Merkland continue to function.[28] However, the Corsock School Board refused to accept this, and appealed against Grant's recommendation.[29] Though their appeal was initially refused, and though Maxwell offered an endowment to forestall future closure, his offer was turned down and the school did shut. Nothing daunted, he set apart a site and had plans for a school to be erected and supported at his own expense on his estate; but failing health prevented this happening.[30] The former Merkland school, schoolhouse, and garden, was bought by Mrs Dunlop of Corsock House in 1879.[31]

---

27 JCM, *Letter to Campbell,* 3 Apr 1873, (C&G1, 386; Harman, II/2, 840).

28 *Glasgow Herald,* 18 Dec 1873.

29 *Dundee and Galloway Standard and Advertiser,* 14 Jan 1874.

30 C&G1, 371.

31 *Dumfries & Galloway Saturday Standard,* 7 June 1879.

4 *Trusteeship.* One issue, which eagle-eyed officials from the Presbytery insisted Corsock Kirk put right, related to trusteeship for the church and congregation. Though most of the kirk's trustees were also its elders, the two bodies were not identical. After 1863 a sloppy habit had developed of the minister and Kirk Session taking decisions which technically were for the trustees.

In 1877 the Presbytery demanded that things be done correctly. The trustees then held a meeting in the Douglas Arms Hotel, Castle Douglas, and elected Maxwell as the chair.[32] Though the trustees found 'everything connected with the Trustees hitherto managed by the Minister and Kirk Session in a satisfactory condition' they resolved to begin *de novo* on the first day of January 1878. In the meantime, they clarified that their specific role was to deal with property, finances, the appointments of precentor and church officer and setting their salaries. A few days later they met again, appointing John Heron as precentor on a salary of £6 per annum, and David Scott as church officer with a salary of £2 and five shilling per annum. The trustees also allocated sittings in the church, setting apart more pews as free sittings, and reserving pew 32 for the officiating elder and pew 37 for the manse family.[33]

With irregularities rectified, the Presbytery approved the new arrangements at Corsock. When James died in November 1879, his cousin Andrew Wedderburn Maxwell of Glenlair, was elected trustee and chairman in his place.[34]

The Mixed Parish

Not everything at Corsock was idyllic. Following the 1843 Disruption, a local Free Church congregation was established in nearby Kirkpatrick Durham, taking several of the Corsock families. Then, in 1851, a Free Church building was erected in Corsock village itself, through the generosity of the formidable Alexander Colquhoun Stirling Murray Dunlop, who, since 1850, owned Corsock House where the Fletcher family once resided. Alexander Dunlop was a Member of Parliament and skilled lawyer, figuring prominently in the events leading up

32 Corsock Kirk Trustees, 1 Oct 1877, (NRS/CH2/446/1/81).

33 Ibid., 5 Oct 1877, (NRS/CH2/446/1/82).

34 Ibid., 11 Nov 1880, (NRS/CH2/446/1/83).

to 1843. He is the subject of a whole chapter in Wylie's *Disruption Worthies,* which is the definitive Roll of Honour for the Free Church Fathers. It was Dunlop's unflagging, energetic, and detailed exertions which lay behind the *Claim of Right* of 1842, plus the *Protest* and the *Act of Separation* of 1843, which set out the legal foundations of the new denomination. Following 1843, Dunlop continued as a tireless defender of the Free Church, both at the bar of the Court of Session and through articles, pamphlets, and books.

Dunlop not only financed the erection of the Corsock Free Church building, but also supported a local Free Church missionary at his own expense. Initially, the place of worship was categorised by the Free Church as a preaching station under the auspices of its Home Mission Committee, but in 1867 the congregation was recognised as a full-status charge within the denomination. Though the building was originally the personal property of the Dunlop family, it was made over to the Free Church in 1875 when a manse was also added.

As often with new groups, the Free Church congregation developed a robust sense of its identity and distinctiveness. In the early years the parish was beset by ecclesiastical divisions. Both Kirks—Established and Free—vied for support and loyalty. Each saw itself as the true Presbyterian presence in the parish. A trivial example of the tension was that, immediately after 1843, only adherents to the Free Church were allowed to curl on the ice at Corsock Loch. In time these petty differences faded, and the Corsock Kirk minister and his parishioners curled alongside their Free Church counterparts.

This happier relationship was aided by the positive spirit of the Revd George Sturrock of the Established Kirk, and that of the Revd Dr Robert Smith (1816–1894), his Free Kirk counterpart. They cooperated in parish life. They worked together in the Corsock Mutual Improvement Society. They gave joint support for a new village hall.[35] There were also shared events, such as an annual tea-party, held for children attending the Free and Established churches

35 *Galloway News & Kirkcudbrightshire Advertiser*, 3 Jan 1890: 'Sunday opening service of Corsock New Hall [which was] filled with a very attentive audience gathered from a wide district. Both Dr Smith, of the Free Church, and the Rev. George Sturrock, of the Established Church, were present. The singing by a select choir was very hearty and well sustained throughout.'

in Corsock, which took place on the lawn of Corsock House, Alexander Dunlop's own home. In 1860 about 150 children were 'plentifully supplied with the usual liberality of Mrs M. Dunlop, whose exertions for their temporal and eternal welfare have been unabated since she came to the district.'[36]

Robert Smith was the Free Church minister whom the Maxwells knew best. He arrived in 1867, when the Free Church congregation was given full-status, and he ministered in Corsock until his death in June 1894. Smith pastored a congregation which, though never large, grew steadily. It numbered 113 in 1868, and by 1900 its roll stood at 128.[37]

One of Alexander Dunlop's grandsons was Sandy Bell, whose *Reminiscences of Corsock* describe local church life of that era. Bell remembered Smith as a kindly, gentle, scholarly man who, in his youth, had been a missionary in Hungary and had written a book on Pesth (Budapest) which was read aloud to the Bell children by their grandmother on Sunday evenings. Bell commented: 'We called it, I fear, Dr Smith's *pest*.'[38] Bell recalled shepherds' collies, which came with their masters to church, remaining quietly outside or in the church porch during the services, which could last two and a half hours. When the concluding praise was sung, the collies started moving about, sometimes venturing up the aisle to their masters' pew during the last verse 'in expectation of the benediction.' This phenomenon, of sheepdogs knowing when a service was ending, was noted in other parishes.

Bell thought Smith's sermons and prayers to be longer than those of any other minister he knew! Visitors, warned about the length of Smith's sermons, were taken by surprise when he concluded his homily much sooner than expected. Each Sunday, after the Scripture reading, Smith embarked on the *exposition*, which was a discourse of only twenty-minutes' duration, explaining the meaning of the text. At the close of the exposition, Bell recalled unwary visitors saying, 'Nothing

---

36 Ibid., 13 July 1860: Brief addresses were given by the Revd J. Adam (1820–1861), Glasgow; and by the Revd J.H. Wilson (1829–1903), Fountain-Bridge, Edinburgh. Both were Free Church ministers.

37 Ewing, *Annals of the Free Church of Scotland, 1843-1900*, Vol. 2, 43.

38 Robert Smith, *Early days of the Mission to the Jews at Pesth.* (Edinburgh: 1893).

like so long as I feared'. But the exposition was merely a prelude to the *sermon*, which would be surprisingly short if it ended in three-quarters-of-an-hour. There were also two long prayers, plus a third very long prayer interposed between the second psalm and the exposition.

In 1947, following various unions of Presbyterian churches in Scotland, ownership of the Corsock Free Church building transferred to the Church of Scotland. It is within this building that the various memorials can now be viewed. There is the stained-glass window featuring 'The Good Shepherd' in memory of Alexander Dunlop and his wife; there is the James Clerk Maxwell memorial plaque; and there is the James Clerk Maxwell window, which depicts the wise men following the star to Bethlehem. The window was transferred from the Corsock Kirk which James knew.

Not all neighbours were Presbyterian. There was a residual Roman Catholic presence in Galloway, and there was also a small, though not insignificant, Episcopalian community. Intense animosity had existed between Presbyterians and Episcopalians during the covenanting times of the seventeenth century when Government dragoons hunted down and killed Covenanters, but Maxwell viewed both traditions as part of his Christian heritage. In an 1857 letter to Richard Litchfield, he mentioned a local Episcopalian:

> The valley seems deserted of its gentry, but we have one gentleman from Dumfriesshire, who is living in a hired house, and building with great magnificence an Episcopal Chapel in Castle-Douglas at his own expense. His own house is twenty miles off, a capital place, and this is perhaps the least Episcopal part of Scotland by reason of the memory of the dragoons.[39]

Maxwell was in two minds about the gentleman, who seemed to be pursuing a personal fixation. Robert Walker, who was privy to some of Maxwell's opinions, wrote of him referring charitably but far from approvingly about the said gentleman's enterprise.[40] Apart from him, there were other local Episcopalians. In Maxwell's letter to Litchfield, he added:

39 JCM, *Letter to Litchfield,* 29 May 1857, (C&G1, 268–269).

40 Robert Walker, 'James Clerk Maxwell', 206.

> One old family of the Stewartry is of that persuasion, and most of the persecutors' families are now Presbyterian and Whig, so that the congregation is but feeble. It is very different at Aberdeen, where the Presbyterians persecuted far more than the Prelatists; so there I actually found a true Jacobite (female: I could not undertake to produce a male specimen), and there are three distinct Episcopal religions in Aberdeen, all pretty lively.

Here, as elsewhere, Maxwell displayed a feeling for the covenanting era beyond that of popular folklore. He recognised that in the north-east of Scotland (the Aberdeen area), the persecutors had become the persecuted after the Glorious Revolution of 1688/89, and though the harrying of Episcopalian clergy was never of the bloody type endured by Presbyterians in Galloway earlier that century, many Episcopalian families had been driven from their homes into lives of destitution and homelessness. Later the same year he wrote also to Campbell about the 'little Episcopal chaplain, a Celt, whom I see often, but do not quite fathom.'[41]

Maxwell was instinctively open and warm-hearted towards other Christian denominations. This was as true in rural Galloway as in London. But the hub of his spiritual life was his beloved parish church.

41 JCM, *Letter to Campbell*, 7 Aug 1857, (C&G1, 271).

Chapter 23
That Diabolical Being!

After the Maxwells returned from Italy in the autumn of 1867, Peter Tait asked James to read draft chapters of his book, *Sketch of Thermodynamics,* which was to be published the following year. In it, Tait discussed a wide range of topics including: the laws of thermodynamics, reversible and irreversible processes, and what he termed Newton's 'impregnable' laws of motion.[1] Intrigued by his friend's approach, Maxwell set out a thought-experiment.[2] This concerned an imaginary 'finite being' whose activity appeared to call in question the Second Law of Thermodynamics.

Maxwell's aim was modest. His thought-experiment was framed simply to demonstrate that the Second Law attributed *statistical probability* to outcomes, rather than *certainty.* As such it differed from other laws of physics such as Newton's laws of motion, which were mechanically deterministic with unwavering consequences. Maxwell's goal was not to destroy the Second Law, but to show its limitations.[3]

He restated his thought-experiment in an 1870 letter to John William Strutt (1842–1919).[4] The concept was then made public in Maxwell's own book on thermodynamics, *Theory of Heat* (1871), in a section sub-titled, 'Limitation of the Second Law of Thermodynamics.'[5] He used the term 'limitation' deliberately,

1 Tait, *Sketch of Thermodynamics,* v.

2 JCM, *Letter to Tait,* 11 Dec 1867, (Harman, II/1, 328–334).

3 Whittaker, "Maxwell's Famous (or Infamous) Demon', 174f.

4 Also known as Third Baron Rayleigh: Strutt succeeded Maxwell at the Cavendish Laboratory. JCM, *Letter to Strutt,* 6 Dec 1870, (Harman, II/2, 582f). See also: C&G1, 571ff, 'The principle of the dissipation of energy, sometimes called the Second Law of Thermodynamics, states that it is impossible by means of inanimate material agency to obtain work at the expense of heat by cooling a body below the temperature of the coldest body in the neighbourhood. This principle was first distinctly given by Sir William Thomson. Maxwell showed that it obtains only in consequence of the coarseness of our faculties not allowing us to grapple with individual molecules.'

5 JCM, *Theory of Heat,* 308f.

highlighting his main point: that the Second Law was not being abolished, but that it described a probability rather than a certainty.

In interpreting the Maxwell/Tait correspondence, we note that they were long-term friends, who teased one another frequently. On one level Maxwell's imaginary 'finite being' was part of a clever jest. Yet, notwithstanding the humour of the original interchange, in the century and a half since Maxwell first shared his idea, it has spawned diverse interpretations with its possible wider implications much debated. In the late twentieth century it played a part in developing information theory.[6]

The Imaginary Finite Being

In his imagined scenario, Maxwell considered the properties of an enclosed quantity of gas at a given pressure and temperature. Though the totality had an average pressure and average temperature, the gas consisted of molecules of widely varying energies.[7] In his thought-experiment he divided the volume into two portions (A and B) which were separated by a partition with small doors. The doors could be opened or closed by means of a slide, and without the expenditure of any energy from the gas. He then introduced an imaginary 'finite being' with 'sharpened' faculties. This 'finite being' not only 'knew' the paths and velocities of all the molecules but oversaw the sliding doors.

In the thought-experiment, the 'finite being' allowed any molecule in B which had a velocity significantly *lower* than the average velocity of molecules in A, to enter A if it were already moving in that direction. Similarly, the 'finite being' allowed any molecule in A, which had a velocity significantly *higher* than the average velocity in B, to enter B if it were already moving in that direction. The 'finite being' kept the slide doors shut for any other molecules, and the number of molecules permitted to move from B to A were the same as those permitted to move from A to B.

Maxwell pointed out that, in such a state, the total energy in B would *increase*, and the total energy in A would *decrease*. The hot

6 Whittaker, 'Maxwell's Famous (or Infamous) Demon', 164.

7 The Maxwell–Boltzmann distribution gives the probable fraction of gas molecules moving at a specified velocity at a given temperature.

system got hotter and the cold colder, yet no work had been done. That was important: no fresh energy had been pumped into the system: 'only the intelligence of a very observant and neat fingered being has been employed.' This seemed to contradict the Second Law of Thermodynamics which, in simple form, involves heat dissipating and always moving from hotter objects to colder objects, *unless energy is supplied* to reverse the dissipation and the direction of flow.[8]

Central to the thought-experiment was the fact that no work was done on the gas by the 'finite being', as work was understood by both Maxwell and Tait, who defined it as the mechanical effect of energy, measured in foot-pounds.[9] Maxwell's 'finite being' was so surefooted and nimble that it never collided with any of the gas molecules or altered their direction: hence, as Maxwell saw it, no 'work' was done within the system. Maxwell had thus created a situation in which the Second Law appeared to be thwarted. It did not matter that an agent was involved. The actions of the imaginary agent were irrelevant with respect to the movement of the molecules.

What was relevant was that a situation could be envisaged whereby, *despite no work being done*, the hot became hotter as the cold

---

8 JCM, *Letter to Strutt,* 6 Dec 1870, (Harman, II/2, 582f): 'Appoint a doorkeeper, very intelligent and exceedingly quick, with microscopic eyes but still an essentially finite being. Whenever he sees a molecule of great velocity coming against the door from A into B he is to let it through, but if the molecule happens to be going slow he is to keep the door shut. He is also to let slow molecules pass from B to A, but not fast ones. (This may be done if necessary by another doorkeeper at a second door.) Of course he must be quick, for the molecules are continually changing both their courses and their velocities. In this way the temperature of B may be raised, and that of A lowered without any expenditure of work, but only by the intelligent action of a mere guiding agent. (Like a pointsman on a railway with perfectly acting switches who should send the express along one line and the goods along another.) I do not see why even intelligence might not be dispensed with and the thing be made self-acting.'

9 JCM, *Theory of Heat,* 87: 'Work is done when resistance is overcome, and the quantity of work done is measured by the product of the resisting force and the distance through which that force is overcome. Thus, if a body whose mass is one pound is lifted one-foot-high in opposition to the force of gravity, a certain amount of work is done, and this quantity is known among engineers as a foot-pound.' Cf. Tait, *Sketch of Thermodynamics,* 9.

became colder. Thus, it was possible to envisage a situation, however unlikely, in which the Second Law did not apply. Consequently, the Second Law did not describe a *certainty*. It merely described an observed situation whose *likelihood* was so high that this was treated *as if it were* a certainty. But it was an '*as if it were*'. It was still a likelihood. Exceptions (however improbable) were possible. Hence the Second Law was a statistical law which held, on average, on a macroscopic scale, and was not a mechanical law as Newton's were.

Maxwell's goal was unassuming. He emphasised to Tait that the sole purpose of the thought-experiment was to demonstrate that the Second Law of Thermodynamics only had a *statistical* certainty.[10] Moreover, though he had shown that it was *theoretically* possible for a system to become hotter (contradicting widespread assumptions about the Second Law), the statistical probability was so low that the event *could be regarded as* impossible, giving the *appearance* of a mechanical certainty. In the 'real world', the ability of a system to move from cold to hot without expenditure of energy, was regarded as impossible.

Agent, Demon, Imp

It was not Maxwell, but others, who used terms such as 'imp' and 'demon' to describe his 'finite being'. William Thomson preferred 'demon' because its activity, if put into practice, would create diabolical confusion across the whole of physics! Maxwell was unhappy with the term, and in an 1875 letter to Tait he stressed that the word was not his but Thomson's. Intriguingly, though the thought-experiment required the 'intelligent action' of a 'guiding agent', Maxwell at first played down the theological implications of his thought-experiment. And he may have deliberately described it as 'finite', not only to make it so small that it could avoid physical collisions with the gas molecules, but to exclude any premature appeal to a divine or infinite being.

But the theological implications were there for all to see. If the Second Law ruled supreme, how could nature ever produce highly complex structures, since contradicting the Second Law seemed to require an intelligent being? Or did it? This, at least, was the

10 JCM, *Letter to Tait*, c. 1875, (Harman, III/1, 185).

assumption, despite Maxwell protesting to Strutt: 'I do not see why even intelligence might not be dispensed with and the thing be made self-acting'?[11] Maxwell's particular suggestion, that intelligence could be dispensed with, is invalid, since the 'finite being' could only be 'self-acting' (in the terms of Maxwell's thought-experiment) if an intelligence had programmed it previously.[12]

What everyone 'knew' was that the Second Law implied that any system left to itself increased in randomness. For theists—including Maxwell, despite his protest to Strutt—this meant that only an intelligent agent could bring the orderliness of the universe into being or produce structured complexities in nature. At the same time, Maxwell was aware that because the Second Law was statistical, then a natural increase in complexity was always theoretically possible: the mathematics allowed for that.[13] The related concept of irreversible processes in nature (which we discuss later) also had theological implications.

For Maxwell, the immediate relevance of the thought-experiment was in demonstrating that the Second Law described a statistical distribution in which the *overall* behaviour of molecules was explained. In Newtonian mechanics every object obeyed a mathematical equation

---

11 JCM, *Letter to Strutt,* 6 Dec 1870, (Harman, II/2, 583).

12 Whitaker, 'Maxwell's Famous (or Infamous) Demon', 180, argues that the demon did not need intelligence, only an ability to take in information, act on it, and discard it if necessary. But Whitaker's 'demon' only works if programmed by an intelligent mind. Whitaker insists that increased order is possible despite the Second Law, but Maxwell's view was that for the demon to function it had to have either inherent intelligence or programmed intelligence.

13 Bronowski, *The Ascent of Man,* 348, held that since the Second Law is a statistical law then the development of complexity in nature was in fact not only possible but likely: 'By and large any orderly arrangement will run down. But "by and large" is not "always". It is not true that orderly states *constantly* run down to disorder. It is a statistical law, which means that order will *tend* to vanish. But statistics do not say "always". Statistics allow order to be built up in some islands of the universe (here on earth, in you, in me, in the stars, in all sorts of places) while disorder takes over in others.' Building from the notion that in the stars simple hydrogen atoms are fused to become *more* complex helium atoms etc, Bronowski argued that the statistical nature of the Second Law allows for more the more complex evolving from the more simple, with progress achieved in slow steps up a ladder of increasing complexity through what he termed *stratified stability.*

without exception, and each object could, in theory, be tracked. But in thermodynamics two factors intruded on this neat model. First, because a volume of gas was not a single object but an amalgam of millions upon millions of molecules, each with differing values, then individual tracking was impossible. Second, in theory at least, in the gas of Maxwell's set-up, pure chance (in an infinite universe?) could result in fast/hot molecules migrating to B, and slow/cold molecules migrating to A: highly unlikely, but not impossible mathematically. But what the Second Law appeared to imply was that such an outcome was unlikely *to such a degree* that its non-occurrence *could be regarded* as a *law* rather than as simply a *possibility*. Consequently, the Second Law had been interpreted *as if* it described a mechanical inevitability on the same level as Newton's Laws. But that was incorrect.

For Maxwell this underlined the fact that there were areas in physics where it was impossible to formulate strict mechanical laws, something which he emphasised in his 1871 Cambridge Inaugural Lecture.[14] At the time, describing physical processes by a statistical function was an innovation.[15]

What may have started as a tongue-in-cheek humorous comment from one friend to another, had sparked deeper musings. Even as a thought-experiment, it envisaged an actual situation in which entropy could decrease and order come out of chaos, *without* additional work or energy being introduced. Yet, for the theist at least, that was *only* through the aid of an intelligent being, finite or otherwise.

Irreversibility, Time, and Disorder

There was a twist. The thought-experiment required a 'finite being', without which the Second Law appeared to imply that any system left

---

14 JCM, *Introductory Lecture on Experimental Physics*, 20: 'The limitation of our faculties forces us to abandon the attempt to express the exact history of each atom, and to be content with estimating the average condition of a group of atoms large enough to be visible. This method of dealing with groups of atoms, which I may call the statistical method, and which in the present state of our knowledge is the only available method of studying the properties of real bodies, involves an abandonment of strict dynamical principles, and an adoption of the mathematical methods belonging to the theory of probability.'

15 Harman, II/1, 12.

to itself had to increase in entropy. What is 'entropy'? The concept was introduced by Rudolf Clausius. It had long been thought that the overall energy of the universe was running down, raising the question as to where the energy went? Was there a pool of inaccessible energy into which usable energy always runs, but from which it cannot be recovered? Clausius proposed that there is energy which is available, and there is also energy which is inaccessible. The inaccessible energy he called 'entropy', and, in Clausius' terms, the Second Law states that entropy is always increasing. In other words, the pool of available energy always *decreases,* and the residue of inaccessible energy always *increases,* making reversibility of a system impossible.

The notion that energy became inaccessible and irretrievable was what Maxwell took his cue from. Later, instead of energy draining into a sort of hidden lake, Ludwig Boltzmann interpreted the degradation of energy differently, as atoms assuming a more disorderly state, with entropy really a measure of disorder.[16] And Boltzmann offered a measure of that disorder through his celebrated equation S=KlogW, formulated in 1877. Boltzmann saw entropy as disorder or randomness, and in systems subject to the Second Law the tendency was for ordered systems to become randomised though he did not stress irretrievability to the same extent.[17]

Significantly, when Maxwell put forward his thought-experiment, the notion of an inaccessible pool was current, linking the Second Law to the one-way arrow of time. As Tait wrote: 'No known natural process is exactly reversible'.[18] This contrasted with Newton's Laws. The mechanical effects of Newton's Laws are all reversible, whereas the statistical systems described by the Second Law are not, because they are constrained by characteristics of natural processes.

---

16 Bronowski, *The Ascent of Man,* 348f.

17 Harman, II/1, 18. Cf. 19: "Maxwell's contrast between the reversible laws of mechanics and the irreversibility of natural processes bore on the problems of materialism and determinism that exercised his circle in this period . . . while the perfect reversibility of particle motions was consistent with the doctrine of materialism, 'in the present dispensation there remain a number of irreversible processes.' [No. 286] The Second Law of Thermodynamics, an irreducibly statistical law, was inconsistent with the supposition of a wholly mechanical universe."

18 Tait, *Sketch of Thermodynamics,* 71.

For Maxwell, this inevitable and universal increase in entropy resulted not only in unstoppable disorder but made the heat-death of the universe inevitable. The concept of heat-death was that the uiniverse would become functionally dead, even though it might continue to exist for ever in some form. It would have no movement, no activity, no meaningful structure, because the amount of *available* energy was insignificant.

That was the implication of irreversibility for the *future* of the universe. But there was also an implication for its *past*. The one-way arrow of time meant that the universe—if it had an infinite past—must have had infinite energy. But that surely was impossible. This conundrum was exploited by Maxwell's friends Stewart and Tait in their 1875 book *The Unseen Universe*, in which they argued that thermodynamical theory demanded a *finite* past, because of the infinite energy requirement of an infinite past.[19] Infinite energy was deemed impossible, therefore the universe began in finite time, and *that* raised huge problems for non-theistic scientists.

Maxwell was aware of these implications. As far as he was concerned, they meant that a purely materialist explanation of the origin of the universe was inadequate. In 1868, in correspondence with Mark Pattison (1813–1884), Rector of Lincoln College, Oxford, he argued that any truly consistent materialist who eschewed the notion of a beginning to the universe—since that presupposed the materialist's *bête-noir*, a Creator—had a major problem with irreversibility.[20] In Maxwell's opinion, a stubborn materialist was forced to reject irreversibility, and had to embrace alternative suppositions such as cyclical processes. As Maxwell saw it, such a dyed-in-the-wool materialist had to believe that every motion great

19 The logic is not foolproof. Even if irreversibility necessitates available energy increasing with each step taken into the past, that does not necessarily imply that an infinite regressive sequence would produce infinite energy. It is possible to construct a sequence in which, for example, one step into the past increases energy by one unit; two steps increases energy by an additional half a unit; three steps increases energy by an additional quarter of a unit, etc. In such a sequence the available energy would increase with every step taken into the past; yet in the infinite past would only amount to two units, not an infinite value. Thus, purely mathematically, an infinite regression does not necessarily imply an infinite value.

20 JCM, *Letter to Pattison*, 7 Apr 1868, (Harman, II/1, 360f).

and small was somehow reversible—otherwise the materialist hit a brick wall—and that if the world were left to itself again, everything could happen backwards. But, this did not happen. Embedded irreversibility presupposed and demanded a beginning and an end.

The Second Law seemed not to allow the scenario of eternal cyclical progressions. Maxwell elaborated on this in a second letter to Pattison, clarifying that though energy is never destroyed, some of its transformations are not reversible.[21] He gave further illustrations:

> Suppose so many black balls put at the bottom of a box and so many white above them. Then let them be jumbled together. If there is no physical difference between the white and black balls, it is exceedingly improbable that any amount of shaking will bring all the black balls to the bottom and all the white to the top again; so that the operation of mixing is irreversible unless either the black balls are heavier than the white or *a person who knows white from black picks them and sorts them.* Thus, if you put a drop of water into a vessel of water no chemist can take out that identical drop again, though he could take out a drop of any other liquid . . . I think you are right in thinking that we are likely to arrive at physical indications of a beginning and an end. That end is not a destruction of matter or of energy, but such a distribution of energy that no further change is possible *without the intervention of an agent who need not create either matter or energy but only direct the energy into new channels.*[22] (Italics ours)

Two years later, in his 1870 lecture to the British Association, Maxwell reflected on the cosmic implications of the irreversible dissipation of energy. He explained that, for him, this resulted in a state of matter which could not have arisen as the consequence of a previous state of matter (hence a finite past), and that 'this critical condition actually existed at an epoch, not in the utmost depths of a past eternity but separated from the present time by a finite interval.'[23] He continued: 'This idea of a beginning is one which the physical researches of recent times have brought home to us, more

21 JCM, *Letter to Pattison,* 13 Apr 1868, (Harman, II/1, 366).

22 Ibid., (Harman, II/1, 367).

23 JCM, 'Address as President of Mathematical and Physical Section', *Report to the 40th Meeting of the BAAS* (1870), 7.

than any observer of the course of scientific thought in former times would have had reason to expect.' For Maxwell, contemporary physics demanded a beginning to the universe.

Two principles fed Maxwell's thinking. First, the Judaeo-Christian teaching that there was a beginning to the universe, and that God had created all things out of nothing. The doctrine of *creatio ex nihilo* asserted that space, time, matter and energy, were brought into existence by the Creator from nothing.[24] There were no pre-existent materials from which they were fashioned, and, going more deeply, neither were there any pre-existent dimensions of space and time in which they were placed. For theists such as Maxwell, *creatio ex nihilo* was buttressed by nineteenth-century thermodynamical theory. It required the universe to have had a beginning, and it provided an example of science pointing inexorably to what theology had long maintained.

Second, Christian theology envisaged a highly structured universe when it was first created, but which was now disintegrating because of the Fall. God had made the universe perfect, with the creation narrative of Genesis 1 repeating the phrase 'God saw that it was good'. But the Fall resulted not only in humanity losing moral perfection, but in the created order losing its physical flawlessness. For theists, the Fall and its consequences dovetailed with the notion of the original perfect order of the universe dissipating as the universe ran towards heat-death. The doctrine of the Fall, and of perfection being lost in a spiral of catastrophe, also militated against the idea that life was evolving through natural selection to something better, and another reason why Darwinism was viewed unfavourably.

From another angle, the doctrine of *creatio ex nihilo* contradicted some classical views of how the universe operated, with Harman suggesting that around this time Maxwell read Lewis Campbell's book, *The Sophistes and Politicus of Plato* (1867), in which several passages represent Greek notions of the reversibility of the universe, which they thought to be possible, but which Maxwell did not.[25] Crucially, as Maxwell insisted in his letters to Pattison, not only was there a 'beginning', but, for some events to occur, there also had to

24 Cf. *Shorter Catechism*, Q&A 9, 'What is the work of creation? The work of creation is, God making all things of nothing, by the word of his power.'

25 Harman, II/1, 361, note 12.

be a 'being' who had knowledge, will, and the ability to influence the environment. Though Maxwell refrained from spelling it out, for entropy to be reversed, and for complex life to occur, the involvement of a sentient being was required.

In his 1870 letter to J.W. Strutt, Maxwell stated, 'The Second Law of Thermodynamics has the same degree of truth as the statement that if you throw a tumblerful of water into the sea you cannot get the same tumblerful of water out again.'[26] He developed this in his anonymous review of Balfour Stewart's, *Conservation of Energy: An Elementary Treatise on Energy and its Laws*, focussing again on the cosmological implications of nineteenth-century thermodynamics, especially the finite duration of the universe which irreversibility demanded.[27] If the universe had a finite prehistory, then it required a Creator; and if irreversibility were denied, then the materialist had to allow for ridiculous scenarios, such as rivers reversing![28]

Maxwell's logic of irreversibility and its implications had its critics, even among friends. William Thomson was strongly wedded to systems of mechanical forces, and he highlighted what he saw as a contradiction in Maxwell's thinking.[29] Thomson argued that in any volume of gas with a finite number of molecules (however large that number may be) each individual molecule obeyed dynamical laws, and therefore all actions of each individual molecule were reversible. But, if true for one, then true for all, and true for the volume of gas as a whole. Computation of molecular movements might be too much for the intellect, but that was a human problem. In essence: reversibility was possible, therefore irreversibility was a false conclusion.

Maxwell was aware of the argument which Thomson put forward. In a note, written as early as 1870, Maxwell stated that the Second Law

> is undoubtedly true as long as we can deal with bodies only in mass and have no power of perceiving or handling the separate molecules of which they are made up; however, if we conceive of a being whose

26 JCM, *Letter to Strutt,* 6 Dec 1870, (Harman, II/2, 582f).

27 Anon, 'Review of *Conservation of Energy: An Elementary Treatise on Energy and its Laws*', *Nature,* Vol. 9/220 (15 Jan 1874), 198–200.

28 JCM, *Letter to Mark Pattison,* 7 Apr 1868, (Harman, II/1, 360).

29 Thomson, 'The Kinetic Theory of the Dissipation of Energy', *Proceedings of the RSE,* Vol. 8, 325-34, (1874). Reprinted in: Thomson, *Mathematical and Physical Papers,* Vol. V, 11–20, esp. 15f.

> faculties are so sharpened that he can follow every molecule in its course, then such a being—whose attributes are still as essentially finite as our own—would be able to do what is at present impossible to us.[30]

But Maxwell was adamant that an 'intelligent being' was necessary for reversal of processes in nature to actually occur. Dissipation of energy introduced non-reversible processes into nature, with nature's ability to reverse entropy made impossible without the aid of such a being.

In an Infinite Universe can anything Happen?

Maxwell assumed that the Second Law of Thermodynamics demanded a finite time-span for the universe. The past was time-finite, and though the future may be time-infinite the universe was only 'active' for some of that future. As he explained to M Pattison in 1868:

> I think we are likely to arrive at physical indicators of a beginning and an end. This end is not a destruction of matter or of energy, but such a distribution of energy that no further change is possible without an intervention of an agent who need not create either matter or energy but only direct the energy into new channels.[31]

But if finitude were *not* an attribute of the universe and if it were eternal with an infinite past, would that infinitude make it inevitable that pure chance could create order?

Some cosmologists have claimed that any event which is at all mathematically *possible*—however inconceivably small that possibility may be—*must* at some point eventually occur if the universe is infinite in space and time. Their argument is that, in an infinite universe, even the rarest of events *must* occur, therefore all events which have a probability greater than zero (however microscopically low that probability is) *must* occur at least once. Under this proposition, Maxwell's assumption that his unlikely event could never happen without the help of an intelligent agent would be false. At some point in the infinite history of an infinite universe the slow-moving and fast-moving molecules of Maxwell's experiment would transfer

30 JCM, *Appendix: From the Manuscript of the 'Theory of Heat'*, Late 1870, (Harman, II/2, 584).

31 JCM, *Letter to M. Pattison,* 13 Apr 1868, (Harman, II/1, 362).

themselves by chance, with no door-keeper necessary. Matter, left to itself, without an interfering agency, would organise itself and could *reverse* itself. Order could come from chaos, without God.

The argument that in an infinite universe everything theoretically possible must happen at least once has a problem. This is because any set of 'all possibilities' includes the possibility of something *not* happening. Thus, even if the universe were infinite in time and space, it cannot be that every theoretical possibility becomes an actuality at some point. This is because the possibility that an event *never* occurs (even in infinite time and space) is itself an option; and if infinite time and space allow *every* possibility to occur at least once, then that option must also become a certainty: which appears to be a contradiction.

The Cavendish Laboratory (1871–1879)

Chapter 24
Professor in Cambridge

Cambridge did not forget Maxwell. During his retirement at Glenlair it called on his services in 1866, 1867, 1869, and 1870, as either moderator or examiner in the Mathematical Tripos, with his input helping to reshape and modernise the syllabus. In 1871, the university finally enticed him to take up the newly created post of professor of experimental physics. William Thomson of Glasgow and Hermann Helmholtz of Heidelberg had already been approached but declined. Though third choice, Maxwell was still viewed as a major coup, and was duly elected on the eighth of March 1871. Three years earlier he had been tempted by the principalship of St Andrews University when James Forbes retired, but that came to nothing. At that point he was happy to stay at Glenlair.

The Cambridge offer proved irresistible, and in the 1871 national census, Maxwell proudly styled himself 'Professor of Experimental Physics at Cambridge', even though he had not yet taken up the post. The position involved teaching; illustrating the laws of heat, electricity, and magnetism; advancing the knowledge of these subjects; and promoting their study in the university.[1] Cambridge was aware it had ground to make up. Strange as it may seem, the university had decided in 1849 that the mathematical theories of electricity, magnetism, and heat, should not be subjects of examination.[2] Admittedly, in 1849 the field had been full of uncertainties, but such a policy was always questionable.

The Maxwells moved south, taking up residence at 11 Scroope Terrace, Cambridge, about half a mile from the site of the future Cavendish Laboratory. Katherine's health was fragile, as it had been for some years, and her widowed sister, Ann McCunn, stayed with them from time to time to give support. James, at this stage of his life, was described as a man of middle height, with a compact, strong frame, who had a sparkle in his eyes and a spring in his step, and who dressed

1 *Cambridge Independent Press*, 8 Nov 1879.

2 Harman, II/1, 33.

for comfortable ease rather than for elegance.[3] His eyes were dark and glowing and his hair and beard perfectly black, greying only in his final years. His daily habit was to walk to work accompanied by his dog. By the time they arrived at the laboratory he was ready for action, with his mind already engaging with the day's issues. At the laboratory he whistled when thinking over a problem, and he had the remarkable ability of maintaining his train of thought no matter the distractions.

The Professorship

Maxwell's Inaugural Lecture was delivered on the twenty-fifth of October 1871. It was so poorly advertised that the senior members of the university missed it and came instead to Maxwell's opening regular class lecture, which they attended in some force, thinking it to be his inaugural address.[4] The students sat in the back rows, whilst Maxwell gave an elementary discourse on heat to the vice-chancellor and other august professorial colleagues, all clad in their formal academic robes. With a twinkle in his eye, and aware of the absurdity of the situation, he expounded basics such as the difference between the Centigrade and Fahrenheit scales!

In his actual inaugural address, which had been heard by only a few students, he set out his future vision: 'The familiar apparatus of pen, ink and paper will no longer be sufficient for us, and we shall require more room than that afforded by a seat and desk, and a wider area than that of a blackboard.'[5] This was Maxwell declaring that the laboratory rather than the lecture room would rule: 'We should begin in the lecture room, with a course of lectures on some branch of physics, aided by experiments of illustration; and conclude in the laboratory, with a course of experiments of research.' The new laboratory was to be a place for original research as much as, if not more than, teaching, so he emphasised that experiments of research, rather than experiments of illustration, were 'the proper work of a physical laboratory.'[6]

3 C&G1, 368f.

4 Jones, 'The Complete Physicist: James Clerk Maxwell', 18.

5 JCM, *Introductory Lecture on Experimental Physics,* 5.

6 Ibid., 8.

Unlike at Aberdeen and London, Maxwell made no direct reference to the deity in his inaugural discourse, but what he did do was to repeat his insistence that physics could never ignore metaphysics, since fundamental questions about the nature of existence were inescapable:

> It has been asserted that metaphysical speculation is a thing of the past, and that physical science has extirpated it. The discussion of the categories of existence, however, does not appear to be in danger of coming to an end in our time, and the exercise of speculation continues as fascinating to every fresh mind as it was in the days of Thales.[7]

His regular lectures in the October term were on thermodynamics, and in the Lent and May terms they were on electricity. His reputation as a poor class lecturer continued, though on a one-to-one level he was all that a keen student would want from a mentor.

During the Maxwell centenary commemorations of 1931, Ambrose Fleming, who studied under Maxwell from 1877 to 1879, described his experience.[8] Fleming came up to the university, chiefly with the aim of benefiting from Maxwell's lectures and laboratory teaching. He was somewhat surprised to find a man, who was already regarded as the greatest living authority on his subjects, lecturing to small groups rather than to a hundred or more. Fleming commented that it was characteristic of Maxwell that he lectured to two or three students with the same care as if his lecture room had been filled to capacity. In fact, given the specialised and advanced nature of Maxwell's work, small classes were inevitable. The laboratory students were exclusively male: that was university policy.[9]

The Cavendish Laboratory

The appointment included supervising the construction of a modern experimental facility. This was known initially as the Devonshire

---

7 Ibid, 23.

8 Fleming, 'Some Memories of Professor James Clerk Maxwell', *James Clerk Maxwell: A Commemoration Volume*, 116–124.

9 JCM, *Letter to Sedley Taylor*, 31 May 1877, (Harman, III/2, 465) plus Harman's instructive footnote 3.

Laboratory but more lastingly as the Cavendish Laboratory.[10] Henry Cavendish (1731–1810) had been an amateur experimenter in chemistry and physics, and it was his later relative—William Cavendish the seventh Duke of Devonshire, and Chancellor of Cambridge University from 1861 until 1891—who was the moving spirit behind the new facility. The Duke underwrote the laboratory's construction plus the cost of most of its equipment, though Maxwell spent considerable sums of his own to ensure that all was first class.

Maxwell wanted the most modern research laboratory possible, especially one which could give measurements of extremely high precision. He visited Thomson's laboratory in Glasgow, Tait's in Edinburgh, and the Clarendon Laboratory in Oxford designed by Robert Clifton (1836–1921). The hard work paid off, and by April 1873 he was able to tell Lewis Campbell that the roof of the laboratory was being put in place, that some floors would be in by May, and that the contractors might be out by October.[11] As he had hoped, the laboratory started use in October 1873, with its official opening by the Duke of Devonshire taking place on the sixteenth of June 1874.[12] William Garnett was appointed as demonstrator.

---

10 Cf. Falconer, 'Cambridge and building the Cavendish Laboratory', [Flood (ed), *JCM: Perspectives on his Life and Work*], 67–98.

11 JCM, *Letter to Campbell,* 3 Apr 1873, (C&G1, 386; Harman, II/2, 840).

12 *Sheffield Daily Telegraph,* 17 June 1874: 'This laboratory, in which every facility is furnished for the prosecution of physical research, is the gift of the Duke of Devonshire, who presents it completed to the University. . . . The ground floor contains a set of rooms for operations requiring great steadiness, such the measurements of length, time, and mass, and of heat electricity, and magnetism. The first floor contains spacious lecture-room, with preparation-room, a large apparatus-room, private room for the Professor, and large working laboratory, fitted with tables standing on beams of their own, so to be independent the vibrations of the floor. All the tables in the building are supported in the same way, and there are in every floor small trap doors, by means of which bodies may be suspended over tbe tables in the room beneath, and through which electric and other communications may be made. The upper rooms are intended for acoustics, radiant heat, optics, electricity, and the graphic reduction of observations. There is also a dark room for photographic preparations. . . . The laboratory is open daily, under the superintendence the Professor of Experimental Physics (Mr J. Clerk Maxwell), for the use of any members of the University who may desire to acquire a knowledge of experimental methods, and so take part in physical researches.'

It was in this facility that Maxwell cemented the custom whereby even undergraduates did actual laboratory work as part of their normal study. He had introduced the practice at King's College, London, and, in the opinion of the quantum theorist C.A. Coulson (1910–1974), it is to Maxwell that British science owes this important aspect of university training in physics.[13]

There were teething problems. Delicate apparatus could not be installed immediately in the rooms on the ground-floor, owing to the dampness of the new walls.[14] One of the external windowsills cracked during the frost and had to be replaced by the contractor. The hot-water pipe in the ceiling of the heat-room became leaky and had to be repaired. More seriously, the lodge-house of the laboratory was poorly constructed, resulting, in early 1879, in the tragedy of one of the attendant's children dying of typhoid.[15]

Despite these setbacks there was a positive beginning with a vast amount of equipment being fitted. Over the next century, the laboratory was the location for some historic achievements by exceptional physicists. J.J. Thomson (1856–1940) identified the electron; John Sealy Townsend (1868–1957) discovered the Townsend Discharge; C.T. Wilson (1869–1959) developed the Cloud Chamber; James Chadwick (1891–1974) detected the neutron; and John Crockford (1897–1967) along with Ernest Walton (1903–1995) made the first experiment to split the nucleus. One of its most distinguished Directors was Ernest Rutherford (1871–1937), who was in charge during the early twentieth century. In the list of Nobel prizewinners from 1895 until 2019, thirty members of the Cavendish Laboratory are honoured. In 1974 the laboratory moved from its original site to a new location in the western side of Cambridge.

Research and Writing

Maxwell undertook the task of editing, annotating, and publishing Henry Cavendish's research. Some biographers have viewed this as

13 Coulson, 'Interatomic Forces', *Clerk Maxwell and Modern Science,* 43.

14 JCM, *Report on the Cavendish Laboratory for 1874,* 14 Apr 1875, (Harman, III/1, 208).

15 Cf. Marston, 'Maxwell, Faith and Physics', 281. The attendant was William Pullin (1834–1880).

an unwanted duty imposed on Maxwell by the university, yet there is some evidence he sought the task himself.[16] His interest may have been sparked years earlier by his Edinburgh professor, George Wilson, who published a biography of Cavendish in 1851. Falconer's opinion is that the work on Cavendish actually enhanced Maxwell's science and helped him gain a critical evaluation of method. This was due to Maxwell not only reading Cavendish's original papers, but seeking out eighteenth-century instruments to test the experiments and their results. He examined Cavendish's work on the discovery of hydrogen, the composition of water, the density of the Earth, and his early experiments in electricity.

The resultant book, *The Electrical Researches of the Hon. Henry Cavendish F.R.S. between 1771 and 1781,* was published in October 1879, a month before Maxwell's death. In repeating many of Cavendish's experiments, Maxwell frequently called in Richard, his laboratory assistant, to take electrical shocks![17] Others volunteered, with Maxwell finding that boating crews had high resistance owing to the toughness of their hands! His work resulted in Cavendish being looked on with more respect, with Peter Tait, in his *Encyclopaedia Britannica* article on Maxwell, noting that Cavendish now had to be regarded as a man of Maxwell's own stamp as a theorist, and as an experimenter of the first rank.[18] Nonetheless, Maxwell did express some regret at having spent so much time on Cavendish to the neglect of his own research.

Despite the time and effort spent on the Cavendish papers, Maxwell was also able to produce fresh work of lasting importance.[19] In 1871 he gave a second lecture on colour at the Royal Institution in London; and in 1873 Oxford University Press published his *Treatise on Electricity and Magnetism*—monumental both in size and significance—with much of it written during his extended sojourn at Glenlair.

---

16 Falconer, 'Cambridge and building the Cavendish Laboratory', [Flood (ed), *JCM: Perspectives on his Life and Work*], 75.

17 Fleming, 'Some Memories of Professor James Clerk Maxwell,' *James Clerk Maxwell: A Commemoration Volume,* 122.

18 Tait, 'James Clerk Maxwell', *Encyclopaedia Britannica* (9th ed.), Vol. 15, 647.

19 NVN1, xviii. Niven points to short articles and reviews for *Nature* and the *Encyclopaedia Britannica,* etc.

In the *Treatise,* Maxwell set out his electromagnetic theory in new form, re-presenting his earlier work more elegantly.[20] One of his new findings concerned the pressure which light exerted on a surface. The notion that light could produce a force came originally from Newton's corpuscular/particle theory of light, but Maxwell showed that light produced pressure, even in his electromagnetic model.[21] He predicted that full daylight would produce a force of about one-pound weight on a half mile square. Twenty-six years later, the Russian physicist Pyotr Lebedew (1866–1912) confirmed that calculation experimentally, publishing his results at the World Physics Congress in Paris in 1900, and providing one of the first quantitative confirmations of Maxwell's theory of electromagnetism. In modern astronomy the radiation pressure of light is important in understanding the internal structure of stars, and Einstein used it to help persuade his contemporaries of the validity of his $E=MC^2$ equation.[22]

In 1876, there was an exhibition of scientific instruments at South Kensington, for which Maxwell wrote the introductory chapter of the exhibition's *Handbook*, focussing on the principles of instrument design. His tireless emphasis on precisely calibrated equipment emerged from his lifelong commitment to empirical research, as well as from his professional responsibility of ensuring that the Cavendish Laboratory was equipped to the highest specifications.[23] The following year, 1877, saw the appearance of his 'small book on a great subject' *Matter and Motion.*

---

20 Cf. Jones, 'The Complete Physicist: James Clerk Maxwell', 18ff.

21 JCM, *Treatise on Electricity and Magnetism,* Vol. 2, 441, Articles 792/793: 'If in strong sunlight the energy of the light which falls on one square foot is 83.4 pounds per second, the mean energy in one cubic foot of sunlight is about 0.0000000882 of a foot pound, and the mean pressure on a square foot is 0.0000000882 of a pound weight. A flat body exposed to sunlight would experience this pressure on its illuminated side only, and would therefore be repelled from the side on which the light falls.'

22 Jones, 'The Complete Physicist: James Clerk Maxwell', 18. Cf. Einstein, *Annals of Physics,* Vol. 20, 627.

23 JCM, 'General Considerations regarding Scientific Apparatus', *South Kensington Museum: Handbook to the Special Loan Collection of Scientific Apparatus, 1876,* 1–21.

The 1878 Rede Lecture

In 1878 Maxwell was appointed as that year's Rede Lecturer. The post was named after Sir Robert Rede (d.1519), who had been Chief Justice of the Common Pleas under Henry VIII. Maxwell gave his speech in the Cambridge Senate House, taking the telephone as his subject.[24] It had been invented two years earlier by Alexander Graham Bell (1847–1922), and Maxwell started the lecture by joking about his own broad Gallovidian accent:

> Professor Graham Bell is the son of a very remarkable man, Alexander Melville Bell, author of a book called *Visible Speech*, and of other works relating to pronunciation. In fact, his whole life has been employed in teaching people to speak. He brought the art to such perfection that—though a Scotchman—he taught himself in six months to speak English! And I regret extremely that, when I had the opportunity in Edinburgh, I did not take lessons from him![25]

Maxwell arranged for a harp to be played in the Geological Museum during the course of the lecture, with the music transmitted to three large telephones in the Senate House, one of which was an original instrument belonging to Bell himself.

Maxwell referred to Faraday's concept of lines of magnetic force in order to explain the undulatory currents in the telephone. He pointed out that Peter Tait had recently measured the strength of these currents, finding that a current of about a thousand million times less than what was used in ordinary telegraphs could—if renewed five hundred times per second—produce audible sound. He further described how, in Bell's telephone, these currents derived their energy only from the voice of thc speaker, and therefore the sound they were able to produce was much weaker than the original vocalisation. However, this limitation vanished when the current was supplied by a battery, with the voice of the speaker merely used to modulate its intensity. With a Hughes microphone, the sound

24 JCM, *Rede Lecture on the Telephone,* 24 May 1878, (C&G1, extracts 361–364; Harman, III/2, 651–664).

25 Ibid., (C&G1 363; Harman, III/2, 661). Concerning Maxwell's accent, in his poetry certain words rhymed perfectly in Maxwell's Gallovidian brogue, though not in standard pronunciation. This is overlooked by Pratt-Smith in her essay: 'Boundaries of Perception', 236.

produced by the receiver could even be stronger than that received by the microphone.[26]

Fleming described Maxwell's presentation as a brilliant discourse, illustrated by flashes of wit, apt analogies, and much learning, though perhaps not conveying to unscientific hearers how a telephone actually operated![27] The lecture was attended by local enthusiasts as well as by dons and students, with newspapers reporting a huge audience.[28] One attendee was the diarist J.W.E. Conybeare, vicar in nearby Barrington, who noted the event in his journal.[29] Despite Fleming's concerns about the technical nature of the lecture, the audience, including the Revd Conybeare, enjoyed one of Maxwell's most popular and well-presented addresses.

The Revd Harvey Goodwin was no longer the preacher at St Edward's. He had moved to Ely in 1858, before becoming Bishop of Carlisle. As a professor, Maxwell was expected to fulfil formal church duties, and on most Sundays, along with heads of colleges and fellow academics, he went to the university sermon in the restored Great St Mary's Church. He also made time for his local Church of England church of St Mary's the Less, where he and Katherine attended the Communion Services along with colleagues such as George Stokes and George Liveing (1827–1924) who were fellow parishioners.[30]

The Vicar of St Mary's the Less was William Henry Guillemard (1815–1887), who later told Lewis Campbell that Maxwell had been constant and assiduous at his church and interested in all church matters.[31] During Maxwell's final illness it was Guillemard who gave

---

26 Cf. JCM, *Letter to Campbell,* 5 Jan 1878, (C&G1, 402; Harman, III/2, 580), 'We have all been conversing on the telephone. Garnett recognised the voice of a man who called by chance. But the phonograph will preserve to posterity the voices of our best speakers and singers.' Also, *Nature,* 5 Jan 1878.

27 Fleming, 'Some Memories of Professor James Clerk Maxwell': *James Clerk Maxwell: A Commemoration Volume,* 121.

28 *Northern Whig,* 27 May 1878.

29 Colloms, *Victorian Country Parsons,* 260.

30 Robert Walker, 'James Clerk Maxwell', 206.

31 Guillemard, *Letter to Campbell,* 19 May 1881, (C&G1, 414).

spiritual support. Guillemard was different from most of the clergy to whom Maxwell was drawn. He was high church rather than low church, and he took a leading part in bringing the Oxford Movement to Cambridge. The Oxford Movement was of a quite different stamp from the evangelicalism which Maxwell sought out and enjoyed in London. That being said, Guillemard was regarded as a moderate high churchman. Importantly for Maxwell, he affirmed traditional Christian doctrine.

Chapter 25
Orderliness and Design

During the period 1862 to 1873, Maxwell grappled with some of the philosophical problems generated by his physics.[1] As well as issues arising from the statistical nature of the Second Law of Thermodynamics and from the reversibility or otherwise of processes in nature, he deliberated on atomism, the orderliness of the universe, determinism and free-will.

It rankled confirmed materialists, such as T.H. Huxley (1825–1895) and John Tyndall (1820–1893), that, in the second half of the nineteenth century, scientific lectures could still include reference to the Deity. William Thomson, for example, in his regular course for his Glasgow students had an annual lecture on the scope and methods of physical science, which included a brief meditation on Psalm 8.[2] Thomson's degree of reference was admittedly unusual, with most religious allusions in scientific papers normally limited to a brief comment that the workings of nature had a divine origin.

It was exasperating to Huxley and Tyndall when religious events such as prayer meetings took place, seemingly under the auspices of the British Association. When the Association met in Bradford in 1873 in the refurbished St George's Hall, a devotional service in connection with the gathering was held in the same location.[3] The event was mostly conducted by laymen who were members of the Association, and, though no sermon was preached, several engaged in prayer and read portions of Scripture with the whole congregation joining in singing hymns. At its conclusion it was agreed that throughout the duration of the Association's gathering a service should be held every morning from 9 am until 10 am in the Ebenezer Chapel at the foot of Horton Road.

---

1 Harman, II/1, 2.

2 Thomson, W, 'Introductory Lecture to the Course of Natural Philosophy', [S.P. Thomson, *The Life of William Thomson: Baron Kelvin of Largs,* Vol. 1], 239–251; especially 249–251.

3 *Bradford Observer,* 22 Sep 1873.

On the day following the service, and in the same St George's Hall, Maxwell read a paper which gained instant renown. This was his 'Discourse on Molecules'.[4] It was distinct from his highly mathematical lecture on molecules, which was one of his official submissions to the Association that year.[5] Maxwell's 'Discourse on Molecules' was an evening lecture under the patronage of the Association, and intended as a popular presentation for the general public. The event was chaired by the Association's President, Professor A.W. Williamson (1824–1904). In his paper, Maxwell made comments which have been interpreted by some (incorrectly) as rejecting Evolution, and by others (correctly) as supporting a modified version of Intelligent Design.[6]

Maxwell was well aware of the exclusively materialist view of science and life championed by Huxley and Tyndall. That position would be set out forcefully by Tyndall in his Presidential Address the following year, partly in response to what Maxwell said in his 'Discourse on Molecules'. But, whereas Huxley and Tyndall demanded a clear separation between science and religion, Maxwell saw that as impossible. In Maxwell's view, any adequate reflection on the deep questions of physics demanded consideration of what lay behind the origin and the rationality of the universe. He could not limit questions raised by physics any more than he could limit enquiry in the field of faith, and he was specifically interested in issues concerning the boundaries of existence. How did matter come to be? How did the rational structure of the universe come to be? How did the laws of physics come to be? Scientists had to discuss these questions.

As a boy in Aunt Isabella's house, he had read William Paley's books, and in 1853—twenty years before his 'Discourse on Molecules'—he had presented an essay 'What is the Nature of Evidence of Design' to the Cambridge Apostles in which he developed Paley's 'watchmaker argument'.[7] What he emphasised in

---

4 Published as: JCM, *'Discourse on Molecules', Philosophical Magazine,* Vol. 46/308 (1873), *453–469.*

5 JCM, 'On the Final State of a System of Molecules subject to Forces of any Kind', *Report of the 43rd Meeting of the BAAS* (1873), 29–32.

6 Cf. Petzold, 'Maxwell, Molecules, and Evolution'.

7 JCM, 'What is the nature of Evidence of Design?', (C&G1, 225–227; Harman, I, 227–228).

that earlier 1853 paper was that, when considering an intricate object such as a watch, we actually become aware of two forms of apparent design, not just one. The first is *complexity*: i.e. the intricate physical mechanism (wheels, springs, etc.). The second is *purpose*: i.e. that a functioning watch indicates the agreed units of civil time. Thus, evidence of design is not in the complexity of the device alone but in its purpose and function; and if an object exhibits strong evidence of complex design *plus* a specific identifiable purpose in human society, then it must have had a designer who had both *personal being* and *intelligent being*. In 1873, it was this thinking which lay behind his address.

'Discourse on Molecules' (1873)

Maxwell started his Bradford lecture by stating the classic position: that an atom was indivisible; that it was a body which could not be cut in two; and that a molecule was the smallest possible portion of a particular substance.[8] He went on to emphasise that no one had ever seen or handled a single molecule, therefore molecular science was one of those branches of science which dealt with things invisible and imperceptible by our senses, and which could not be subjected to direct experiment.

A few years later, Maxwell began his 'Atom' article in the *Encyclopaedia Britannica* in similar fashion, stating that the atom is indivisible, and though space and time may be infinitely divisible the truly elementary units of matter were not.[9] Maxwell's view of space and time was, of course, pre-quantum. On molecules, what he had in mind were molecules of the same element, rather than molecules made up of different elements.

Our modern understanding of atoms is vastly different from Maxwell's. We know about sub-atomic particles, isotopes, radioactive decay, nuclear fusion, etc., which he did not. But the essential question which Maxwell was asking is still valid: if, in the make-up of

8 JCM, 'Discourse on Molecules', *Philosophical Magazine*, Vol. 46/308 (1873), 453. That atoms were *not* indivisible was finally settled by J.J. Thomson in 1897 in Cambridge when he discovered the electron.

9 JCM, 'Atom', *Encyclopaedia Britannica*, (9th ed.), Vol. 3, 36f. Coulson, 'Interatomic Forces', *Clerk Maxwell and Modern Science*, 45, points out that Maxwell was vague as to what these individual units actually are.

the universe, there are irreducible and fundamental entities (*whatever these may be*) which underpin everything else, how did these irreducible and fundamental entities come to exist, and how did their properties, characteristics, laws, and structure come to exist? He argued that there *must* be irreducible entities, otherwise we are faced with an infinite regression into an infinite past, which was impossible according to the thermodynamical theory of the time. Maxwell said: 'We have reached the utmost limits of our thinking faculties when we have admitted that because matter cannot be eternal and self-existent it must have been created.'[10] Maxwell's argument was that any truly fundamental and irreducible entity could not come into being as the result of a process, otherwise it would itself be composed of prior constituent parts, and hence, by definition, would be neither truly fundamental or irreducible.

At the core of his argument was the conviction that no developmental (evolutionary) process could—again by definition—produce a fundamental *entity*, nor could any process produce the fundamental *logic* of the universe. These had to be fully-developed from their moment of existence: 'None of the processes of Nature, since the time when Nature began, have produced the slightest difference in the properties of any molecule. We are therefore unable to ascribe either: the existence of the molecules, or the identity of their properties to any of the causes which we call natural.'[11]

By 'molecules' Maxwell was referring to what, at the time, he understood to be the basic building blocks of nature. In a modern context we might refer to something on the sub-atomic level, but Maxwell's essential point remains. In nature—at its irreducible fundamental level—there must exist entities and structures which have never emerged through a process, but which came into existence in fully-developed fashion.

A large part of Maxwell's lecture was devoted to properties of molecules and atoms in relation to the diffusion of gases, and though the lecture was meant to be a simple presentation of science for the public, much of it was hard to follow. Its subsequent fame arose from a section towards its close, when Maxwell reflected on how spectroscopic analysis had shown that molecules (e.g., of hydrogen)

10 JCM, *Discourse on Molecules*, (C&G1, 359–360). (Italics ours).

11 Ibid., (C&G1, 359).

were exactly the same on earth or as observed in the light from distant stars:

> In the heavens we discover by their light, and by their light alone, stars so distant from each other that no material thing can ever have passed from one to another; and yet this light, which is to us the sole evidence of the existence of these distant worlds, tells us also that each of them is built up of molecules of the same kinds as those which we find on earth. A molecule of hydrogen, for example, whether in Sirius or in Arcturus, executes its vibrations in precisely the same time. Each molecule, therefore, throughout the universe, bears impressed on it the stamp of a metric system as distinctly as does the metre of the Archives at Paris, or the double royal cubit of the Temple of Karnac.

Maxwell interpreted this uniformity of the characteristics of hydrogen throughout the universe as validating his point that fundamental entities had constant characteristics everywhere and always. The science of 1873 also provided Maxwell with no awareness of intra-atomic activity causing radiation, and so he thought of atoms or molecules as creating radiation by vibrating in the ether. He concluded:

> No theory of evolution can be formed to account for the similarity of molecules: for evolution necessarily implies continuous change, and the molecule is incapable of growth or decay, of generation or destruction.[12] None of the processes of Nature, since the time when Nature began, have produced the slightest difference in the properties of any molecule. We are therefore unable to ascribe either (a) the existence of the molecules or (b) the identity of their properties to any of the causes which we call natural.[13]

This echoed what he had written in his *Theory of Heat* (1872).[14] In using the word 'evolution' Maxwell was not discussing biological

12 For Maxwell this meant that the logic of fundamental entities had to be full-grown from the first moment of existence.

13 JCM, 'Discourse on Molecules', (C&G1, 359). Cf. Ellicott, *Modern Unbelief: Its Principles and Characteristics,* 104.

14 JCM, *Theory of Heat* (1872), 310: 'In the case of the molecules, however, each individual is permanent; there is no generation or destruction, and no variation, or rather no difference, between the individuals of each species. Hence the kind of speculation with which we have become so familiar under the name of theories of evolution is quite inapplicable to the case of molecules.'

evolution. Darwin was not in his mind. He was *contrasting* rather than *comparing* the biological and molecular worlds.[15] He was explaining why truly fundamental entities could not arise from prior processes, and that was why, in his 'Discourse on Molecules', he preferred Herschel's term 'manufactured' rather than the term 'evolved'.

Maxwell accepted that the universe, and anything within it, could change and mutate through natural processes. Anything, that is, apart from truly fundamental entities. These, and these alone, were immune—by definition—and even though every object in the natural world is made from molecules, and even though natural processes may alter and reshape such objects, the constituent molecules themselves remain the same everywhere for all time. He described them as the 'foundation stones of the material universe.'[16] That was his point:

> The exact equality of each molecule to all others of the same kind gives it, as Sir John Herschel has well said, the essential character of a *manufactured* article, and precludes the idea of its being *eternal* and *self-existent*.[17]

Truly fundamental entities have the appearance of being 'manufactured' because they are phenomena which have no processes (evolution) leading to what they are now. In that respect, they differ radically from phenomena produced by processes which have led to the present form of an object, and whose *subsidiary* component parts are capable of examination and analysis. Truly fundamental entities can have *no subsidiary* constituent parts, nor can they have previous versions of themselves.

Though Maxwell knew nothing of electrons, protons, neutrons, quasars, positrons, or radioactive decay, his argument can be modified to apply to whatever might now be considered to be an irreducible entity in nature. One of Maxwell's major points was that though science could investigate how such fundamental entities relate to one another in nature, science could not go behind the fundamental entities themselves:

---

15 Petzold, 'Maxwell, Molecules, and Evolution'.

16 JCM, *Discourse on Molecules,* (C&G1, 360).

17 Ibid., (C&G1, 359). Cf. Herschel, *Preliminary Discourse on the Study of Natural Philosophy*, Ch.3, Art 28.

> We have been led, along a strictly scientific path, very near to the point at which Science must stop . . . Science is incompetent to reason upon the creation of matter itself out of nothing. We have reached the utmost limits of our thinking faculties when we have admitted that because matter cannot be eternal and self-existent it must have been created.[18]

Mary Hesse (1924–2016), well capable of giving theists a hard time if she questioned their reasoning, would agree with Maxwell on this issue. She wrote: 'There are certain metaphysical issues that are not going to be settled by the methods of science alone. These are issues such as determinism, the contingency or self-sufficiency of laws, the beginning and end of the natural universe, and the significance of mind and spirit in nature.'[19]

For Maxwell, it was because no natural process could produce any truly fundamental entity that he adopted the term 'manufactured', though he employed it in relation to the *origin* and *sameness* of basic entities, rather than to their *complexity*. It was in this vein that he referred to molecules of pure hydrogen, which had the exact same features and spectral lines wherever they were observed in the universe (though today we know of isotopic variations). For him, these molecules were the absolute, unchanging, and stable backstops of material existence.[20] This echoed thoughts expressed in his 1870 address to the British Association, when his faith implications were more explicit:

> When we find that here, and in the starry heavens, there are innumerable multitudes of little bodies of exactly the same mass, so many, and no more, to the grain, and vibrating in exactly the same time, so many times, and no more, in a second, and when we reflect that no power in nature can now alter in the least either the mass or the period of any one of them, we seem to have advanced along the path of natural knowledge to one of those points at which we must accept the guidance of that faith by which we understand that 'that which is seen was not made of things which do appear' (Hebrews 11:3).[21]

---

18 Ibid., (C&G1, 359–360).

19 Hesse, 'Retrospect', [Peacocke (ed), *The Sciences and Theology in the Twentieth Century*], 286.

20 Marston, 'Maxwell, Faith and Physics', 274, points out that, in contrast to Descartes' theory of the evolution of molecules, Maxwell found molecules to be 'unalterable by any of the processes which go on in the present state of things.'

21 JCM, 'Address as President of Mathematical and Physical Section', *Report to the 40th Meeting of the BAAS* (1870), 7.

Maxwell cited Scripture, which was the type of religious allusion which infuriated Tyndall and Huxley. Maxwell's point was that in science we eventually arrive at—not gaps in knowledge—but irreducible and unchanging phenomena which are the fundamental axioms of reality. He was intrigued by issues on the boundaries of science, recognising that at these points science looked into a region beyond her domain.[22]

'Foundation Stones of the Material Universe'

In Maxwell's faith and in Maxwell's physics the universe was created out of nothing. Consequently, the irreducible atomic and molecular constituents of nature were neither 'eternal' nor 'self-existent'. They had not been formed by natural processes, though—as 'foundation stones of the material universe'—they were endowed with an absolute stability.[23] They had characteristics for which science could not account since their characteristics had emerged from the initial conditions of their formation, which was outside the scope of human science.[24] Science could investigate processes of nature, but it could not investigate the coming into existence of nature. Science was 'incompetent to reason upon the creation of matter out of nothing.'[25]

Sub-atomic theory had still to be developed. Maxwell himself, in his 1871 Cambridge Introductory Lecture on Experimental Physics, had admitted that the current molecular hypothesis might

---

22 Cf. Gaal, *The Faith of a Scientist,* 34.

23 JCM, 'Discourse on Molecules', 1873, (C&G1, 360).

24 Torrance, 'Christian Faith and Physical Science', 223.

25 JCM, *Discourse on Molecules,* (C&G1, 359). Coulson, 'Interatomic Forces', 52, comments: 'With knowledge as it was at the time 'the uniformity and similarity of molecules] was bound to be an insoluble problem. With our greater insight now we can understand that without some form of quantum theory there could be no answers to questions about the identical nature of all atoms and molecules in the same substance. . . . Maxwell had got himself to the position where he was wanting to ask one of the most important questions physics ever has had to ask . . . Why do we get quantization in certain phenomena; why are there quantum effects, showing discrete and not continuous variations?' Coulson, himself a Methodist lay-preacher, saw Maxwell's religious convictions obtruding in his molecules speech.

be ousted by some new theory of the constitution of matter.[26] Later physicists, sympathetic to Maxwell's conclusions, needed to reset his argument, as acknowledged by Sir John Arthur Thomson in his 1916 Gifford Lectures: 'For *molecules* a modern chemist would read *atoms* and even then he would remind us of the apparent disintegration of the atom in radioactive substances. But, in the main, what Clerk-Maxwell said remains true.'[27] Twenty-first century science has retuned Maxwell's reasoning even further.

What Maxwell focussed on were the fundamental and irreducible building blocks of nature—whatever these may be. Truly fundamental entities could not evolve: that was why they were 'fundamental'. Yet, the origin of such building blocks was a question which could not be ignored by science, even if it introduced metaphysical issues. This lay behind Maxwell's oft-quoted statements on the technical limits of science: 'Thus we have been led, along a strictly scientific path, very near to the point at which science must stop', and 'Science is arrested when she assures herself, on the one hand, that the molecule has been made, and, on the other, that it has not been made by any of the processes we call natural.'[28]

For Maxwell the implications were clear: 'Because matter cannot be eternal and self-existent it must have been created. It is only when we contemplate, not matter in itself, but the form in which it actually exists, that our mind finds something on which it can lay hold.'[29] Science can examine the *form* in which matter actually exists, but not its *existence* as such:

> That matter, as such, should have certain fundamental properties, that it should exist in space and be capable of motion, that its motion should be persistent, and so on, are truths which may, for anything we know, be of the kind which metaphysicians call necessary. We may use our knowledge of such truths for purposes of deduction, but we have no data for speculating as to their origin. But that there should be exactly so much matter and no more in every molecule of hydrogen is a fact of a very different order.

26 JCM 'Introductory Lecture on Experimental Physics', 22.

27 J.A. Thomson, *The System of Animate Nature*, Vol. 1, 65.

28 JCM, 'Discourse on Molecules', (C&G1, 359).

29 Ibid.

Why are fundamental entities as they are? Is the form of existence of fundamental entities a necessary form? Are their characteristics necessary characteristics? Could they have had different characteristics? Is their structural logic (embedded in nature from the moment of creation) unique and necessary? Are the laws of nature necessary laws? Do the laws of nature have a necessity independent of any cosmos existing at all? All of these were questions which Stephen Hawking realised were issues which the physicist could not ignore.[30]

Maxwell himself, within his own scientific and philosophical context, alluded to the terminology (*collocation*) and argument used by Thomas Chalmers in his 1830s Bridgewater essay. Chalmers argued that it was the *comprehensibility* of nature, rather than its *complexity,* which pointed to a Creator, with nature's *comprehensibility* implying that the *rationality* of the universe was itself something created and *sui generis.*[31] Maxwell agreed.

The Theological Significance of Science

Toward the close of his lecture, Maxwell did link his observations to his theism though, careful as ever, he did not state directly that God had made these 'manufactured articles', though that was implied. Everyone knew it was implied.

What he did voice was the thought that, because humanity was confronted with such ultimate realities, then the role of science in human society was essential and had theological significance:

> [Molecules] continue this day as they were created, perfect in number and measure and weight; and from the ineffaceable characters impressed on them we may learn that those [human] aspirations after accuracy in measurement, and justice in action—which we reckon among our noblest attributes as men—are ours because they are essential constituents of the image of Him who in the beginning created, not only the heaven and the earth, but the materials of which heaven and earth consist.[32]

30 Hertog, *On the Origin of Time,* esp. 254-266.

31 JCM, 'Discourse on Molecules', (C&G1, 360). Cf. Chalmers, 'Treatise 1', *The Bridgewater Treatises,* 21.

32 Ibid.

Why is science important? Maxwell looked beyond its functional value. He believed that the human urge to gain scientific truth existed because humanity was made in the image of the God who had created all things and who had given the universe its rationality, orderliness, and intelligibility. The desire to understand the universe was a God-given gift. More than that, it was an integral aspect of being human. To seek to understand nature was a consequence of the *imago dei.* For Maxwell, the scientific task was a Christian vocation as an act of joyful obedience to his Lord and Maker.

What he found truly remarkable was nature's uniform design, which appeared to point to a plan and a purpose. *Complexity* by itself may prompt thoughts of a designer, but it was the added evidence of *purpose*—with the watch indicating a time which harmonised with society's agreed value for time—which gave force to the watchmaker argument. As he later wrote to Bishop Ellicott:

> What I thought of was not so much that uniformity of result which is due to uniformity in the process of formation; as a uniformity intended and accomplished by the same wisdom and power of which uniformity, accuracy, symmetry, consistency, and continuity of plan are as important attributes as the contrivance of the special utility of each individual thing.[33]

In his Bradford lecture of 1873, Maxwell's theism was on display. Traditional arguments from design had seemed undermined by Darwin's *Origin of Species*, as Darwin appeared to offer explanations for the complex structures found in nature without having to invoke the supernatural. But Maxwell recast the *genre* and gave it fresh life. The core thesis of his lecture was that, at a much deeper level—and underlying every process of nature—the existence and properties of truly fundamental entities still cried out for a creator or manufacturer.

When Maxwell finished his lecture, it fell to Francis Sharp Powell (1827–1911), the local Member of Parliament for Bradford, and a life-member of the Association, to propose the vote of thanks. Politicians are rarely lost for words, and though Powell may have struggled with the intricacies of Maxwell's paper, he followed it enough to express gratification that the poet Lucretius (a personal favourite of

33 JCM, *Letter to Ellicott,* 22 Nov 1876, (C&G1, 393: Harman, III/1, 417).

Powell's during his younger days) had been quoted by the lecturer![34] Concerning the closing section of Maxwell's address, Powell was 'thankful for the manly manner in which Professor Maxwell had acknowledged the existence of a great primary cause.' In Powell's view, if the scientific world acknowledged and proclaimed the existence of a final cause, and if religion and science joined together, 'both might safely advance from triumph to triumph without halting or doubting.'

Powell's vote of thanks was seconded by the Oxford mathematician, Professor H.J.S. Smith (1826–1883). Smith was far more scientifically qualified than Powell, and he was more hesitant in his evaluation. He confessed he had found the paper difficult in some parts, and believed that was the case with many. He hoped that all of them would read the lecture carefully when it was published. And he was confident that the more they looked at it, the more they would be struck—not only with its scientific truth—but with the beauty of the language and the clearness of the illustration which had been put before them. The meeting ended with prolonged applause.

34 *Bradford Observer,* 23 Sep 1873.

Chapter 26

Reaction and Response

The following day, the botanist Joseph Dalton Hooker (1817–1911) wrote to Charles Darwin, decrying Maxwell's speech: 'Clark Maxwell's lecture on Molecules last night was simply dull, dry, and singularly unintelligible as delivered: without a point or significance to the outside world.'[1] Within a few days the fair-minded Darwin had read the lecture for himself, and suggested to Hooker that he should reconsider: 'I think if you read Clerk Maxwell's lecture you will alter your opinion.'[2] Darwin respected Maxwell. A few months later, he sent a copy of Maxwell's work on the Rings of Saturn to the Russian academic and palaeontologist V. O. Kovalevsky (1842–1883).[3]

Mixed verdicts continued. Maxwell's old friend, C.J. Monro, questioned the notion of a 'manufactured' article, and asked *Nature's* readership to assist him. Tongue-in-cheek, he appealed to anyone who knows that molecules possess the essential character of a manufactured article to kindly 'explain how he knows a manufactured article when he sees it (in his mind's eye or elsewhere).'[4] Some of Monro's questions were answered in Maxwell's article on the 'Atom' for the *Encyclopaedia Britannica*. Other critics of Maxwell included the young mathematician, William Kingdom Clifford of University College, London.[5]

In the wider world, Maxwell's contribution was embraced with delight by a public for whom recent scientific theories (e.g., Darwinism) had been disturbing, and whose confidence in the ability of religion to live alongside modern science had been shaken. Within a few weeks, the *South London Chronicle* reported:

---

1 Hooker, *Letter to Charles Darwin*, 23 Sep 1873, (DCP).

2 Darwin, *Letter to Hooker*, 27 Sep 1873.

3 Darwin, *Letter to Kovalevsky*, 24 Jan 1874 (DCP).

4 Monro, *Letter to Nature*, 15 Oct 1874: *Nature*, Vol. 10/259 (15 Oct 1874), 481.

5 Marston, 'Maxwell, Faith and Physics', 275.

> The great speech recently delivered by Professor Clerk-Maxwell at Bradford, is causing considerable excitement in what may be called scientific circles, and attracting great attention in London. The fact that the learned Professor prefers to allow the existence of the Almighty, and is unwilling to agree to the theory that in some mysterious, unexplained, and fanciful manner, matter created itself and then tumbled together as to form the universe and all it contains, has been quite enough to stir up plenty of enemies, and we may expect to see some strong criticism of what was said in the course of a very short time.[6]

The newspaper continued in sermonic tones, deprecating the growing trend to downgrade and question matters of religious faith and belief in God. It deplored the sight of physicists and scientists competing to say the 'boldest and most impious things in respect to the Almighty', commenting: 'No one can read any of the scientific papers without noticing this unpleasant fact.' But the *South London Chronicle* could now point with satisfaction to two great scientific figures of the age—John Burdon-Sanderson and James Clerk Maxwell—who showed that impiety was not the inevitable consequence of scientific thinking.[7]

The newspaper rejoiced that such men had declared their belief in the necessity for some over-ruling intelligence, despite meeting with determined opposition from some of their fellows. Yet it warned that, in this battle for faith, the opposing forces would be untiring and unrelenting:

> On one point their opponents are agreed; that it would detract from the lustre of their own discoveries if it could be proved that there was an intelligence superior to their own, and their knowledge was but derived, and not original with themselves. So they fight with all the ardour of fanaticism, exposing themselves to the contempt of all who question the right of specialists of any kind to expel the Almighty from the universe.[8]

---

6 *South London Chronicle*, 4 Oct 1873.

7 John Burdon-Sanderson (1828–1905) was a respected national medical figure: official investigator of the 1858 diphtheria outbreak in England, and anticipator of Sir Alexander Fleming in being aware that *penicillium* inhibited the growth of bacteria. On Burdon Sanderson's faith see: Lady Burdon Sanderson, *Sir John Burdon Sanderson: A Memoir*, 18f, 57, 160, 300.

8 *South London Chronicle*, 4 Oct 1873.

The Tyndall Critique

As the *South London Chronicle* predicted, big guns were soon trained on Maxwell. A year later his ideas were subjected to sustained heavy bombardment at the 1874 British Association meeting in Belfast. At that meeting, John Tyndall, the Association's new President, delivered one of the nineteenth-century's strongest expositions of the materialist position, with Tyndall's broadside followed by T.H. Huxley's questioning of long-held views on the nature of humanity.

John Tyndall succeeded Michael Faraday as professor of natural philosophy at the Royal Institution in London. Credited today as discovering the Greenhouse Effect, Tyndall was an exciting and effective communicator. Along with Huxley, he formed the X-Club which aimed to shape the future of British science, convinced that only natural causes should be permitted in science, and that no theistic explanations should be discussed, even on the question of ultimate origins.

Maxwell knew and respected Tyndall, and they had many amicable exchanges. In 1864, Tyndall had invited Maxwell to become a member of the Philosophical Club, an invitation which Maxwell accepted.[9] It was also Tyndall who suggested to Maxwell alternative ways of thinking through some of Faraday's ideas. On Maxwell's part, he evaluated several of Tyndall's Royal Society papers before they were published in the Society's *Transactions*; and in Maxwell's 1870 address to the British Association he spoke admiringly of Tyndall's penetrating insight.[10] He also enjoyed Tyndall as a dynamic speaker, who could light up any conference, commenting: 'The only man I know who can make everything the subject of discussion is Dr Tyndall. Secure his attendance, and that of somebody to differ from him, and you are all right for a meeting!'[11]

Tyndall admired Faraday tremendously, though the two men were far apart on matters of faith. Though the two men had a long acquaintance, they rarely discussed spiritual matters. Tyndall wrote of Faraday: 'Never once during an intimacy of fifteen years did he

---

9 JCM, *Letter to Tyndall,* 20 Apr 1864, (Harman, II/1, 147). The Philosophical Club was founded in 1847.

10 JCM, 'Mathematics and Physics, *Report of the 40th Meeting of the BAAS (1870),* 1–9. T.H. Huxley was BAAS president.

11 JCM, *Letter to H.R. Droop,* 19 July 1865, (C&G1, 343; Harman, II/1, 226).

mention religion to me, save when I drew him on the subject. He then spoke to me without hesitation or reluctance; not with any apparent desire to "improve the occasion" but to give me such information as I sought.'[12] Tyndall interpreted Faraday as believing the human heart to be swayed by a power beyond any science or logic, commenting: 'Right or wrong, this faith, held in perfect tolerance of the faiths of others, strengthened and beautified his life.' Whereas a theist might close an account of Faraday's life by writing that God had given him great gifts, Tyndall only looked to nature: 'By some natural process in the formation of this man, beauty and nobleness coalesced, to the exclusion of everything vulgar and law.'[13] This was Tyndall putting his materialist philosophy into practice when paying homage to a man whom he respected immensely.

Tyndall died tragically through accidental poisoning, when his wife gave him an overdose of sleeping medication. The Inquest into his death was told that he had been in delicate health for about three years, and was taking two medications, magnesia and syrup of choral.[14] When a new supply of choral was obtained at the local chemist it was placed close to the magnesia on his bedside table. Tyndall usually took his medicines at 8.30 each morning, and, on the fateful day, his wife measured a tablespoonful of magnesia—as she thought—without checking the label. Tyndall gulped it down, but realised straightaway that something was wrong. A stomach pump was applied immediately, and medical help came quickly. Despite all efforts, Tyndall was dead by 6.30 in the evening. Louisa Tyndall was haunted by the mistake for the rest of her life.

In 1874, Tyndall was at the height of his powers, and, as the new president of the British Association, he had a platform to set out his scientific vision. Maxwell and like-minded friends such as William Thomson, Balfour Stewart, and Peter Tait, believed that acknowledgement of the Deity as the ultimate cause and source of all things was proper in scientific discussion. Tyndall was of a different stamp. In his Belfast address, delivered with all the skill of expression

12 Tyndall, *Faraday as a Discoverer,* 151. 'To improve an occasion' meant 'to drive home an application'.

13 Ibid., 170.

14 *Westminster Gazette,* 7 Dec 1893, 'Inquest on Professor Tyndall'.

for which he was famous, he set out the exclusively materialist approach as the only credible stance for a scientist to take:[15]

> We claim—and we shall wrest from theology—the entire domain of cosmological theory. All schemes and systems which thus infringe upon the domain of science must . . . submit to its control and relinquish all thought of controlling it.[16]

Tyndall argued forcibly for a clear divide between what he saw as the objectivity of science (based on evidence and logic) and the subjectivity of religion (based on feelings and faith). He extolled Darwin's evolutionary theory, and proposed that primitive thinking had invented the notion of 'gods' to account for gaps in the explanation of phenomena. For Tyndall, science was committed to explaining natural phenomena by purely natural causes, and any reference to the theological had to be banished entirely from the natural sciences. He also suggested that life could arise from inanimate origins.

His speech made its mark. The highly religious Belfast audience was shocked. They had come, expecting that the President of the British Association would give due honour to the Almighty at some point in his speech, but what had they heard? Rampant, undiluted, materialistic atheism! Newspapers in Britain, Ireland, Europe, and North America carried reports of Tyndall's *tour de force* on their front pages.[17] Local reaction was strong, with Robert Watts (1820–1895), professor of systematic theology at Belfast's Theological College, challenging Tyndall to a public debate, and asking the Association for an opportunity within the programme to answer Tyndall. When both were denied, and when Huxley followed up Tyndall's speech with an even more provocative address, Watts initiated a city-wide series of meetings to give an alternative view.[18]

Tyndall was not only setting out his personal convictions. He was responding to Maxwell's discourse on molecules of the previous

15 Tyndall, 'Presidential Address', *Report of the 44th Meeting of the BAAS,* (1874), lxvi–xcvii.

16 Ibid., xcv.

17 Tyndall's speech did not halt references to the Deity at the British Association immediately. In 1876 Professor Thomas Andrews of Belfast referred to the 'Supreme Author and Upholder of the Universe.' (Andrews, 'Presidential Address', *Report of the 46th Meeting of the BAAS* (1876), lxxxv.).

18 Rodgers, *The Life and Work of Robert Watts,* 27f.

year, and he was reacting to theistic statements made by prominent Association members in the recent past, such as those put forward by William Thomson at the Association's 1871 meeting in Edinburgh. Thomson's Edinburgh address was delivered after initial difficulties in getting into the venue as he forgot his ticket! [19] Thomson had stated that though scientists should not invoke an abnormal act of divine creative power unnecessarily, yet 'overpoweringly strong proofs of intelligent and benevolent design lie all around us.'[20] This was exactly what Huxley and Tyndall wanted to banish. In their minds, Thomson, Maxwell, and others, failed to draw a clear line between science and superstition. It had been particularly galling for Huxley to have to listen to Thomson in 1871, as he had just vacated the presidency.

Tyndall's Belfast speech sought to root science in a purely material universe, excluding all religious aspects. With Maxwell's 'manufactured' articles in mind, Tyndall criticised the inferring of a first cause. He rejected Maxwell's methodology which necessitated a Creator, arguing that Kant had pronounced inductive logic invalid, and consequently he had to 'doubt the legitimacy of Maxwell's logic.'[21] Given that inductive reasoning was at the core of Maxwell's thinking, Tyndall aimed to torpedo Maxwell's whole thesis, holing

---

19 *Scotsman,* 3 Aug 1871.

20 Thomson, 'Presidential Address', cv: 'Overpoweringly strong proofs of intelligent and benevolent design lie all around us, and if ever perplexities, whether metaphysical or scientific, turn us away from them for a time, they come back upon us with irresistible force, showing to us through Nature the influence of a free-will, and teaching us that all living beings depend on one ever-acting Creator and Ruler.' The *Scotsman* (3 Aug 1871) reported the BAAS proceedings, noting another speaker's comment that Thomson had used Huxley's argument that all life must come from life to prove (in the second speaker's view) that in Huxley's philosophy life had to be eternal as Aristotle had argued. The apparent necessity (under materialist assumptions) for there to be infinite regression for biological causes as well as for physical ones is addressed by Stewart and Tait's *The Unseen Universe.* Stewart and Tait may have had this interchange in mind. See also: Kelvin, *Popular Lectures and Addresses,* Vol. II, 132–205.

21 Cf. Tyndall, 'Presidential Address', lxxviii: 'In his "manufactured articles" as he calls the atoms, Professor Maxwell finds the basis of an induction which enables him to scale philosophic heights considered inaccessible by Kant, and to take the logical step from the atoms to their Maker. Accepting here the leadership of Kant, I doubt the legitimacy of Maxwell's logic.'

it below the waterline! As a polite nod to a respected fellow scientist whose friendship he enjoyed, Tyndall conceded that it had been impossible not to feel the ethic glow with which Maxwell's 1873 lecture had concluded—even though he rejected Maxwell's entire proposition.[22]

The Ellicott Correspondence

If John Tyndall disagreed with Maxwell's 'Discourse on Molecules', the opposite was the case with John Ellicott, the Bishop of Gloucester and Bristol. Maxwell had read some of Ellicott's early books during his student years in the 1850s, and now, in the 1870s, Ellicott was busy preparing a series of lectures in defence of the faith. His aim was to narrow the widening gap between faith and science in the minds of parishioners, and he wanted to be sure he had understood Maxwell correctly. He wrote to Maxwell, seeking clarification of what he meant by 'manufactured' articles. In passing, Ellicott referred to a theory which was in vogue in Christian circles:

> Are you, as a scientific man, able to accept the statement that is often made on the theological side, viz. that the creation of the sun posterior to light involves no serious difficulty, the creation of light being the establishment of the primal vibrations, generally; the creation of the sun, the primal formation of an origin, whence vibrations would be propagated earthward? My own mind, far from a scientific one, is not clear on this point.[23]

Ellicott was referring to the creation narrative of Genesis 1, which describes light as existing *before* the sun was created. He wondered if the biblical narrative was hinting that the first mention of light was a reference to the universal ether, which could then 'carry' the light which streamed from the sun and stars. Though Maxwell did not dismiss Ellicott's ideas, in his reply he warned against theologians fixing on particular scientific theories, and for two reasons. First, because scientific theories change. Second, because the adoption of a particular scientific theory by religion tends to fix that theory as

22 On Maxwell's arguments in seeking to refute the claims of Tyndall's 'scientific materialism' see Harman, *The Natural Philosophy of James Clerk Maxwell*, 199–208.

23 C.J. Ellicott, *Letter to JCM*, 21 Nov 1876, (C&G1, 392).

an article of faith, even when it is no longer accepted by science.[24] Years later, when George Walker of Castle Douglas reflected on his conversations with Maxwell, he recalled the same emphasis: 'He distinctly refused to bind himself, or to hold others bound, to the results of *present* attempts to harmonise science with Christianity, which might afterwards have to be reviewed.'[25] Ellicott did mention the theory in his lectures, but only as an interesting hypothesis.[26]

1 *Manufactured Articles.* On the specific question of atoms or molecules being 'manufactured' articles, Maxwell reminded Ellicott that the comparison had originally been made by John Herschel in his 1830 book on natural philosophy (republished in 1851).[27] Maxwell also sent Ellicott several papers in which he had written about 'certain kinds of equality among all molecules of the same substance', with reflections on how this might relate to their origin. By 'certain kinds of equality' Maxwell was pointing to fundamental characteristics and consistent forms of structure among all molecules—he referred to molecules rather than to atoms, because a 'molecule' was the first basic *arrangement* of fundamental being. An 'atom' was fundamental being itself, whereas molecules were atoms in relationship.

When George Walker wrote a Maxwell tribute in his local newspaper, he described him as having viewed molecular physics as 'the science which treats of the *ultimate* constitution of material

---

24 JCM, *Letter to Ellicott,* 22 Nov 1876, (C&G1, 393–395; Harman, III/1, 416–418): 'I should be very sorry if an interpretation founded on a most conjectural scientific hypothesis were to get fastened to the text in Genesis, even if by so doing it got rid of the old statement of the commentators which has long ceased to be intelligible. The rate of change of scientific hypothesis is naturally much more rapid than that of Biblical interpretations, so that if an interpretation is founded on such an hypothesis, it may help to keep the hypothesis above ground long after it ought to be buried and forgotten.'

25 George Walker, 'James Clerk Maxwell', *Galloway News & Kirkcudbright Advertiser,* 18 May 1883.

26 Ellicott, *Modern Unbelief: Its Principles and Characteristics,* 106.

27 Herschel, *Preliminary Discourse on the Study of Natural Philosophy,* Ch.3, Art 28: 'The discoveries alluded to effectually destroy the idea of an eternal self-existent matter, by giving to each of its atoms the essential characters, at once, of a manufactured article, and a subordinate agent.' Herschel discussed the discovery that atoms of a particular type are always exactly the same, and asserted that these discoveries destroyed the idea of an *eternal self-existent matter,* implicit in primitive atomistic theories.

things.'[28] That was how Maxwell had explained things to Walker in conversation. On Ellicott's part, he was delighted with Maxwell's reply, and sent a letter of thanks:

> Allow me not to lose a post in thanking you most warmly for your most kind letter and for the packet of pamphlets, for which I hardly know how enough to express my gratitude. They are exactly what I needed, yet I fear I may be taking from your stock more than I ought to take.[29]

Volume Three of the ninth edition of the *Encyclopaedia Britannica* (Athens–Boissonade) had been published in 1875, and included Maxwell's article on the 'Atom'. It was an off-print of that article which Maxwell sent to Ellicott, prompting the bishop's concern that Maxwell may have depleted his stock unnecessarily. The letter of thanks also touched on a range of other scientific issues: attraction, gravitation, the ether. Importantly, Ellicott accepted Maxwell's advice on not tying religion to a particular scientific theory, since jumping to premature conclusions helped neither science nor faith:

> It seems to me that we want for several things, e.g., light, the conception of an ether-beach all-round the visible universe from which waves might be reflexively started, and at which the particles might be more closely packed; but then again I see not what it is that keeps up the beach. . . . I cordially agree with you as to the light question. *Theologians are a great deal too fond of using up the last scientific hypothesis they can get hold of.* (Italics ours)

After Ellicott delivered his lectures, they were published as *Modern Unbelief: Its Principles and Characteristics* (1877), in which he referred directly to Maxwell, especially Maxwell's 'manufactured' articles, i.e. entities which were not the product of a process: 'If the molecule is "incapable of growth or decay, of generation or destruction" how can we reconcile such characteristics with the operation of those purely

28 *Galloway News & Kirkcudbrightshire Advertiser,* 18 May 1883.

29 Ellicott, *Letter to JCM,* 24 Nov 1876, (C&G1, 395). Ellicott's book, *Modern Unbelief: Its Principles and Characteristics,* 76, 80, 141, echoes Maxwell's warning concerning theologians adopting scientific theories which might soon be reversed. He quoted almost word for word from Maxwell's letters: 'If we take our stand on the science of 1876 . . . remember that the science of 1896 may be further advanced in these very particulars than the science of today' (107); 'Theology becomes clogged with antiquated and presumably incorrect explanations, and so the truth greatly suffers' (141).

natural causes [evolution] which are now so persistently claimed to be the constructive principles of the universe?'[30] Like Maxwell, Ellicott was not referring to evolutionary processes in biology. What he was rejecting was the notion that *every* natural phenomenon was the result of an evolutionary process.

Following Maxwell, Ellicott held that truly fundamental entities—whatever they may be—were unchanged and unchanging. They had to be: otherwise there could never be any *truly fundamental entities*. 'Manufactured' articles pointed unerringly to a Maker. Science demanded a divine Artificer.[31]

2 *The Logic of Being*. There was a deeper issue. This concerned a possible Christological interpretation of the logic of the cosmos. Maxwell wrote to Ellicott:

> At the same time, I think that each individual man should do all he can to impress his own mind with the extent, the order, and the unity of the universe, and should carry these ideas with him as he reads such passages as the 1st Chap, of the Ep. to Colossians (see Lightfoot on Colossians, p. 182), just as enlarged conceptions of the extent and unity of the world of life may be of service to us in reading Psalm viii, Heb. ii. 6, etc.[32]

Maxwell was referring to chapter one of Paul's *Letter to the Colossians* in the New Testament. The Christian doctrine of creation is rooted in numerous biblical passages, not only Genesis 1. When taken together, these gave Maxwell an impression of 'the extent, the order, and the unity of the universe.' He viewed Colossians 1 in particular as pointing to the ultimate logic of the universe, since it seemed to hint that its intelligibility was rooted in Christ and existed because of Christ. The critical verses were Colossians 1:15-17:

> [Christ] is the image of the invisible God, the firstborn of every creature: for by him were all things created, that are in heaven, and that are in earth, visible and invisible, whether they be thrones, or dominions, or principalities, or powers: all things were created by him, and for him: and he is before all things, and by him all things consist.'

30 Ellicott, *Modern Unbelief: Its Principles and Characteristics,* 36.

31 Ibid., 63.

32 JCM, *Letter to Ellicott,* 22 Nov 1876, (C&G1, 394f; Harman, III/1, 417f).

Crucial to Maxwell's interpretation of Colossians 1 was J.B. Lightfoot's commentary on the Greek text of the epistle. Maxwell knew J.B. Lightfoot (1828–1889) through the Eranus Club, founded in 1872 by the biblical scholar B.F. Westcott.[33] The group promoted scholarly discussion and debate on a variety of questions, and its active members included rising academics such as F.J.A. Hort (1828–1892) as well as Lightfoot and Westcott. Lightfoot had published his commentary in 1875, a year before the Maxwell and Ellicott correspondence. Since Maxwell was solidly grounded in the classical languages, he was well able to appreciate his friend's work.

Lightfoot interpreted Colossians as a rebuttal of Gnostic influences on Christian thought. Gnostics viewed Christ as one of several spiritual beings, or demiurges, who occupied a place between God and the created order. Lightfoot understood verse 15 of chapter 1 as Paul asserting that Christ was much more than a demiurge, 'claiming for Him the absolute supremacy, both in relation to the *universe* (the natural creation), and in relation to the *Church* (the new moral creation)'.[34] In Lightfoot's view, Paul combined these and explained this twofold sovereignty by 'the absolute indwelling of the *pleroma* [the entirety of *all* spiritual things] in Christ'.[35]

Significantly—and relevant to Maxwell's thinking about the source of the logic of the cosmos—Lightfoot stated that the philosophical idea of the *Logos* underlay the whole of this section of Colossians (even though the term *Logos* was not present) with

33 A possible meaning of Eranus is 'hungry', and the society was sometimes called the Picnic Club, with its followers hungry for knowledge! A.F. Hort (ed.), *Life and Letters of Fenton John Anthony Hort,* Vol. 2, 185.

34 Lightfoot, *Commentary on Colossians,* 126, 143. (Italics ours).

35 Ibid., 102: 'On the one hand, in relation to Deity, He is the visible image of the invisible God. He is not only the chief manifestation of the Divine nature: He exhausts the Godhead manifested. In Him resides the totality of the Divine powers and attributes. For this totality Gnostic teachers had a technical term, the *pleroma* or *plenitude*. . . . But in all cases this *pleroma* was distributed, diluted, transformed and darkened by foreign admixture. They were only partial and blurred images, often deceptive caricatures, of their original, broken lights of the great central Light. . . . In contrast to [Gnostic] doctrine, [Paul] asserts and repeats the assertion, that the pleroma abides absolutely and wholly in Christ as the Word of God.'

Logos denoting 'both reason and speech'.[36] As God's *reason,* the *Logos* denoted God's purpose or design. As God's *speech,* it implied his revelation. For Christians, that *Logos,* Word, Reason, or Speech, was a divine person who became incarnate in Jesus Christ.

In Lightfoot's view, though Christian usage of the term *Logos* was not identical with its usage by non-Christians, such as the Jewish scholar Philo of Alexandria, many aspects of *Logos* philosophy had been carried over into Christian thought. Included in this was the notion that the *Logos* was the source of the original conception and archetypal idea of creation which was afterwards realised in the form taken by the material world.[37]

That was what Lightfoot proposed in the main body of his detailed verse by verse comments on the text of Colossians 1. Earlier in his book, he had been even more explicit, with the page number (182), cited in Maxwell's letter to Ellicott, relating directly to Lightfoot's comments in that introductory section where Lightfoot wrote:

> All things were created through Him, are sustained in Him, are tending towards Him. Thus, He is the beginning, middle, and end, of creation. This He is, because in Him dwells the plenitude of Deity. *This creative and administrative work of Christ the Word in the natural order of things* is always emphasised in the writings of the Apostles when they touch upon the doctrine of His Person.' (Italics ours).

Lightfoot explained that in Pauline thought Christ was not a demiurge or subordinate deity, as in some Gnostic thinking, but was the fullness of God himself.[38] What Maxwell took from this was that the 'extent and unity' of the world *came from* Christ, was *sustained in* Christ, and had *meaning in* Christ. In other words: all things exist in him, all things exist because of him, and he is the one who gives logos, reason, and rationality to the cosmos. That gave Maxwell a theological underpinning for the orderliness and logic of the natural world. It was Christ the *eternal logos* who gave the created order its own *logos.* Ultimate rationality derived from Christ, in whom and from whom that rationality undergirds and permeates the created order.

---

36 Not all New Testament scholars would agree that Paul had a *Logos* theology in mind, but Maxwell followed Lightfoot.

37 Lightfoot, *Commentary on Colossians,* 144.

38 Cf. Marston, 'Maxwell, Faith and Physics', 279.

In his monograph *The Faith of a Scientist* Botond Gaal leaned heavily on Maxwell's appeal to Colossians 1. Gaal understood Maxwell, in his letter to Ellicott, to be saying that the ultimate unifying principle in the created universe was its maker, the Son of God, Jesus Christ, and that when Maxwell read Colossians 1 he 'read the heart of Christian teaching about the order and unity of the created world, which is Jesus Christ':

> That is, the unity of the universe is not in the human mind as Hamilton taught it, neither brought about by the human mind as Hume thought, but 'all things hold together' in Jesus Christ. That was Maxwell's firm conviction, and not mysticism, as Sir William Thomson thought anxiously about him.[39]

Whereas positivism can inhibit a scientist in his or her attempts to progress beyond scattered evidence and fragmentary data, Maxwell's rooting of the logic of reality in Christ gave him confidence to go further. He worked on the principle that in every creative scientific investigation there occurs at some point a leap of understanding in which the mind grasps the 'whole' and not just the 'parts', and that it is this grasp of the 'whole' which enables the parts to be viewed, and interpreted, and inter-connected in their proper logic. Maxwell believed that the source of created order was Jesus Christ, who was, as Botond Gaal expressed it, 'the absolute reality, through whom, by whom, and in whom, Clerk Maxwell could see the perfect wholeness.'[40]

When Maxwell sought a basis for the ultimate order and unity of the universe, he had one. Unlike some others, he was not trapped in an endless cycle of systems behind systems behind yet other systems, but was able to break out of that spiralling dead-end to the ultimate foundational order and unity given in Jesus Christ.[41]

Maxwell was a theist. God was his starting point. He viewed Colossians 1 as pointing beyond a simple doctrine of God making all things, to a sophisticated doctrine in which the structure and logic of the universe came from Christ and was rooted in Christ. The ultimate source of the logic of the universe was God himself, in and through Christ.

39 Gaal, *The Faith of a Scientist,* 24.

40 Ibid., 39.

41 Ibid., 40. Gaal (42) elaborately argues that this implied for Maxwell that the unity of God was based on God's triunity.

Yet Maxwell's assured belief in an ultimate cause never tempted him to abandon his search for natural causes. As far as he was concerned, God, in his artistry, had created a universe of endless connections, and so he always assumed there was never a situation in science where there were no more links to be found in the God-given chain of natural causation. Consequently, God should not be invoked simply because science could yet explain why such and such occurs.

At the same time, though the important notion of a primary or first cause cannot be *deduced* from the chain of natural causes, Maxwell held that *inductive* reasoning, plus the conviction that there had to be a beginning to the cosmos, pointed to this as necessary. It was this methodology which Tyndall questioned but which Ellicott endorsed in his book: 'All true science is now pointing to a beginning, and to an intelligent and all-powerful Will [which] made that beginning, and has since taken part in its progress and development.'[42]

42 Ellicott, *Modern Unbelief: Its Principles and Characteristics*, 81.

Chapter 27
Automata or Free?

Tyndall's 1874 Belfast lecture was Part One of a combined assault on Maxwell's position. A few days later, T.H. Huxley followed up with Part Two. Tyndall's speech attempted to counter Maxwell's 1873 'Discourse on Molecules' through attacking the validity of any proper connection between religion and science. Huxley's address had a radical interpretation of what it meant to be human, questioning the ability of humans to exercise true free-will. Huxley had the honour of giving that year's prestigious British Association Evening Lecture, which he delivered to a crowded gathering in the Ulster Hall. The audience included Maxwell. Huxley had the prime public slot which Maxwell had occupied in 1873.

Huxley's Hypothesis

The title of Huxley's address was: 'The Hypothesis that Animals are Automata, and its History'. Huxley argued that all animals—including humans—are automata whose actions are predetermined by mechanical, chemical, and biological causes.[1] The sheer complexity of inter-related causes lying behind human actions

---

1 Huxley's talk is referred to, but not included, in the official *Report of the 44th Meeting of the BAAS* (1874), (*Report,* lvi). It can be found in: Huxley, *Collected Essays.* Vol. 1, 199–250. Huxley stated (243f): 'It is quite true that, to the best of my judgment, the argumentation which applies to brutes holds equally good of men; and therefore, that all states of consciousness in us, as in them, are immediately caused by molecular changes of the brain-substance . . . it follows that our mental conditions are simply the symbols in consciousness of the changes which takes place automatically in the organism; and . . . the feeling we call volition is not the cause of a voluntary act, but the symbol of that state of the brain which is the immediate cause of that act. We are conscious automata, endowed with free-will in the only intelligible sense of that much abused term—inasmuch as in many respects we are able to do as we like—but none the less part of the great series of causes and effects which, in unbroken continuity, composes that which is, and has been, and shall be, the sum of existence.'

might give the illusion of free-will, but in reality there was none. He drew on Descartes' view that all events in nature come to pass purely because of physical material causes, admitting that total determinism was the logical outcome of his paper. Though Huxley was reluctant to abandon the notion that men and women could make truly free decisions, he was determined to go where his argument led. He declared defiantly:

> As to the logical consequences of this conviction of mine, I may be permitted to remark that logical consequences are the scarecrows of fools and the beacons of wise men. The only question which any wise man can ask himself, and which any honest man will ask himself, is whether a doctrine is true or false. Consequences will take care of themselves.[2]

His conclusion was inescapable. It was a verdict demanded by all who allowed only material causes for all phenomena. All of the great forces of nature, including natural selection, geology, and conservation of energy, were under the sway of determinism. Nature had no place for free-will.[3] Free-will was illusory.

Huxley spoke five days after his friend John Tyndall delivered his Presidential Address, in which he had argued for a purely materialist methodology in natural science and that science should make no reference to the spiritual or the divine. Over the weekend, preachers had thundered from Belfast pulpits against this 'impious philosophy'.[4] When Huxley stood up on the Monday evening, his aim was not only to set out his own thesis, but to defend his and Tyndall's shared position from what he regarded as the prejudiced fulminations of religious bigots.

Towards the close of his remarks, and aware he was in Protestant Ulster, Huxley mischievously cited St Augustine, John Calvin, and Jonathan Edwards in support of his argument, though without quoting directly from any of them. These theologians were all well-known proponents of divine predestination, and Huxley impishly allied their predestinarian doctrines with his argument for deterministic

2 Huxley, 'The Hypothesis that Animals are Automata', *Collected Essays*, Vol. 1, 244.

3 Cf. Whitaker, 'Maxwell's Famous (or Infamous) Demon', 178.

4 *Tablet*, 29 Aug 1874.

causality. He also took the opportunity to take aim at the religious establishment:

> May I say, in conclusion, that it really would be well if ecclesiastical persons would reflect that [religious] ordination—whatever deep-seated graces it may confer—has never been observed to be followed by any visible increase in the learning or logic of its subject. Making a man a Bishop, or entrusting him with the office of ministering to even the largest of Presbyterian congregations, or setting him up to lecture to a Church congress, really does not in the smallest degree augment such title to respect [any more than] his opinions may intrinsically possess.[5]

Huxley's barnstorming finale, delivered with his famed eloquence and panache, meant that he sat down to loud and prolonged applause.[6] According to *The Newtonards Chronicle,* at least one person did question Huxley's argument.[7] This was Dr W.B. Carpenter of the University of London, one of the British Association's vice-presidents, who had been appointed to second the vote of thanks. Carpenter congratulated Huxley on his talk, but taxed him for omitting any reference to the moral 'ought' in the 'dictionary of man'. Was it not the case that Huxley's deterministic causality undermined any basis for moral choices?

For many ordinary members of the public in the Ulster Hall, the rapturous ovation accorded to Huxley's lecture was as disturbing as the content of his speech. The north of Ireland was intensely religious, highly Protestant, and strongly church going. They interpreted the applause as signifying agreement with Huxley's views, rather than as expressing appreciation for a gripping lecture. Local Christian believers had yet again been shocked by what they had heard. Prominent scientists visiting their city had not only failed to exalt the dignity of the Creator, but had expounded non-theistic materialism, and had reduced humanity from the *imago dei* to automata.

Belfast reacted as strongly to Huxley as it had to Tyndall. When the monthly meeting of the Belfast Presbytery took place in early September, concern was expressed that Huxley had ignored man's

5 Huxley, 'The Hypothesis that Animals are Automata', *Collected Essays.* Vol. 1, 249.

6 *Belfast Newsletter,* 25 Aug 1874.

7 *Newtonards Chronicle,* 29 Aug 1874.

moral responsibility and had taught that man was nothing more or less than a mere conscious machine.[8] The Revd William Johnston (1818–1894) referred to the fact that though Belfast had been honoured by the Association coming to the city, and though all had looked forward with interest to the visit of such illustrious savants, they had not been prepared for the 'onslaught which had been made upon the principles of Christianity by the leading members of that Association.'[9] The Presbytery's counterattack in defence of Christian faith was led by Robert Watts at his series of rallies which attracted huge audiences.

Watts' speeches had the title: 'Professor Huxley's Address before the British Association Examined and Refuted!'[10] Watts was troubled. He knew that any paper read to the British Association automatically carried weight and influence. As such, Huxley's lecture had to be answered. In his response Watts focussed on his own conviction that nothing so far discovered in science added to or took away from Paley's arguments that there had to be an intelligent designer: 'If nothing out of nothing come, how can a conscious intellect and will come forth from the womb of an unconscious, unintelligent, purposeless thing called force?'

Turning to Huxley's *automata* hypothesis, Watts interpreted Huxley's statement 'that animal organisms are mere conscious automata' as implying, 'that all the actions of animals, from the mollusc to the man, may be accounted for by referring them to the molecular changes in the nervous system.'

Huxley had embraced Descartes' hypothesis that changes in the molecular condition of the nervous system are sufficient to account 'for all the movements of those wondrous organic structures', and Watts was prepared to allow some of Huxley's points. Watts was perfectly happy to describe the body as a wondrous machine. He also conceded that the immediate cause of muscular movement involved chemical and molecular changes in the nervous system. But he moved on to an analogy with the recently laid trans-Atlantic cable and the signals going through it. What or who gave these signals? How came there to

8 *Londonderry Sentinel*, 3 Sep 1874.

9 *Freeman's Journal*, 3 Sep 1874.

10 *Belfast Weekly News*, 29 Aug 1874; *Freeman's Journal*, 17 Sep 1874; *Newry Reporter*, 19 Sep 1874.

be cables and wires at all? For Watts, such developments in modern science strengthened the need for a designer rather than diminishing it, since the initiation of signals—whether to the telegraph system or to the human nervous system—required a thinking agent. Watts insisted that 'the constitution of the human mind is such that it must ask these questions', voicing a criticism which was levelled at Tyndall and Huxley by many theists: that they ignored ultimate questions.

The *Westminster Confession of Faith* was Irish Presbyterianism's doctrinal basis, as it was of Maxwell's Church of Scotland. In its Articles, the *Confession* affirmed both the freedom of God and human responsibility. It described God as 'most wise, most holy, *most free*, most absolute'.[11] The phrase '*most free*' was critical: on the one hand, it countered the Aristotelian determinism of some medieval theology; on the other hand, it emphasised the biblical notion of God as the living God who is free to act, always in accordance with his nature. The *Confession* also stated that God had endued the will of man with, 'that natural liberty, that it is neither forced, nor by any absolute necessity of nature determined, to do good or evil.'[12] Over against the deterministic straightjacket of materialists such as Huxley, the *Confession* taught that the will and its freedom was a gift of God as part of the image of God in humanity, and distinct from the physical constitution of humanity.

This was a different view of what it meant to be human from that of Huxley, and was what Watts and Maxwell—both well acquainted with the *Confession*—believed. Consequently, Watts' argument against Huxley focussed on moral implications, insisting that if the doctrine that animals are mere automata were to be adopted, then human society would become an impossibility since all responsibility would be at an end: 'For such a theory not only puts an end to all ideas such as justice, rewards, and punishments, but

11 *Westminster Confession of Faith,* II/1

12 *Westminster Confession of Faith,* IX:1. It is probable that Maxwell's understanding of 'will' was influenced by the *Confession of Faith,* which, as an elder of the Church of Scotland, he studied closely. A longer quotation from the *Confession's* description of the nature of God (Chapter II/1) is: 'There is but one only, living, and true God, who is infinite in being and perfection, a most pure spirit, invisible, without body, parts, or passions; immutable, immense, eternal, incomprehensible, almighty, most wise, most holy, *most free*, most absolute; . . .' (Italics ours).

puts an end to any necessity for the laws enacted by the Legislature for the protection of society, as, according to [Huxley's] theory men would be no more responsible for their acts than the Atlantic cable for the messages it received.'[13] Watts did not prove Huxley's logic false, but he showed its consequence.

Huxley may have been Peter Tait's golfing partner, but the Belfast lecture showed how far apart they were on key issues. Tait's response came the following year in his co-authored work *The Unseen Universe* (1875). As for Maxwell, he was well acquainted with the materialist position and in fact agreed with Huxley and Tyndall some of the way.

Maxwell accepted that all examinable material phenomena had traceable material causes, and that for every natural phenomenon a natural cause could be discovered. In that respect he thought similarly to Huxley and Tyndall, though this did not contradict his Christian faith. As a Christian who was a scientist, he believed that God had given the universe a rationality which could be sought within itself, and he believed that it was precisely this God-given integrity of the created order which meant that nature was explicable. It was a *God-given* internal coherence which enabled nature—apart from questions concerning its ultimate origin and its rational order—to be understood. However, for Maxwell not all phenomena and not all causes were *purely* natural. He disputed the materialist's assumption that the non-material or the spiritual had no reality, and that physical deterministic causes were the *only* type of causes in nature.

As an aside, we note that it is inherently impossible to prove free-will as the cause of an event. This is because any formal proof can only deal with arguments in which, at each stage, an 'A' *necessarily* implies a 'B'. But if an event occurs outside a matrix of *necessary* sequences (i.e. as an act of true free-will) then no argument which presumes causal necessity can prove that the cause came from outside that matrix. 'Freedom' and 'free-will' are unconditioned entities—to use Sir William Hamilton's terminology. Consequently, no system of proof—operating purely within a conditioned causal system—can ever indicate a cause outside that system. It is impossible to prove freedom.

13 *Downpatrick Recorder*, 19 Sep 1874.

Maxwell was aware that true freedom of the will is something which can only be posited as a God-given constituent of the self.

Freedom and the Will

Establishing 'freedom' was not straightforward. The issue of 'freedom' versus 'necessity' was a well-discussed topic in Scottish intellectual circles well before Maxwell's time though it was more often analysed from a theological standpoint than from that of physics.[14]

Maxwell's approach started with all reality being contingent on the mind of God. That 'all' included both the universe's material origin and its underlying structural logic. He was no deist, for whom God originated the 'stuff' of the universe, gave it its order, and then let it run its course without interfering. Instead, Maxwell saw an ongoing dynamic interaction between the Creator and the created order in which God upheld the universe in Christ by his Word. This was what he had expressed in his correspondence with Bishop Ellicott. And because Maxwell never based faith in a 'God of the gaps' argument, he would have been uncomfortable with those who appeal to enigmas in science—such as the mystique of indeterminacy or the puzzle of the quantum—as being inexplicable phenomena which might provide a locus for the religious dimension or allow for gaps of causation in nature. Such phenomena may be difficult to understand and express, but Maxwell would regard them as part of the created order and therefore capable of explanation from within nature, at least theoretically.

Given the interaction between the Creator and the created order, what was critically important to Maxwell was that personal

---

14 One argument was that not only was God's omnipotence compromised if any truly free human action took place (since if truly free, such human actions were outside his control thus contradicting his omnipotence) but so were his majesty and dignity. The possibility that human freedom was no more than an illusion troubled James Boswell who refers to the issue several times in his *Journal*. Cf. Boswell, *Journal*, (entry written 17 Feb 1781): 'I read in Lord Monboddo's *Ancient Metaphysics* that there *could be no such thing as contingency*, and that every action of man was absolutely fixed and comprehended in a series of causes and effects from all eternity; so that there was an universal Necessity.' [Reed & Pottle, *Boswell: Laird of Auchinleck 1778–1782*, 283]. The Law Lord Monboddo was James Burnett (1714–1799). Burnett's *Ancient Metaphysics: Or, The Science of Universals* was published in 1779. Monboddo focussed on Aristotle's concept of God as the 'first mover' and Newton's laws of motion.

being—whether divine or human—had true freedom of will. But such a free-will faculty was not to be found in some physical part of the body. It was an entity in its own right, and neither reducible to the material nor found within it.

In referencing 'free-will', what do we mean by 'will'? Within the context of our discussion, the 'divine will' is not the same as irresistible divine omnipotence. It is not a reference to unfettered absolute power. Nor does it indicate the triumph of a totalitarian deity. Instead, the 'divine will' is simply the freedom of God to act. It is the liberty of God to act as he chooses, though always within a faithfulness to his covenant love. The *Westminster Confession of Faith* states that an essential attribute of God is that he is '*most free*'.[15] Thus, for Christians such as Maxwell, the divine 'will' was the divine 'freedom to act' unconstrained by any necessity internal or external, apart from God's own character. There is a divine freedom to love, a divine freedom to create, a divine freedom to interact with humanity, and a divine freedom to redeem. That freedom can even involve God choosing to curtail, restrict, limit, curb his omnipotence in 'emptying himself' (cf. Philippians 2:1-11). The divine 'will' is therefore that which chooses to act in the freedom of his covenant love. Such qualities are also *mutatis mutandis* qualities of the human will: men and women have the ability to exercise 'will', not in the sense of exercising naked power, but in the sense of exercising a freedom to act and a freedom to choose, which is not wholly predetermined by environment, or by inherited characteristics, or by biochemical activities in the brain. Though this human ability to 'will' or to 'choose' may be influenced by a multitude of factors, it is still a true freedom, granted by God to a humanity which he has created in his image. That was Maxwell's position.[16]

15 *Westminster Confession of Faith,* II/1.

16 Sometimes termed 'voluntarism'. Cf. John Henry, 'Voluntarist Theology at the Origins of Modern Science: A Response to Peter Harrison', *History of Science,* 47/1, (2009), 79–113, who analyses historical and contemporary discussions of divine omnipotence, reason, freedom, and will. He points out that for some medieval theologians there was the fear that Aristotelian logic would 'lead to conclusions about God which either diminished his omnipotence or, more subtly, implied that he was not fully in control of his own actions because he always had to follow the dictates of reason, and acted, therefore, out of necessity' (80). It is because of this debate that it is important to clarify our use of 'will'.

A year before Huxley's *automata* speech, Maxwell had addressed issues of determinism and free-will at the Eranus Club.[17] The title of his 1873 paper was: 'Does the Progress of Physical Science tend to give any advantage to the opinion of Necessity (or Determinism) over that of the Contingency of Events and the Freedom of the Will?'[18] It impressed W.E. Gladstone who, like Carpenter and Watts, was concerned that the materialist denial of free-will undermined Christian teaching on morality and responsibility.[19] But, before exploring Maxwell's 1873 paper, we need to refer to some of his even earlier thinking in an 1858 letter to Richard Litchfield.

The 1858 Letter to Richard Litchfield

Maxwell's lengthy epistle to Litchfield, in the spring of 1858, first asked a general question: 'How do we pursue truth in any academic discipline?'[20] His answer was that the physical sciences give the proper methodology for all areas of human enquiry (including that of freedom) since they provide the analogies of thought which other disciplines build on.

On human 'freedom', Maxwell was anxious to distinguish between: (i) irresponsible acts of human free-will which brought disorder and caprice; and (ii) a liberty which was not disorder, but which was no less free than the other, 'indeed more free.' This distinction, between *false-freedom* which resulted in chaos, and *true-freedom* which resulted in 'order, certainty, and trustworthiness' was something to which he returned during his career, and which he saw as fundamental for both science and theology.

True-freedom was not chaos or randomness. Instead, *true-freedom* was contingent upon order, and in turn generated more order. Maxwell extended this thought analogically, contending that in human behaviour *true-freedom* of the will was freedom to do or think whatever was right and good in any and every field of human activity. In contrast,

17 Cf. C&G1, 366. Cf. JCM, *Letter to B.F. Westcott,* 14 May 1877, (Harman, III/2, 450).

18 JCM, 'Does the progress of Physical Science tend to give any advantage to the opinion of Necessity (or Determinism) over that of the Contingency of Events and the Freedom of the Will?', 11 Feb 1873, (C&G1, 434f; Harman, II/2, 814–823).

19 Marston, 'Maxwell, Faith and Physics', 285.

20 JCM, *Letter to Litchfield,* 5 Mar 1858, (C&G1, 304–307; Harman, I, 588f).

false-freedom was wilfulness, with decisions controlled by something alien to our true nature: humanity deceived itself when it thought that an ability to choose what is wrong was part of true freedom. He stated to Litchfield that the wilfulness of our wills '*feels to us like liberty*' though it is in reality *necessity:* and 'wilfulness is seen to be *anything but free-will,* since it is merely a submission to the strongest attraction.' To put it another way: wilfulness means that the human sense of free-will can be an illusion, driven by bondage to alien and unhealthy forces impacting upon a man or woman. Maxwell explained:

> Nevertheless, there is another phase still, in which there appears a possibility of . . . an abandonment of wilfulness without extinction of will; but rather, by means of a great development of will, whereby, instead of being consciously free [but] really in subjection to unknown laws, it becomes consciously acting by law, and really free from the interference of unrecognised laws.

In other words, true freedom occurs when the human mind acts in accordance with objective truth, rationality, and goodness. Maxwell's notion of men and women exercising 'true freedom'—in both the moral and physical domains—was of them voluntarily acting in accordance with the laws by which they were formed.[21] Hence, true freedom occurs when mind and will align themselves—not with subjective desires of what we *want* to be the truth—but with what is actually the truth. True freedom occurs when we *will* to do the right because of our submission to the good.

For the theistic Maxwell as much as for the materialistic Huxley, 'free-will' could be an illusion, though Maxwell's understanding of its illusory nature was quite different to that of Huxley's. For Huxley, free-will was an illusion because all human actions (including choices) were predetermined. For Maxwell, free-will was an illusion when the will (which *could* make truly free decisions) was in captivity to forces stronger than it. On this point he may have been influenced by the *Confession of Faith's* assertion that true human freedom (i.e. a freedom to only do the good) is an eschatological event rather than something experienced here and now. The *Confession* states: 'The will of man is made perfectly and immutably free to do good alone in the state of glory only.'[22]

21 Torrance, 'Christian Faith and Physical Science', 222.

22 *Westminster Confession of Faith,* IX:5.

Maxwell strove to apply his metaphysical insight concerning freedom to both science and life. In science, true freedom of thought occurred when thought aligned itself with the rational and the logical. In life, true freedom of decision occurred when the will aligned itself with the good. He may have had a biblical text in mind: '*You shall know the truth, and the truth shall make you free*' (John 8:32). True freedom occurs, not when we select from unlimited possibilities, but when we think, will, choose, and act in conformity to what is right. True freedom did not consist in having every possibility, good or bad, open to a person, but in having the ability to think and do the good.

That was why Maxwell wrote to Litchfield that all want of order, all caprice, and all unaccountableness, resulted from interference with liberty; whereas true liberty (if unimpeded) brought order, certainty, and trustworthiness. Moreover, though what we commonly call free-will (often a broken form of freedom) may produce caprice and disorder, 'there is a liberty which is not disorder, and this is by no means less free than the other, but more.'[23]

The human will has a *false* feeling of freedom when it chooses evil. In fact, in such a situation the human will acts as the *result of law* and is in bondage to a determinism: 'The wilfulness of our wills feels to us like liberty, being in reality necessity'—in the sense that it is at the mercy of forces it cannot control. Thus, wilfulness is anything but true free-will, since it is merely a 'submission to the strongest attraction, after the fashion of material things.'

If the will is so easily swayed by evil, should a Christian mortify the will 'till nothing of himself remains and the man and his selfishness disappear together?' Maxwell referred to Buddhist thinking, noting that some Christian traditions had advocated something similar.[24] He rejected this, arguing instead that there was a genuine *true-freedom* apart from the *false-freedom* which many confused it with. There was a freedom to think rationally. There was a freedom to do the right. There was no need for men and women to abandon usage of that God-given gift. It was perfectly possible to have, 'an abandonment of wilfulness without extinction of will'.

---

23 JCM, *Letter to Litchfield,* 5 Mar 1858, (C&G1, 305; Harman, I, 588).

24 Ibid., (C&G1, 306; Harman, I, 588f); with reference to, Max Muller (1823–1900), *Ueber den Buddhistischen Nihilismus, etc.,* (Keil: 1869).

And there could be 'a great development of will whereby instead of being consciously free—and really in subjection to unknown laws—it becomes consciously acting by law, and really free from the interference of unrecognized laws.'[25]

The 1873 Eranus Club Paper

In Maxwell's 1873 Eranus Club submission—'Does the Progress of Physical Science tend to give any advantage to the opinion of Necessity (or Determinism) over that of the Contingency of Events and the Freedom of the Will?'—he insisted that his focus was determinism in the physical sciences rather than in theology or metaphysics, though, with his roots in Scottish Calvinism, the problem of freedom/determinism was as much a theological issue as a scientific one. But determinism was not the same as predestination—despite Huxley's attempt to conflate them—and Maxwell's paper was not on that issue. At least not directly.[26]

Taking aim at continental positivists, Maxwell emphasised that science was about more than simply observing and measuring material phenomena. Science involved putting forward theories (doctrines) which explained phenomena in a logical, consistent, and reasonable form.[27] This was rooted in his belief that good science revealed the true nature of things, and that predictions were possible because science could give a knowledge of what reality is truly like. For Maxwell, scientific predictions were not based solely on statistical likelihoods derived from past occurrences—though in some areas scientists had to use probabilistic tools—but on an understanding of the inner structure of reality.

---

25 Ibid., (C&G1, 306; Harman, I, 589).

26 For a discussion of aspects of Maxwell's paper see: Marij van Strien, 'Vital Instability: Life and Free Will in Physics and Physiology, 1860-1880', *Annals of Science*, 72/3, (2015), 381-400. Strien describes Maxwell's position as dualistic since 'since it involves an immaterial entity (mind, soul, or vital principle) that intervenes in the body' (383); and that Maxwell 'argued for an immaterial mind or soul that could cause changes in the body (where the body is treated as a mechanical system), and thus regarded an antimaterialist and dualistic metaphysics as essential for free will.' (398). In our view, Maxwell was proposing *twin* entities rather than *dualistic* entities.

27 JCM, 'Determinism and the Freedom of the Will', (C&G1, 442).

Turning to determinism—the central issue of his paper—he pointed out that in the physical sciences it was assumed that the same conditions always produced the same consequences. But he had two caveats. First, it could never be guaranteed that *exactly* and *precisely* the same conditions applied on two separate occasions. It might even be that the particular *time* and particular *place* of an event were themselves factors affecting the event. In other words, *time* and *space* may not be constant absolutes. Hence, even for physical events governed by apparently identical mechanical forces, there were a multitude of contingencies, known and unknown, which affected outcomes. Second, looking further afield than the physical sciences, human conduct was even less predictable than physical events, though some predictions of human conduct were possible. For example, when dealing with people of *weak character* (those who tended to follow the crowd) behaviour could be predicted on a statistical basis; and when dealing with individuals of *confirmed character* (strong-minded persons), their behaviour could also be predicted to some extent.

His cautions touched on two distinct considerations. First, the difficulty in *knowing* whether exactly the same conditions were in fact ever operating. Could a scientist ever be aware of absolutely every factor involved? Under this caveat, any uncertainty concerning outcomes derived from limitations in knowledge rather than from something indeterminate in the nature of things.[28] Second, and more importantly,

28 JCM's, 'Presidential Address to Mathematics and Physics Section', *Report of the 40th Meeting of the BAAS* (1870), 1–9), mused on the incompleteness of knowledge, though incompleteness was not the same as objective indeterminacy. Convinced that mathematical theory can correspond to the embedded mathematical structures of nature (thought weds fact), Maxwell stated: 'I shall only make one more remark on the relation between mathematics and physics. In themselves, one is an operation of the mind, the other is a dance of molecules. The molecules have laws of their own, some of which we select as most intelligible to us and most amenable to our calculation. We form a theory from these partial data, and we ascribe any deviation of the actual phenomena from this theory to disturbing causes. At the same time, we confess that what we call disturbing causes are simply those parts of the true circumstances which we do not know or have neglected, and we endeavour in future to take account of them. We thus acknowledge that the so-called disturbance is a mere figment of the mind, not a fact of nature, and that in natural action there is not disturbance.' See also, Coulson, 'Interatomic Forces', *Clerk Maxwell and Modern Science*, 49.

he allowed the existence of a non-mechanical factor in human behaviour: i.e. the will, and the mind behind the will. But what was the will? Was it a God-given entity (the Ego) superadded to all the physical things which make us human?[29] And did it free humanity from the Cartesian matrix of deterministic causes and effects to which Huxley appealed? If not—and on this point Maxwell in 1873 anticipated Huxley in 1874—was the human feeling of freedom of the will only an illusion?

> The doctrine of free-will asserts that in some such cases the Ego alone is the determining cause. [Whereas] the doctrine of determinism asserts that in every case, without exception, the result is determined by the previous conditions of the subject, whether bodily or mental, and that Ego is mistaken in supposing himself in any way the cause of the actual result, as both what he is pleased to call decisions and the resultant action are corresponding events due to the same fixed laws.[30]

Maxwell saw 'the Ego' as a God-given component of the personality, which participated in the conscious decision-making process. It vaguely corresponded to 'soul'. But was it free? Within his understanding of *true-freedom* over against *false-freedom*, and in line with the *Confession of Faith's* teaching on human responsibility, Maxwell believed that it was. It was an added gift of God which governed human actions. Once a decision was made, then the physical body acted in accordance with it and in line with physical causes and principles such as the conservation of energy; but these physical entities (cause, energy, etc.) did not affect the 'soul' itself.[31] Importantly, determinism did not rule. Humans were more than automata.

Maxwell's position has not always been represented clearly. Marston and Van Strien interpreted Maxwell's main concern as: given

29 JCM, *Essay for the Eranus Club on Psychophysick*, 5 Feb 1878, (C&G1, 452–463; Harman, III/2, 598–607).

30 JCM, 'Does the progress of Physical Science…?', 11 Feb 1873, (C&G1, 441; Harman, II/2, 819).

31 Cf. JCM, *Letter to Campbell*, 21 Apr 1862, (C&G1, 336; Harman, I, 711): "The soul is not the mere mover. Food is the mover, and perishes in the using, which the soul does not. There is action and reaction between body and soul, but it is not of a kind in which energy passes from the one to the other—as when a man pulls a trigger it is the gunpowder that projects the bullet, or when a pointsman shunts a train it is the rails that bear the thrust.'

the principle of conservation of energy, how was it possible for the soul to alter behaviour in the absence of a source of energy?[32] But that assumes that the 'soul' or 'mind' or 'will' is something constituted physically, and though that was Huxley's position it was not Maxwell's.

Similar remarks can be made on Harman's comment that Maxwell argued for the universe being fundamentally *causal* yet not *deterministic*, citing the swerve of Lucretian atoms as an illustration of the instability of a dynamical system at singular points.[33] But in that example Maxwell was *illustrating* the concept of free-will. He was not arguing that free-will was *caused* or *permitted* by unpredictable atomic swerves. His concern was to counter a popular mystical notion: namely, that humans, by a remote act of will, or telepathy, could affect events:

> No leopard can change its spots, nor can anyone by merely wishing it, or, as some say, *willing* it, introduce discontinuity into his course of existence. Our free-will at the best is like that of Lucretius's atoms—which at quite uncertain times and places deviate in an uncertain manner from their course. In the course of this our

32 Marston, 'Maxwell, Faith and Physics', 284. Van Strien ('Vital Instability', 384, cf. 396) has implied that Maxwell and others were primarily attempting 'to make energy conservation compatible with the intervention of mental and vital causes, through an appeal to instability and indeterminism in physics'. However, when Maxwell stated, 'There is action and reaction between body and soul, but it is not of a kind in which energy passes from the one to the other' JCM, *Letter to Campbell,* 21 Apr 1862), he was placing the act of decision-making *apart* from physical structures and as something which was *not* part of a matrix of physical causes and effects. As such, there was no need for Maxwell to seek instability or indeterminism as a 'causal gap' in physical reality. Van Strien has suggested that Maxwell developed his ideas as an explicit reaction to how the law of conservation of energy was being interpreted, and the claimed implications that 'free action' could not occur in a world governed by the conservation of energy ('Vital Instability', 391). But does a mental action of the 'soul' or 'will' itself affect energy? Van Strien (392) has suggested that there could be a 'very small' effect, which she believed Maxwell was aware of (395), and which could compromise Maxwell's argument. Whether or not that is be the case, Maxwell saw no issue. What Van Strien's paper omits is appreciation of the doctrinal convictions lying behind Maxwell's thinking on 'freedom', the 'will' etc.

33 Harman, II/1, 19.

> mortal life we more or less frequently find ourselves on a physical or moral watershed, where an imperceptible deviation is sufficient to determine into which of two valleys we shall descend. The doctrine of free-will asserts that in some such cases the Ego alone is the determining cause.[34]

Maxwell was clear what a doctrine of determinism implied:

> The doctrine of determinism asserts that in every case, without exception, the result is determined by the previous conditions of the subject, whether bodily or mental, and that Ego is mistaken in supposing himself in any way the cause of the actual result, as both what he is pleased to call decisions and the resultant actions are corresponding events due to the same fixed laws.

What Maxwell was expounding was the notion of small variations producing large consequences; and because such instabilities were incalculable, there were limits to the perfect predictability of a deterministic universe.[35] But he never confused limitations of knowledge with a proof for freedom of the will. Difficulty in determining outcomes was not the same as proving that the will is truly free, or proving that the universe is indeterminate as such. But in Maxwell's thinking the will *was* free; and it was so because it was a God-given entity in its own right, not imprisoned by physical causes and effects.

The 1878 Eranus Club Paper

Four years after Huxley's lecture, Maxwell shared more of his thinking in another Eranus Club paper 'Psychophysik' (February 1878), in which he discussed the nature of the human person, touching on issues of consciousness, memory, mood, and will.[36] Consistent with his previous position, he maintained that there was true free-will, and he rejected Emil Du Bois Reymond's Huxlean thesis that human behaviour was totally controlled by mechanical determinism.[37]

---

34 JCM, 'Does the Progress of Science', (C&G1, 441; Harman, II/2, 820).

35 Harman wrongly suggests Maxwell regarded activity on the molecular level as giving room for indeterminacy. (Harman, III/1, 15).

36 JCM, 'Psychophysik', Paper read to Eranus Club, 5 Feb 1878, (C&G1, 452–463; Harman, III/2, 598–607).

37 Ibid., (C&G1, 462; Harman, III/2, 606).

Du Bois Reymond (1818–1896) was the co-discoverer of nerve action potential and a developer of experimental electrophysiology. Maxwell rejected his hypothesis that human behaviour was merely the effect of mechanical or electrical stimuli. In his view, the 'will' was independent of these material phenomena and able to make truly free decisions.

Maxwell also took issue with Karl Wilhelm von Nägeli (1817–1891) of the University of Munich, who argued that human persons were not unique in terms of self-awareness. Nägeli's theory of consciousness proposed that perhaps every object, including molecules, had qualities of sensation and could experience feelings of pleasure and thought. Balfour Stewart and Peter Tait slammed Nägeli in their 1875 book *The Unseen Universe,* and although Maxwell did not agree with Stewart and Tait in everything, he was at one with them on this point. Maxwell stated that for Nägeli to attribute life, sensation, and thought to objects—even though these attributes had not been established by evidence—was nothing more than the 'good old figure of personification.'[38] Atoms did not think. Molecules did not think. And men and women did not think just because each of their atoms could think! The whole was more than the sum of the parts. The answer to the question 'How am I what I am?' had to be found elsewhere than in our component parts.[39]

Another whom Maxwell discussed was the Liverpool physician John Drysdale (1816–1892), categorised by Maxwell as a Christian materialist.[40] Maxwell was uncomfortable with splitting the human person into mind, body, and soul, since that overlooked the essential integrity of a person. As Maxwell saw it, Drysdale, in order to preserve his particular theory of life and mind, had found himself proposing that, because the soul or spirit of a person in this life cannot be identified by science, perhaps it came into being after

---

38 JCM, 'Psychophysik', (C&G1, 458; Harman, III/2, 603). JCM repeated the same thoughts (and words) in his 1878 review of Stewart and Tait's *Paradoxical Philosophy.* (JCM, 'Review of *Paradoxical Philosophy*': *Nature,* Vol. 19/477 (19 Dec 1878), 141–143).

39 Ibid., (C&G1, 460; Harman, III/2, 604.) Also discussed by Maxwell

40 Drysdale rejected Stewart and Tait's, *The Unseen Universe.* He also dismissed Maxwell's concerns about the size and number of gemmules required for mechanisms of heredity to operate.

death.[41] For Maxwell, this was woefully inadequate. He believed that the soul and will were intimately connected; they were active in this life; and they were evidenced by the ability of the will to make free decisions.

Maxwell was of the view that it may be impossible from within science to ever answer the question of human identity as free-agents. One reason was that the starting-point of consciousness—the proposition 'I am'—cannot 'be used in the same sense by any two of us; and, therefore, it can never become science at all'.[42] Such subjectivity made impossible the normal methodology of natural science which requires repeat experiments with identical outcomes.

Since Maxwell's era, psychoanalysts have delved further into questions of human *behaviour,* but the fundamental issues which Maxwell identified—*identity* and *will*—are as elusive as ever. Maxwell was a cautious and careful thinker. He was reluctant to conjecture. What he never did was to argue from the notion of a free-will to that of a non-determinist structure to the physical universe. Free-will was a gift of God to men and women (perhaps also to animals): the system of causation which applied in the physical universe was quite different.

41 JCM, 'Psychophysik', (C&G1, 463; Harman, III/2, 607).

42 Ibid.

Chapter 28
The Unseen Universe

In the spring of 1875, Balfour Stewart and Peter Tait published their jointly-authored book *The Unseen Universe: Or Physical Speculations on a Future State.* Stewart was professor of physics at Manchester and Tait held the chair of natural philosophy at Edinburgh. Both were respected scientists. Both held key positions in the academic world. Both were long-time friends of Maxwell. Maxwell had known Tait from his schooldays, and he had been acquainted with Stewart since the 1850s when Stewart assisted James Forbes at Edinburgh University. Maxwell and Stewart had also worked together on standardising measurements of electrical resistance.

Though the book was issued anonymously, its authorship was soon known.[1] Within its pages Stewart and Tait sought to demonstrate that there were men of science (themselves) who rejected a purely materialist interpretation of nature and of life, and who believed in the possibility of a future existence.

The book was a sensation. It was welcomed as a push-back against the growing materialist approach to science and life which had gathered momentum through the eighteenth and nineteenth centuries, and which had accelerated with Darwinism. In most church circles Stewart and Tait's work was embraced enthusiastically as a robust response to materialist presuppositions; even more so when it was confirmed that its authors were top level scientists who carried weight and authority. The drawback was that its argument was complex and convoluted. So Stewart and Tait followed up with a second book, *Paradoxical Philosophy*, adopting an easier-to-follow, story-telling, dialogue technique.

---

1 Knott, *Life and Scientific Work of Peter Guthrie Tait,* 237. The fourth edition of *The Unseen Universe* (April 1876) had the authors' names on the title-page. The one conspicuously new feature was an introduction setting out the motive of the book, which the authors felt had been misunderstood by some of the earlier critics.

The title, *Unseen Universe*, had biblical overtones. Though the word 'unseen' does not occur in the Authorised Version of the Bible, the concept does. The opening page of *The Unseen Universe* quoted 2 Corinthians 4:18: *'The things which are seen are temporal, but the things which are not seen are eternal'*. In context, the verse describes the new spiritual humanity awaiting believers beyond this life. Such 'unseen things' became incredibly important for Maxwell when he knew he was terminally ill. He talked to his parish minister, George Sturrock, about, 'the probable nearness to us, even in this life, of the unseen universe.'[2] He may have been reflecting on the text from 2 Corinthians, or he may have had his friends' book in mind, or both.

Maxwell and Tait communicated regularly, often discussing draft texts before publication. It was common for Tait's views on a range of subjects to have been chewed over by Maxwell before they appeared in print, and so he would have been aware of Tait's direction of thought even before *The Unseen Universe* was published. Though we do not know the identity of the author who reviewed *The Unseen Universe* for *Nature* in May 1875, we do know that when Stewart and Tait brought out their next volume, *Paradoxical Philosophy*, it was Maxwell who reviewed it as one of his last contributions in print.[3]

A draft copy of *Paradoxical Philosophy,* held by the National Library of Scotland, and marked throughout in pencil by Maxwell, displays not only Maxwell's proof-reading corrections to the text, but some of his comments on what Stewart and Tait had written.[4] The draft copy was posted to Glenlair in October 1878, from whence it was immediately forwarded to Maxwell in Cambridge. What he saw on page one when he opened the package, was an instruction, written above the printed text in heavy pencil: 'Keep it dark and return it!' The book and contents were to be secret until publication. Maxwell sent it back in time for most of his suggested textual

---

2 *Galloway News & Kirkcudbrightshire Advertiser,* 18 May 1883.

3 Anon, 'Review of *The Unseen Universe*', 41–43; & JCM, 'Review of *Paradoxical Philosophy*', 141–143.

4 NLS: Acc.10000/399, 'Balfour Stewart & P.G. Tait, *Paradoxical Philosophy* with annotations by JCM.'

amendments to be included in the first edition, which was on the shelves by mid-December 1878.

The Problem of Belfast

The catalyst for *The Unseen Universe* was John Tyndall's Belfast lecture. Even before that was delivered, Maxwell, Stewart and Tait were aware that scientific papers now had fewer references to the Deity, that natural theology had lost its traditional place in scientific thinking, that there was strong pressure to disconnect all religious notions from scientific enquiry, and that the trend was to exclude any cause apart from forces of nature. All three were present when Tyndall orated his ode to materialism. That Belfast gathering was the last which Maxwell attended, due to a combination of pressure of work at Cambridge and health problems for Katherine and himself.[5] His contribution had been a paper on electricity.[6]

As Maxwell, Stewart, and Tait listened to Tyndall, they were concerned that he not only advocated a purely materialist vision for science but pre-empted alternative scenarios by imposing strict limits to the questions science was allowed to consider. They saw that as wrong. In their view, Tyndall deliberately and unreasonably curtailed the scientist's freedom to explore *every* question connected with the natural sciences, including questions of ultimate origins. In their opinion, Tyndall was erecting an artificial barrier, blocking perfectly valid enquiry, and disallowing perfectly reasonable hypotheses. Stewart and Tait's primary aim, in writing *The Unseen Universe* was to show that was a perverse restriction.

After Belfast, Maxwell returned to Glenlair to work on his 'Atom' article for the *Encyclopaedia Britannica.* His own response to Tyndall's lecture was twofold. During the conference he dashed off some light-hearted and humorous verses which touched on the issues, and at Glenlair he addressed some of Tyndall's concerns in

5 JCM, *Letter to Stokes,* 12 Aug 1878, (Harman, III/2, 707): 'Mrs. Maxwell has not been strong this summer. She has got drives every day through all the fine weather but is still weak.'

6 JCM, 'On the Application of Kirchhoff's Rules for Electric Circuits to the Solution of a Geometrical Problem': *Report of the 44th Meeting of the BAAS* (1874), 18–19.

his 'Atom' article.[7] The poems were printed in Campbell and Garnett's *Life*.[8] It is difficult to quote just a few lines, since the light-hearted logic carries from verse to verse, but Maxwell mainly focussed on the problem of ultimate origins. Can something come from nothing? Can atoms explain intelligent life? And what about the problem of inherited characteristics? A taste of his wit comes towards the end of his poem 'The British Association: 1874':

> Thus the pure elementary atom,
> the unit of mass and of thought,
> By force of mere juxtaposition,
> to life and sensation is brought;
>
> So, down through untold generations,
> transmission of structureless germs
> Enables our race to inherit
> the thoughts of beasts, fishes, and worms.
>
> First, then, let us honour the atom,
> so lively, so wise, and so small;
> The atomists next let us praise,
> Epicurus, Lucretius, and all.
>
> Let us damn with faint praise Bishop Butler,
> in whom many atoms combined
> To form that remarkable structure,
> it pleased him to call his mind.

---

7 Cf. Harman, III/1, 15: 'In his [*Ency. Brit.*] article on 'Atom' Maxwell . . . repeated his view, first expounded in his famous lecture on 'Molecules' in 1873, that the creation of matter lay beyond the realm of scientific explanation. Science was concerned with the form in which matter exists and with the constitution of molecules, but the formation of atoms was not transparent to scientific investigation. Maxwell gave this argument a focus on the problems of materialism and scientific naturalism that flourished in the 1860s and 1870s, fostered by the publication of Darwin's theory of evolution by natural selection in 1859 . . . [but] he did not advocate a form of scientific obscurantism. He maintained that any harmonisation between science and religion was personal.' See also: Harman, *The Natural Philosophy of James Clerk Maxwell*, 199–208.

8 JCM, 'Molecular Evolution', 'Song of the Cub', and 'British Association 1874', (C&G1, 637–641).

Stella Pratt-Smith has proposed that some of Maxwell's Belfast poems expressed exasperated annoyance rather than humour.[9] Though even the benign Maxwell could be irritated, was he expressing anger in these verses? Pratt-Smith interpreted the two poems, 'Molecular Evolution' (1874) and 'Song of the Cub' (1874), as doing just that, arguing that four stanzas of 'Molecular Evolution' reveal the hurt and annoyance which Maxwell felt about the rejection of biblical explanations by scientists and intellectuals. She concluded that, in the 'Song of the Cub', Maxwell was expressing alarm concerning the effect of secularisation on the next generation of scientists, and she suggested that Maxwell was depicting the repercussions of flawed mentoring upon an increasingly disillusioned hypothetical young scientist, laying the blame on the 'jackal' Tyndall. Pratt-Smith's analysis is intriguing, but is weakened by her added verdict that Maxwell's reply in *Nature* to Stewart and Tait in his poem, 'A Paradoxical Ode (1878), was also an expression of ire and annoyance. It was not.

If Maxwell had a generally irenic relationship with Tyndall, that was not so for Tait. In 1873, a year before Belfast, there had been an acrimonious exchange between them in the pages of *Nature*. This sprang from Tait's review of Tyndall's pamphlet *Principal Forbes and his Biographers* which Tait interpreted as impugning the reputation of his distinguished Edinburgh predecessor.[10] Tyndall said it had not. Tait said it had. The atmosphere between them was decidedly unhealthy.

Personal issues aside, Stewart and Tait saw where things were heading if Tyndall were left unanswered, and they resolved to produce a substantial reply. As soon as they returned to their respective universities, they set to work. During the winter of 1874/75, it began to be whispered among Edinburgh students that Tait was engaged on a book which was to overthrow materialism by a purely scientific argument.[11] Given these expectations, when *The Unseen Universe* was released to the public in April 1875 some assumed it would give proofs—based on current scientific knowledge—for the spiritual, the unseen, the immortal, and God. In a later letter to *Nature*, Tait needed to explain that such was never the object of the book, and that the

9 Pratt-Smith, 'Boundaries of Perception', 253.

10 Tait, 'Tyndall and Forbes', *Nature*, Vol. 8/202, (11 Sep 1873), 381–382.

11 Knott, *Life and Scientific Work of Peter Guthrie Tait*, 236ff.

theory developed in its pages was never put forward as *probable*, but only as *possible* though 'not inconsistent with any part of our present knowledge.'[12] A similar declaration came in the Introduction to the fourth and subsequent editions: 'All that we have yet endeavoured to show is that the theory of a future life is not in any way whatever in contradiction to any ascertained facts or principles of science.'[13]

C.G. Knott identified two goals of Stewart and Tait's work.[14] One was to show that there was nothing in physical science which denied the *possibility* of human intelligences existing after death in another universe. The other was to argue that certain aspects of the modern theory of energy suggested—if not proving—the *probability* of such an Unseen Universe. This is important. Stewart and Tait were not using science (especially thermodynamics) to 'prove' the spiritual dimension. Their aim was to expose the inadequacies of the materialist view of science promoted by Huxley and Tyndall. For Stewart and Tait, materialism could not account for all phenomena; materialism should not exclude alternative explanations; and materialism was self-contradictory if the non-material were excluded.

The Principle of Continuity

The Unseen Universe was reviewed anonymously for *Nature* in May 1875. The reviewer began by stating that the book confirmed at least one thing: that scientific men were not necessarily unbelievers, since some accepted frankly and fully the whole scheme of trinitarian Christianity, finding in it the most adequate expression of their physics.[15] The reviewer explained that the authors' argument was built on what they termed the 'principle of continuity'. Stewart and Tait explained the principle at great length (Sections 62 to 77), citing the physical laws governing the motion of celestial bodies as a worked example. They stressed that adoption of the principle meant that scientists assumed that the universe operated according to laws; that

12 Tait, 'On the Formula of Evolution', *Nature*, Vol. 23/578 (25 Nov 1880), 81.

13 Stewart & Tait, *The Unseen Universe*', 10.

14 Knott, *The Life and Scientific Work of Peter Guthrie Tait*, 239f.

15 Anon, 'Review of *The Unseen Universe*', *Nature*, Vol. 12/290 (20 May 1875), 41–43.

these laws applied in a smooth and continuous manner; that there were no random exceptions to these laws; and that God had given the universe such continuity because he was not a devious or trick-inclined deity. In science, both theist and non-theist trusted this principle. For Stewart, Tait, and Maxwell, that trust had a theological basis, rooted in the actions of a God who was neither capricious nor misleading:

> It thus appears that, assuming the existence of a Supreme Governor of the universe, the principle of continuity may be said to be the definite expression in words of a trust that he will not put us to permanent intellectual confusion. (Sec. 77) . . . The power of the Divine Being is surely unlimited, but, nevertheless, we have perfect trust that God will work in such a way as not to put us to permanent intellectual confusion. (Sec. 81)

The principle of continuity was the principle of linear development. It was what geology had adopted when it put aside explanations based on sudden and unpredicable supernatural interventions in nature:

> We dig into the crust of the earth and find therein stratified deposits containing fossil forms: and we may *either* suppose that God created these as they are; or that they came into their place through the operation of natural forces, and represent the relics of an ancient world of life. The latter of these is undoubtedly the scientific hypothesis.The only other hypothesis is that of certain theologians of the Church of Rome who asserted that the devil put the fossils there. (Sect. 85. Italics ours)

For the joint-authors, the whole of natural science was based on the principle of continuity. Critically, that principle was their common-ground with materialists, and it was from that shared basis that Stewart and Tait developed their *reductio ad absurdam* argument that pure materialism was self-contradictory. Given that starting-point, we can understand why some critics thought that Stewart and Tait disallowed any possibility of God intervening in nature. But that was not the case. Their aim was to develop a *reductio ad absurdum* argument. They wanted to show—starting from agreed common-ground—that the purely materialist position of their ideological rivals was inadequate; that physical science itself showed the inadequacy of an exclusively materialist interpretation of the universe; and that nothing in science argued against the possibility of the existence of a future state after death.

One of the simplest summaries of Stewart and Tait's thesis is in a letter which they sent to *Nature* in January 1879, under the pseudonym of Professor Hermann Stoffkraft. This was after they had published their follow-up volume, *Paradoxical Philosophy*:

> The principle of continuity forbids us to imagine that the collocation called the atom has existed as it is from all eternity. This the authors of the *Unseen Universe* have insisted upon, and I need not go further than their title-page to remind Mr. Hallowes that in like manner they do not contemplate a future eternal existence for the atom. But this principle cannot tell us what was the exact nature of the thinkable antecedent of the present universe, nor can it tell us the exact nature of that state which will follow the disappearance of the present system. There are, however, *strong scientific analogies* which lead us to believe that the thinkable antecedent of the present system was a spiritual unseen, which not only developed but which now sustains the present order.[16]

Stewart and Tait's argument was built, in part, on Maxwell's 1873 'Discourse on Molecules', in which Maxwell contended that truly fundamental entities are now as they have always been though they were not eternal.

Stewart and Tait also argued that fundamental physical entities could not have existed eternally, since thermodynamic theory not only inferred a heat-death for the universe in the future, but implied the impossibility of the physical universe having had an infinite past. Thus, the fundamental entities of which the material universe was composed must have 'come into existence' at some point. But how? That question, they insisted, could only be answered—even by a materialist—by positing a non-material (spiritual) universe sustaining the visible material one. Moreover, an origin in the unseen should not be too difficult a concept to swallow, since science already accepted concepts such as electricity, magnetism, and the ether, all of which were *unseen*. There was already an unseen universe of some description within the physical natural order itself.

16 Stewart & Tait, *Letter to Editor, Nature,* Vol 19/481, (16 Jan 1879), 242. (Italics original).

Stewart and Tait insisted that not only was there an unseen *physical* universe, but there had to be an unseen *spiritual* universe. This was because the 'principle of continuity'—which allowed only natural causes to explain natural phenomena—would otherwise create a contradictory situation. In the absence of an infinite regression of natural physical causes (impossible because the material universe had a finite past), another type of cause (non-material/spiritual) was necessary for fundamental entities to begin to exist. They complained that the materialists did not want to consider the big issues, such as questions of origins with its allied problem of an impossible infinite regression.

In Stewart and Tait's view, the principle of continuity led to a paradox which was solvable only by the supposition of an *unseen universe* which could not be physical. Thus, the principle of continuity opened up the reasonable possibility of the spiritual, the immortal, and the divine.

Heat Loss and its Implications

Much of *The Unseen Universe* was concerned with consequences surrounding the continual and irreversible heat loss of the universe. The phenomenon of irreversible heat loss was widely accepted across science, among theists and non-theists alike. This raised two major questions. Where did the energy come from? Where did it go to? The first question was no problem for theists who believed in a Creator, but for non-theists it raised the problem of the apparent necessity of an infinity of energy in the universe's possibly infinite past. On the second question, Stewart and Tait attempted to harmonise the principle of continuity with the doctrine of immortality by considering the apparent waste of energy which, as in Clausius' theory, had become innaccessible. They hypothesised an invisible universe, which had perhaps developed out of another invisible universe, and so on *ad infinitum*.

On this point, the *Nature* reviewer highlighted their notion that our natural bodies are probably accompanied by 'a sort of invisible framework or spiritual body' which has a kind of connection with that unseen universe. The reviewer explained that the transition from the visible universe to the invisible was intelligible—at least to the authors—by the existence of the ether, 'a substance into which energy is continually being passed' and into which it is perpetually and finally absorbed, and that some beings, such as angels, may be capable of communicating between the visible and the invisible.

After expounding the book's main thesis, the reviewer reflected on several issues. First, Stewart and Tait took certain tenets as axiomatic. They assumed the existence of a Creator. They proposed that 'mind' was more than 'matter'. They held that the brain was more than chemical particles, with something superadded to its chemical and physical qualities. And they also believed that mind and consciousness pointed beyond the merely material. Second, though science assumed that the visible physical universe must one day become mere dead matter, in Stewart and Tait's opinion the principle of continuity demanded a dynamic continuance, though perhaps in a form beyond the visible. Third, on Darwinism, they accepted the possible development of new species, but alleged that the origin of life was left unexplained by Darwinism and that life may only be able to come from life: this directly contradicted Tyndall who, at Belfast, proposed that life could emerge from inert matter, basing it on his own interpretation of the continuity.[17]

Overall, Stewart and Tait envisaged a physical universe which was finite in extent, finite in duration, and with a beginning and an end; surrounding this was an invisible immaterial universe stretching infinitely forward and infinitely backwards, which was the forerunner and continuation of the visible.[18]

The reviewer conjectured that Stewart and Tait may have been guilty of unsubstantiated assumptions in stating that whatever energy

---

17 Tyndall, 'Presidential Address', *Report of the 44th Meeting of the BAAS*, (1874), xci–xciii.

18 Stewart & Tait, *The Unseen Universe*, Sects. 84, 85: 'We do not hesitate likewise to assert, that the visible universe must, *certainly in transformable energy, and probably in matter,* come to an end. . . . But the principle of continuity upon which all such arguments are based still demanding a continuance of the universe, we are forced to believe that there is something beyond that which is visible, or that, to use the words of an old writer . . . *'the things which are seen are temporal, but the things which are unseen are eternal.'* Looking back instead of forward—to the origin of this visible universe rather than to its end—we are brought to a similar conclusion. If the visible universe is all that exists, then the first abrupt manifestation of it is as truly a break of continuity as its final overthrow.'

perished in the visible universe passed into the invisible universe. Also, even if their model were correct, surely dissipation of energy had to continue in the invisible universe also, only delaying the inevitable end of all things? Moreover, if the *invisible* universe could be infinite, why could not the *visible* one be so, with an equally infinite store of energy? The reviewer concluded that Stewart and Tait's whole elaborate machinery of invisible universes, piled one on top of the other, 'seems to us to fall like a house of cards, if we can accept the eternal duration of an infinite by-sense-perceptible universe.' Maxwell could not have been the anonymous reviewer, since he would not have constructed an objection based on the 'eternal duration of an infinite by-sense-perceptible universe.' Maxwell did not believe in an eternal universe.

Stewart and Tait responded in a brief letter to *Nature*.[19] They accepted that the review was fair in many respects but protested they had not been fully understood, perhaps because their presentation needed improvement. Concerning the ether, they responded that the 'ether may be looked on as forming part of the invisible universe'. Concerning the possible infinity of the physical universe: they argued that the reviewer's thinking was wrong:

> Even allowing (which we are not disposed to allow) that the visible universe is infinite, this would not affect our argument against its past eternity. Our argument (see p. 127 of the book) is that *the dissipation of the energy of the visible universe proceeds,* pari passu, *with the aggregation of mass, and the very fact therefore that the large masses of the universe are of finite size is sufficient to assure us that the process cannot have been going on for ever.*

Based on the principle of continuity, the conservation of energy, and the phenomenon of irreversible heat loss, Stewart and Tait were satisfied that a purely materialist physics ran into insuperable problems. The elephant in the room for pure materialists was that there *had* to be a beginning, and there also *had* to be something which extended into the future beyond the heat-death of the universe. For Stewart and Tait, that something was the *unseen universe*. But, as Maxwell later pointed out when reviewing *Paradoxical Philosophy*, was that 'something' the *spiritual* and the *divine*?

19 Stewart & Tait, *Letter to the Editor, Nature*, Vol. 12/292 (27 May 1875), 66.

For Stewart and Tait, there was nothing to suggest that the soul did not survive death. As they saw it, the principle of continuity implied there could be no break in nature and, under a purely materialistic interpretation, this resulted in the paradox of infinite regressions. The only way around the paradox was to postulate that the material universe is surrounded by concentric, invisible universes of which *one must be non-physical*, since if *all* were physical then the problem of an infinite physical universe reappears.

They exploited the popular notion of a common religious consciousness of humanity, opening their book with an account of beliefs about the afterlife in ancient Egypt, Greece, Buddhism and Christianity.[20] They followed this preamble by setting out the scientific part of their argument in great detail, reflecting on the nature of matter, the concept of the ether, Newton's laws, and the idea that, through something vaguely analogous to electromagnetism, the soul upon death might transfer from the visible to the invisible universe.[21] The visible universe was only a portion of the whole, yet it had energy: and because energies were never lost then that gave a reasonable basis for immortality. One unfortunate and unintended consequence of their work was that some of their ideas were picked up by the late nineteenth-century fascination with psychic, occult, and spiritualist phenomena.[22]

Stewart and Tait aimed to show that the presuppositions of their opponents allowed and perhaps even necessitated notions of immortality. In spite of them protesting that they did not claim to 'prove' anything, the public read their book as an attempted proof.

A second edition was soon issued, with more in quick succession. All were hugely successful in terms of sales and public reaction. Alexander Fergusson, one of Tait and Maxwell's contemporaries at Edinburgh Academy, read *The Unseen Universe* with proud delight, reckoning its logic unanswerable.[23] He assumed that it harmonised

20 William Robertson Smith offered insights from an Old Testament and Middle Eastern perspective. Cf. Tait, *Letter to Smith,* 5 June 1875: Knott, *Life and Scientific Work of Peter Guthrie Tait,* 238.

21 Cf. P.M. Heimann, 'The "Unseen universe": Physics and the Philosophy of Nature in Victorian Britain', *The British Journal for the History of Science,* Vol. 6, No. 1 (Jun. 1972), 73-79, (Cambridge University Press).

22 Pratt-Smith, 'Boundaries of Perception', 255.

23 Fergusson, *Chronicles of the Cumming Club,* 204.

with Maxwell's position, given the many references to Maxwell in its pages, but in this he was not entirely correct. Despite its success, the book attracted critics from theology, philosophy, and science.

The Unseen Universe: The Theological Response.

The response within ecclesiastical circles was generally positive and one of relief. At last respected scientists had stood up for spiritual realities! Eminent churchmen quoted from the book in articles and sermons.

James Fraser, the popular anti-Darwinian Bishop of Manchester, referred at length to the views held by 'certain scientific men of the present day' when he addressed a fund-raising event in his diocese. The bishop held a copy of *The Unseen Universe* in his hand as he spoke and expressed satisfaction that a scientific book, written by scientific men, if not vindicating Christian doctrine—which could only be known by revelation—at least demonstrated that Christian faith was neither improbable nor absurd.[24] Stewart and Tait were delighted with his enthusiasm which they acknowledged in the preface of the second edition: 'The Bishop of Manchester has very clearly described our position by stating that (*from a purely physical point of view*) we contend for the possibility of immortality and of a personal God'.

Others worried that Stewart and Tait conceded too much. They welcomed Stewart and Tait's rejection of materialist philosophy but felt they discounted important Christian doctrines in doing so, and that they undervalued divine revelation.

One such critic was Robert Watts, who in 1874 had led the response from Belfast Presbytery against Tyndall and Huxley. Watts now turned his attention from alien-doubters (Tyndall and Huxley)

---

24 *County Advertiser & Herald for Staffordshire and Worcestershire*: 29 May 1875: '[The Bishop explained that] The writers said that the outcome of all their speculations led them to regard the Christian system as affording full scope for development in all respects, whether of the universe or the individual; that its law was pre-eminently that of liberty; and that the doctrine of the Trinity, or something analogous to it, formed, as it were, as avenues through which the universe itself led us to the conception of the infinite and eternal One. . . . [The Bishop] welcomed these collateral supports to our Christian faith.'

to fellow-believers (Stewart and Tait). In his opinion they muddied the waters by offering a flawed defence of Christian faith. In October 1877, Watts addressed a packed YMCA meeting in the Kinnaird Hall, Dundee.[25] He described *The Unseen Universe* as an important contribution, but one which made dangerous concessions and included unsafe principles. In his opinion, Stewart and Tait's use of the 'principle of continuity' resulted in them admitting that there could be 'no physical sequence which had not an immediate physical antecedent.' This, Watts argued, 'excluded the scriptural ideas of [direct divine intervention in] the creation, miracles, and the resurrection: or, while admitting these truths, virtually denied them.'

Watts conceded that *The Unseen Universe* did give important testimony to the necessary existence of God as opposed to principles laid down by the materialists, but he exhorted his listeners to examine the book's fundamental principles and not to be carried away by the names of its authors. He confessed that he had spoken strongly, but 'on the question of the resurrection no compromise could be accepted.'

Eighteen months later Watts was back in Scotland, again speaking on *The Unseen Universe*. This time in the Free South Church, Blairgowrie.[26] He explained that he did not reject natural theology, but it had to be conducted properly. For Watts, the starting point for any natural theology was the text: *'For the invisible things of [God] from the creation of the world are clearly seen, being understood by the things that are made, even his eternal power and Godhead.'* (Romans 1:20). He restated his delight that the authors of *The Unseen Universe* had 'rebuked' the materialists, but again singled out what he considered to be errors in their book.

Watts' critique had some validity. In pursuing a *reductio ad absurdum* argument, Stewart and Tait did concede ground, though only in order to establish an agreed starting-point with their opponents, and to show that, even from materialist presuppositions, a purely material universe was impossible, and that the prospect of immortality was not improbable. But this form of dialectic was easily misunderstood.

25 *Dundee Courier*, 5 Oct 1877.

26 *Dundee Evening Telegraph*, 29 Apr 1879.

Watts was wrong to say that Stewart and Tait 'would admit of no physical sequence which had not an immediate physical antecedent.' In fact, Stewart and Tait's overall argument was the exact opposite. The thrust of their book was that the postulation of a non-physical cause was absolutely necessary: this was in order to break free from a never-ending sequence of physical causes stretching back infinitely in time, which was a Pandora's Box of problems for pure materialists.

Where Watts was correct was in stating that Stewart and Tait posited a *spiritual* resurrection rather than a *physical* one. They interpreted New Testament references to physical resurrection as an accommodation to what the original readers of Scripture could understand.[27] Watts identified this as contradicting biblical teaching on bodily resurrection. Stewart and Tait had allowed their abstract concept of an unseen spiritual universe to dictate what the New Testament teaching on resurrection should be, forgetting to move on from their *reductio ad absurdum* mind-set to positive theology. On other issues such as the Trinity, they were more obviously orthodox.

The Unseen Universe: The Philosophical Response.

In November 1875, Henry Calderwood (1830–1897), professor of moral philosophy at Edinburgh University, gave his opening public lecture of the session, making *The Unseen Universe* the central feature of his address. Calderwood knew that one of its authors was Peter Tait, but did not refer to him directly. In a wry circumlocution, he merely suggested that there was every reason to suppose that the University of Edinburgh was 'not unrepresented' among its anonymous authors![28]

---

27 Stewart & Tait, *The Unseen Universe,* Introduction, 26ff.

28 *Edinburgh Evening News,* 2 Nov 1875. '[The professor described physicists' view of metaphysics as] a sort of limbo into which they could shoot all their own defunct theories.' Calderwood was not totally averse to what Stewart and Tait were trying to do. In his *The Philosophy of the Infinite,* (1854), he attacked Hamilton's assumption that there can be no knowledge of the Infinite. Calderwood maintained that such knowledge, though imperfect, was real and ever-increasing; and that faith implies knowledge. He argued that, beside the data of experience, the mind has pure activity of its own whereby it apprehends fundamental realities.

Calderwood outlined the book's arguments before criticising what he regarded as the authors' weak grasp of philosophy, warning his students of its metaphysical deficiencies. Calderwood was no atheist or philosophical materialist; neither was he an unsympathetic dry-as-dust academic. He was an ordained minister who had been involved in the Moody and Sankey evangelistic campaigns of 1873–1875, and was a popular speaker at student events.[29]

A similar verdict on Stewart and Tait's philosophical abilities was expressed in the *Pall Mall Gazette* in its lengthy review.[30] Like Calderwood, the *Gazette* reviewer questioned the authors' competence in metaphysics: 'If it would do much good to many dabblers in philosophy to learn some physics of Professor Tait, it would have done no harm to the authors of *The Unseen Universe* if they had taken some little survey of philosophy before they blew that exquisitely coloured bubble.' In the reviewer's opinion, Stewart and Tait wrote 'nonsense of the most ingenious kind.'

The *Gazette* lambasted the naïve treatment of profound philosophical issues, pointing out that, whereas Stewart and Tait would be the last men in the world to be caught out in any point of natural science, in their book they had made naïve metaphysical assumptions, and had 'committed themselves with the utmost innocence and simplicity to the snares of "common-sense", falsely so called.' The *Gazette* complained that Stewart and Tait were guilty of trying to settle 'offhand in a few lines' questions which serious students of philosophy knew were extremely difficult even to state properly, far less solve. Moreover, and irony of irony: 'In their laudable endeavours to provide man with a convenient spiritual body, they went very near to exposing themselves to the charge of gross materialism.' That was an interesting observation. Others, including Maxwell, concluded similarly. There was a suspicion that Stewart and Tait had turned the spiritual into the material rather than *vice versa*.

The *Gazette's* verdict was that Stewart and Tait had missed an opportunity: 'The *Unseen Universe* might conceivably have been

29 Cf. 'United Prayer for Scotland Statement', *Huntly Express*, 3 Jan 1874; G.A. Smith, *The Life of Henry Drummond*, 187.

30 *Pall Mall Gazette*, 26 June 1879.

worked out on really philosophic lines. As it is, it must remain a mere exercise in imaginary physics.'

The Unseen Universe: Maxwell's Response

The book was ignored by most scientific journals, apart from the influential *Nature* magazine which reviewed not only *The Unseen Universe,* but its follow-up, *Paradoxical Philosophy.*[31] Maxwell, who critiqued the latter publication, was widely read in philosophy, almost certainly more so than Stewart and Tait. He was aware of some of the book's philosophical weaknesses, though it may have been biblical convictions which made him uncomfortable with his friends' reasoning.

Maxwell was reviewing a work penned by two friends, and his sense of fun intruded from the start. He began playfully by asking if the book were at all a scientific text, and if a scientific journal was the place in which to discuss it? But, given that the book claimed tongue-in-cheek to be the proceedings of a 'learned society', and that Professor Herman Stoffkraft—the central character—purported to be a materialist scientist debating with an assortment of individuals in the 'Paradoxical Society', then he, Maxwell, would treat it as such.

Maxwell was hesitant about how Stewart and Tait had made a 'spiritual universe' a simple cause of the 'physical universe' as if the spiritual and the physical could both appear in the same causal framework. Philosophically, Maxwell had been taught by Sir William Hamilton that the finite could not grasp the infinite. And theologically, as alluded to in his *Encyclopaedia Britannica* article on Faraday, he believed that knowledge of the transcendent was only bridgeable by revelation and not by human reasoning.[32] As such, there could be no straightforward 'unending continuity' between the spiritual and the physical as Stewart and Tait implied. Maxwell shared Stewart and Tait's belief in God. He shared their conviction that there was more to life than material existence. He believed with them that there was a life beyond the grave. He was their friend. But he queried their argument.

---

31 JCM, 'Review of *Paradoxical Philosophy*', 141–142. For observations on Maxwell's reaction to *The Unseen Universe* and *Paradoxical Philosophy,* see: Marston, 'Maxwell, Faith and Physics', 258–291.

32 Cf. JCM, 'Michael Faraday', *Encyclopaedia Britannica,* (9th ed.), Vol. 9, 31.

What did Stewart and Tait mean by the 'unseen'? On their title page they quoted the Apostle Paul: *'The things which are seen are temporal, but the things which are not seen are eternal.'* (2 Corinthians 4:18). Maxwell knew that in the original biblical context, 'things not seen' referred to heavenly realities, but in Stewart and Tait's work the things 'not seen' (*unseen universe*) appeared to refer to something created, even if invisible and immaterial and beyond the immediate universe.

When Stewart and Tait had responded to the anonymous review of *The Unseen Universe* they had objected: 'May we not say that when energy is carried from matter into ether it is carried from the visible into the invisible?'[33] But that described energy moving from one form of created existence into another form of created existence, and it envisaged the *unseen* and the *invisible* both existing within the created order. Stewart and Tait's *unseen* involved no break-through to the divine or the heavenly, which was what the Apostle referred to when using similar terminology. That was a major concern for Maxwell. After musing on what Stewart and Tait might mean by the 'unseen universe' (was it the 'ether' or a 'still more subtle medium'?), he reproached them for referring to Scripture:

> What then is the Unseen to which they raise their thoughts? In the first place the luminiferous aether, the tremors of which are the dynamical equivalent of all the energy which has been lost by radiation from the various systems of grosser matter which it surrounds. In the second place a still more subtle medium, imagined by Sir William Thomson as possibly capable of furnishing an explanation of the properties of sensible bodies; on the hypothesis that they are built up of ring vortices set in motion by some supernatural power in a frictionless liquid: beyond which we are to suppose an indefinite succession of media, not hitherto imagined by anyone, each manifoldly more subtle than any of those preceding it. To exercise the mind in speculations on such media may be a most delightful employment for those who are intellectually fitted to indulge in it, though we cannot see why they should on that account appropriate the words of St Paul.[34]

33 Stewart & Tait, *Letter to 'Nature'*, Vol. 12/292 (27 May 1875), 66.

34 JCM, 'Review of *Paradoxical Philosophy*', 142.

Maxwell's perceptive comment was a stab to the heart of Stewart and Tait's argument. At the same time, and in their support, he expressed doubts as to whether modernist views of the soul had actually made the argument *against* immortality any stronger, as some materialists too readily assumed: 'No new discoveries can make the argument against the personal existence of man after death any stronger than it has appeared to be ever since men began to die; and no language can express it more forcibly than the words of the Psalmist, *"His breath goeth forth, he returneth to his earth; in that very day his thoughts perish"*.'

What Maxwell did welcome was that science had helped to clear away 'the haze of materialism which clung so long to men's notions about the soul'. Thankfully, it was no longer thought that the soul could be discovered through dissecting a body!

Maxwell's review of *Paradoxical Philosophy* was only one of his responses to the book. It also inspired his humorous poem: 'To Herman Stoffkraft, PhD: the Hero of a recent work called *Paradoxical Philosophy*' (1878), which picked up several themes discussed by Stewart and Tait.[35] In the following list the italicised phrases are taken from his poem:

> The issue of infinite regression: *'Long avenues of universes'*.
>
> The possibility of a finite universe: *'Klein and Clifford fill the void, with one finite, unbounded homaloid, and think the Infinite is now at last destroyed.'* (Clifford's notion of curved space and a universe which was finite but unbounded was taken up by Einstein).
>
> The problem of heat-death: *'When earth and sun are frozen clods; when, all its energy degraded, matter to ether shall have faded.'*
>
> Confidence in the principle of continuity: *'As I still in thee have trusted, so trusting let me never be confounded!'*
>
> Conviction that the chain of causes could be relied on: *'Oh never may direct Creation break in upon my contemplation; still may thy causal chain ascending, appear unbroken and unending.'*

35 JCM, 'To Herman Stoffkraft, PhD', (C&G1, 649–651).

> Allowance for divine intervention: *'While Residents in the Unseen—Aeons and Emanations—intervene, and from my shrinking soul the Unconditioned screen.'*

The poem climaxed by referring to the notion of the Unconditioned, introduced to Maxwell by Sir William Hamilton. The Unconditioned was the primary cause behind all other causes, but was itself produced or conditioned by no cause. Garnett and Campbell's *Life* footnotes additional words which Maxwell penned in relation to the Unconditioned:

> And where that chain is lost to sight
> Let viewless fancies guide my darkling flight,
> Through atom-haunted worlds in series infinite.[36]

The 'chain' was the linear chain of natural causes. Could that '*causal chain ascending*' ever reach the destination Stewart and Tait wanted? Following Hamilton, Maxwell questioned any argument from the finite to the infinite, from the caused to the uncaused, or from the created to the uncreated. As a 'Hamiltonian' he was perhaps more aware of the problem of the *unconditioned* than were Stewart and Tait. At the same time, as recorded by F.J.A. Hort, he sympathised with the concept of the 'unseen':

> [Maxwell's] strong sense of the vastness of the world . . . and of the partial nature of all human modes of apprehension, seemed to enlarge for him the domain of reasonable belief. Thus in later years it was a favourite thought of his that the relation of parts to wholes pervades the invisible no less than the visible world, and that beneath the individuality which accompanies our personal life there lies hidden a deeper community of being as well as of feeling and action.[37]

Maxwell knew that Stewart and Tait had a major problem: how could they move logically from this universe to another? If all they were arguing was that physics could not explain *this* universe without hinting at something else undergirding it, then that was valid. But, despite protestations about their limited goals, they seemed to claim more.

---

36 Ibid., (C&G1, 651).

37 F.J.A. Hort, *Letter to Campbell*, 4 Feb 1882, (C&G1, 419).

He agreed with some of Stewart and Tait's thesis: the origin of the visible universe by an intelligent agency; implications arising from atoms (fundamental entities) appearing to be 'manufactured articles'; and implications arising from the transformable energy of the visible universe coming to an end.[38] But he was unconvinced by the way they pushed the principle of continuity to 'prove' immortality through the 'necessity' of life continuing in the invisible universe. (Stewart and Tait would argue that if life only comes from life—as theistic scientists maintained, but which materialists disputed—then 'life' and 'consciousness' surely had to be immortal.) For Maxwell, the continuation of the personality beyond death—a major aim for Stewart and Tait—was outside the range of science.

> [Science] has added nothing of importance to what has always been known about the physical consequences of death, but has rather tended to deepen the distinction between the visible part, which perishes before our eyes, that which we are ourselves, and to show that this personality, with respect to its nature as well as to its destiny, lies quite beyond the range of science.[39]

Stewart and Tait responded to their friend's review by writing another letter to *Nature*:

> It is strange to see . . . Professor Clerk-Maxwell, falling into the same *errors of observation* as more ordinary mortals. Neither the authors of the *Unseen Universe*, nor any of the members of the Paradoxical Society, have, so far as I am aware, expressed the notion that the invisible order of things which continuity requires an antecedent to the visible order, is in any sense *material*. They only assume that it must be conditioned. Indeed, the authors of the *Unseen Universe* have expressed this conviction *in the preface to the second edition of their work, in italics, and in language that is not only exceedingly clear, but also extremely strong*.[40] (Italics ours)

Maxwell had known the authors for years. They worked together. They travelled together. They visited each other. They dined together. They discussed ideas together. Yet, despite deep sympathy with their aims, he did not think their argument was solid. They had not been

38 Marston, 'Maxwell, Faith and Physics', 277.

39 JCM, 'Review of *Paradoxical Philosophy*', 143.

40 Stewart & Tait, 'Letter to Editor', *Nature*, Vol. 19/479 (2 Jan 1879), 193f.

able to prove that a physical, material, entity was caused by a spiritual, immaterial, reality. As far as Maxwell was concerned, though the physical may 'cry out' for the spiritual, the spiritual cannot be proved deductively from it. Inductive reasoning might bridge the gap, but Stewart and Tait had not. Maxwell had learned well from Sir William Hamilton. Maxwell's spirituality did not rest in Stewart and Tait's speculations but in his trust in the Christian revelation. That faith soon became an existential issue.

Chapter 29
Evening Shadow

The summer of 1876 was the last which James and Katherine were able to enjoy before James' health issues were no longer ignorable: 'We have been at Glenlair all the summer, and intend to remain until about the 20th October when we return to Cambridge. We have been doing nothing but taking delicious drives and enjoying so great quiet that we hardly know what the busy world is like. Katherine has been improving in health during the summer.'[1] Apart from some family property issues which needed to be straightened out, life was good.

From the spring of 1877, colleagues and students began to notice problems. The initial symptom was a difficulty in swallowing, which James eased by taking bicarbonate of soda. By April 1879 self-medication was inadequate. He mentioned the problem to Dr George Paget (1809–1892), his family physician in Cambridge. Paget was Regius professor of physic, and, as a medical practitioner, had a brutally direct bedside manner. It was he who had given a candid assessment of Charles Mackenzie's life expectancy if he returned to Africa, telling Mackenzie's sister that any Life Assurance Company would estimate her brother's chance of survival at no more than two years![2] When consulted by James, the straight-talking Paget gave a frank prognosis. Despite this, the Maxwells stuck to their usual plans and went to Glenlair for the summer.

At the start of June, and before term finished at Cambridge, two beautiful stained-glass windows had been installed in Corsock church, one on either side of the pulpit. What the Maxwells saw on their return to Galloway was their simple country kirk more handsome than ever.[3] But if the kirk was in good health, both in terms of its outward

---

1 JCM, *Letter to R.D. Cay,* 29 Sep 1876, (Harman, III/1, 396).

2 Goodwin, *Memoir of Bishop Mackenzie,* 219.

3 The gift was acknowledged by the Corsock Kirk Trustees: 'These windows were allowed to have greatly improved the appearance of the Church and Mr. Barbour received a vote of thanks.' (Corsock Kirk Trustees, 11 Nov 1880, NRS/CH2/446/1.) Cf. *Galloway News and Kirkcudbrightshire Advertiser*

appearance and in terms of its membership, James was not. Over the next few months, friends and visitors to Glenlair marked a worsening in his condition. He wrote to George Stokes in August, expressing his hope that the Royal Society would not send him any more papers to review since, 'I could not do it.'[4] When the recently married William and Rebecca Garnett came as guests in September, they were alarmed at what they saw when James came to greet them. There had been a major deterioration. Nevertheless, that evening Maxwell conducted family worship as usual for the assembled household, and each day he tried to be his old self.[5]

Soon the only nourishment which he could take was milk. By October he was in considerable pain, despite the occasional encouraging sign. On the second of October, Dr William Rutherford Sanders (1828–1881), former fellow-student, friend, physician, and professor of pathology in Edinburgh, was summoned to Glenlair.[6] Sanders told James he had only a few weeks to live—perhaps six at the most—and if they intended to return to Cambridge they should do so as quickly as possible.[7] The abdominal cancer which had killed James' mother was now attacking him. Despite the weakness of his condition they journeyed south almost immediately. One source suggests they

(Friday 06 June 1879): 'Corsock Parish Church—Two very chaste and beautifully stained glass windows have been placed in this church, the one bearing the inscription, 'To the memory of James Barbour, Barley, died 1855,' and who was one of the original subscribers to the church. The windows are the gift of his son, the well-known and liberal laird of Barley and Dalshangan, who was the first trustee chosen by the male communicants, in terms of the constitution of the church and parish. On Sabbath last, at the close of the services, Mr Sturrock took occasion to say to his congregation that since they had last met together two windows had been placed before them—a memorial to a good man, who took an interest in the raising up of their house of God, and a memorial of which he was sure all must be proud, and with which they could not help being well pleased. It was the first thing of the kind in Corsock, perhaps in the immediate neighbourhood, but he trusted that it would not be the last.'

4 JCM, *Letter to Stokes,* 21 Aug 1879, (Harman, III/2, 833).

5 JCM, *Letter to Garnett,* 8 Aug 1879, (C&G1, 404; Harman, III/2, 832).

6 Robert Walker, 'James Clerk Maxwell', 210.

7 JCM, *Letter to Garnett,* 3 Oct 1879, (Harman, III/2, 854). Cf. fn. 4: Sanders, *Letter to Tait,* 5 Nov 1879, also footnote 5 (p. 855).

left during the week beginning the sixth of October, though James had hardly the strength to walk from the train to a carriage.[8] In Cambridge, Dr Paget did his best to alleviate the pain during the final weeks.

Around Glenlair it had become public knowledge that the gentle and likeable James Clerk Maxwell was terminally ill. The Irish physicist and mathematician Joseph Larmor (1857–1942) made frequent journeys from Ireland to England, with part of his route going through Galloway. As he returned by rail to Cambridge in October 1879, the train halted in Dalbeattie and a visibly upset country farmer came into the compartment. The farmer remained silent for a time, then remarked with emphasis as something that concerned the world to know: 'Clerk Maxwell has been taken away, mortally stricken: he will never come home again.'[9]

Dealing with Death

There are three accounts of James living with terminal illness. All from clergymen. Two were Scottish Presbyterians, evangelical in outlook. The third was Anglican with high church tendencies. All wanted to speak well of him, and their reports are read with that in mind.

The Revd George Sturrock. Writing a few years after Maxwell's death, Sturrock recalled the final time James attended Corsock Kirk, ten days or so before he and Katherine left for Cambridge. The sermon title was: '*Eye hath not seen nor ear heard*' (1 Cor 2:9), and the text which lay behind the sermon's title was: '*Eye hath not seen, nor ear heard, neither hath entered into the heart of man, the things God hath prepared for them that love Him. But God hath revealed them unto us by His Spirit; for the Spirit searcheth all things, yea, the deep things of God.*' Sturrock recalled Maxwell being more than usually attentive throughout the service.

The next Sunday, James was unable to attend church. Sturrock visited Glenlair that afternoon.[10] James was sitting in a chair, but experiencing considerable pain. Seeing this, Sturrock expressed

8 Riddell, *Clerk Maxwell of Glenlair,* 42. The final Sunday at the Kirk was probably 28 Sep 1879, with the journey on 8 Oct 1879.

9 Larmor, 'James Clerk Maxwell', *Nature,* Vol. 118, Suppl. 2981, (18 Dec 1926), 47.

10 Sturrock, *Corsock Parish Church,* 21.

concern that the Maxwells were set to travel to Cambridge only a few days later: 'You should not go away until you are a little better.' James replied: 'You know that there will be no betterness till the end comes; and I wish to leave my wife in her own house.' Lying behind this remark was the fact that Glenlair was not James' property outright. He and Katherine had no children, and the Middlebie and Glenlair estate would pass elsewhere in the family as soon as he died, whereas in Cambridge the house was leased in their own name. James' cousin Andrew Wedderburn (Aunt Isabella's son) became the new Laird of Glenlair by the end of the year.

A cushion and some spiritual books were left in Corsock Kirk. Pews can be uncomfortable, so James had his own flat cushion, which he rolled up and took back to the house each time he left for the south. He did not know that the final Sunday he attended Corsock Kirk was to be his last in the little church which he loved and which his father had helped to build. The cushion, plus his volumes with his personal jottings, lay in the Glenlair pew for years afterwards.

The Revd George Walker. James and Katherine were also called on by George Walker from nearby Castle Douglas.[11] Like several of Maxwell's friends, Walker was of a strongly evangelical disposition, and had been preacher at the Castle Douglas Mission House for two years before becoming minister of the full charge in 1874. After Campbell and Garnett's *Life of James Clerk Maxwell* was published, Walker reviewed it for the Church of Scotland's *Life and Work* magazine, with his article reprinted in the local *Galloway News & Kirkcudbrightshire Advertiser*.[12] His review incorporated some reminiscences, including an account of a conversation which Maxwell had had with George Sturrock.

Through his pain, James had talked to his minister about the probable nearness, even in this life, of the unseen universe—thoughts first expressed over a decade earlier in a letter to William Huggins.[13] On the 1879 occasion, as reported by Walker, James felt that the unseen universe was 'near indeed'. Compromised health meant he

11 Cf. *FES*, II, 403.

12 George Walker, 'James Clerk Maxwell', *Galloway News & Kirkcudbrightshire Advertiser*, 18 May 1883.

13 Cf. JCM, *Letter to W. Huggins*, 13 Oct 1868, (Harman, II/1, 451).

could only speak with difficulty, but he referred to the invisible world as encompassing everyone, of heaven bordering earth, of the human touching the divine, of the difficulty of fixing a boundary-line between them, of the darkness within ourselves, and of the fact that the light is very great and very near. He mused on the Spirit of God coming and dispelling darkness so that all becomes light, with God filling the soul with a flood of light, for what the eye cannot see is spiritually discerned.

Walker's memories of Maxwell were of a man for whom—in science as in religion—light was central. As a toddler at Glenlair, James delighted in his mirror flashing light all around the room. To his infant mind it brought the sun into the house. As a student, he was fascinated by light's qualities. As a mature scientist, he established principles relating to the electromagnetic spectrum, seeing light as but one segment of it. In his spiritual life, light was equally a central motif: Jesus was the light of the world; the Spirit of God created light out of darkness; and God filled the soul with a flood of light. Walker did not mention Maxwell speaking about the cross or his Saviour, though these may have been part of their intimate conversations. Significantly, Walker identified an emotional dimension to Maxwell's faith:

> He was too deep a thinker to be sceptical . . . the fact is that an emotional religion, such as is supplied by some of the leading features of our common Christianity is as absolutely essential to the comfort and satisfaction of a nature like Maxwell's as was his irrepressible tendency to attempt to find out the 'go' of things. As he wrote himself, 'It is a universal condition of the enjoyable that the mind must believe in the existence of a law, and yet have a mystery to move in.'[14]

Not only was there an emotional dimension to his faith, there was also one to his science, alluded to by Maxwell himself in a letter to William Huggins, where he wrote of, 'deep emotion sharpening the intellect'.[15]

---

14 George Walker, 'James Clerk Maxwell', *Galloway News & Kirkcudbright Advertiser,* 10 Nov 1882, quoting from: JCM, 'Has everything Beautiful in Art its original in Nature?' (1854), (C&G1, 231).

15 JCM, *Letter to W. Huggins,* 13 Oct 1868, (Harman, II/1, 451).

The Revd William Guillemard. During Maxwell's last month in Cambridge, William Guillemard visited with diligent regularity, giving support and spiritual ministry. Years later, Guillemard recalled his visits to 11 Scroope Terrace.[16] He remembered Maxwell as being very low physically, yet joining 'fervently in all acts of prayer' and listening, 'with a most intelligent interest to all I read, either out of the Bible (which he knew well-nigh by heart) or out of any of our great devotional writers in prose or poetry.'

Guillemard noted Maxwell being especially fond of new hymns, and of frequently reciting from memory favourite passages by authors such as George Herbert. Guillemard added that Maxwell's faith 'in the grand cardinal verities was firm, simple, and full; and he avowed it humbly but unhesitatingly, with the deepest gratitude for the revelation of the truth in Jesus.' For Guillemard, James was 'serenely resigned to the will of God and bowed in meek acquiescence before what he believed to be the Word of God.' James' main concern was for Katherine, with her own health problems.

In a letter to the Revd Isaac Bowman (1831–1907), written only a month after Maxwell's death, Guillemard said that his parishioner was fully aware of his condition from the first and knew what it meant: 'He welcomed me at once as visiting him, not only as a friend, but as the parish priest come to assist him and to minister to him; and spoke of our relations with a grave, simple cheerfulness.'[17] He commented that Maxwell was never downcast or overburdened, and was 'the humblest and most diffident of men, with the deepest sense of his own unworthiness, of his many short-comings, of his neglected opportunities.'

Some of Guillemard's words may have been conventional expressions when speaking of the respected departed (*de mortuis nil nisi bene*), but nothing of what we know of James Clerk Maxwell's character suggests they would be untrue.

Guillemard visited almost every day during that last month. Each time he read to him. Each time he prayed with him. He told Bowman that Maxwell knew by heart many of the set prayers of the Church of England, and that these and other devotional literature

16 Guillemard, *Letter to Campbell,* 19 May 1881, (C&G1, 414).

17 Guillemard, *Letter to the Rev Isaac Bowman,* 9 Dec 1879, (C&G1, 415-416).

were important to him. He was astonished at Maxwell's knowledge of the Bible and recalled him asking for his most deeply prized passages. On the Saturday before his death, he received Holy Communion with 'holy, reverent, fervid devotion, and said what strength it gave him.' While Guillemard was donning his surplice, Maxwell recited to him George Herbert's lines on the priest's vestments. The poem, called 'Aaron' and based on a believer's union with Christ, is redolent with significance for a Christian:

Holiness on the head,
Light and perfections on the breast,
Harmonious bells below, raising the dead
To lead them unto life and rest:
Thus are true Aarons drest.

Profaneness in my head,
Defects and darkness in my breast,
A noise of passions ringing me for dead
Unto a place where is no rest:
Poor priest, thus am I drest.

Only another head
I have, another heart and breast,
Another music, making live, not dead,
Without whom I could have no rest:
In Him I am well drest.

Christ is my only head,
My alone only heart and breast,
My only music, striking me e'en dead;
That to the old man I may rest,
And be in him new drest.

So holy in my head,
Perfect and light in my dear breast,
My doctrine tuned by Christ (who is not dead,
But lives in me while I do rest),
Come, people; Aaron's drest.

Guillemard saw him once more, though he was now too weak for sustained conversation. As Guillemard rose from kneeling in prayer, Maxwell was able to say: 'My dear friend, you have been a true under-shepherd to me. Read me, before you go, the beautiful prayer out of the Burial Service, Suffer us not at our last hour.' Guillemard did so, noting, 'his grasp of my hand, as we parted, told me all he felt':

> Man that is born of a woman, hath but a short time to live, and is full of misery. He cometh up, and is cut down, like a flower; he fleeth as it were a shadow, and never continueth in one stay. In the midst of life, we are in death.
>
> Of whom may we seek for succour, but of thee, O Lord, who for our sins art justly displeased? Yet, O Lord most mighty, O holy and most merciful Saviour, deliver us not into the bitter pains of eternal death.
>
> Thou knowest, Lord, the secrets of our hearts; shut not thy merciful ears unto our prayers; but spare us, Lord most holy, O God most mighty, O holy and most merciful Saviour, thou most worthy Judge eternal, suffer us not at our last hour for any pains of death to fall away from thee.[18]

Maxwell had been regular in his attendance at church, rarely failing to join in the monthly celebration of Holy Communion. Even so, Guillemard confessed that he knew little of his inner self before his illness: 'He was singularly reticent; and, though we occasionally discussed a text critically, we rarely got upon doctrine, or anything that touched upon the spiritual life.' In Guillemard's estimation, the illness 'drew out the whole heart and soul and spirit of the man.'

Despite their conversations rarely touching on doctrine, Guillemard stated that Maxwell had a 'firm and undoubting faith in the Incarnation and all its results' including the full sufficing of the atonement, and the work of the Holy Spirit. He reported that his parishioner 'had gauged and fathomed all the schemes and systems of philosophy and had found them utterly empty and unsatisfying: "unworkable" was his own word about them and he turned with simple faith to the Gospel of the Saviour.' This was Maxwell's faith as seen through Guillemard's lens.

As death neared, James reflected on his life: 'I have been thinking how very gently I have always been dealt with. I have never had a

18 Book of Common Prayer, 1662.

violent shove all my life. The only desire which I can have is like David to serve my own generation by the will of God, and then fall asleep.'[19] Another time, when prolonged conversation was impossible, and he had been lying with closed eyes, he looked up and repeated the verse: "Every good gift and every perfect gift is from above, and cometh down from the Father of lights, with whom is no variableness, neither shadow of turning" (James 1:17). He quoted Baxter's hymn frequently:

> Lord, it belongs not to my care,
> Whether I die or live;
> To love and serve Thee is my share,
> And that Thy grace must give.

Towards the end, he said to Colin Mackenzie: 'I have read up many queer religions: there is nothing like the old thing after all'; 'I have looked into most philosophical systems, and I have seen that none will work without a God.'[20] The substance of this statement was printed within days of James' death, in the obituary which Garnett submitted to Nature:

> About three weeks ago he remarked that he had examined every system of Atheism he could lay hands on, and had found, quite independently of any previous knowledge he had of the wants of men, that each system implied a God at the bottom to make it workable. He went on to say that he had been occupied in trying to gain truth, that it is but of truth that man can acquire, but it is something to 'know in whom we have believed.' His simple Christian faith gave him a peace too deep to be ruffled by bodily pain or external circumstances, and left his mind free to the last to contemplate all kinds of questions of general interest.[21]

Even the strongest believer can be anxious about death. Stella Pratt-Smith has suggested that Maxwell's poem 'A Paradoxical Ode' (1878), was composed when he was already feeling the symptoms of the cancer that would kill him, and that it reflected a degree of fearfulness about death and the breakdown of the partition between the physical

19 Quoted in: Hort, Letter to Campbell, 4 Feb 1882, (C&G1, 421).

20 C&G1, 426

21 Garnett, 'James Clerk Maxwell', Nature, Vol. 21/524 (13 Nov 1879), 45.

and non-physical worlds.[22] She cites the lines, '*Oh! Never may direct creation, break in upon my contemplation*', suggesting that the 'direct creation' may refer to James' physical body with its weaknesses. It is more likely that the words indicate problems with any purely material explanation of nature if God were to intrude the created order with an unexpected act of the divine will.

Final Journey

James Clerk Maxwell died at noon on Wednesday, the fifth of November 1879, aged forty-eight. It was the same age his mother had been when she died of the same illness. The following Monday, a service was held in Cambridge's Trinity Chapel.[23] The vice-chancellor of the university, nearly all the heads of houses and professors, a large number of Masters of Arts, plus numerous undergraduates and a few of the general public, assembled in the chapel. At a quarter to five the coffin was met at the Trinity Great Gate by the Revd V.H. Stanton as Senior Dean, the Revd W.F. Kirkpatrick as Junior Dean, and the Revd L. Borrison as precentor, along with the surpliced choir.

The procession entered the chapel with the choir singing the opening sentences of the funeral liturgy. The chief mourners were Colin Mackenzie, Professor Stokes, and Professor Paget, along with Mr Pullen who was the assistant in the Cavendish Laboratory. The coffin was placed on a bier in the middle of the floor, and the choir sang the Ninetieth Psalm: 'Lord, thou hast been our dwelling-place in all generations.' The Dean read 1 Corinthians chapter 15, plus the first part of the Anglican Burial Service. John Goss's anthem, 'For if we believe that Jesus died and rose again' was sung by the choir, after which a prayer was offered by the Senior Dean, and the service concluded. The procession re-formed, and the coffin was preceded to the Great Gate by the choir singing, 'Jesus lives! No longer now, can thy terrors, Death, appal me.' On the night air, the words echoed over the great court of Trinity. The coffin was placed in the hearse and taken to the railway station for conveyance to Scotland and Glenlair.

22 Pratt Smith, 'Boundaries of Perception: James Clerk Maxwell's Poetry of Self, Senses, and Science', 255f.

23 *Galloway News & Kirkcudbrightshire Advertiser*, 14 Nov 1879. Cf. Garnett, 'James Clerk Maxwell', Nature, Vol. 21/524 (13 Nov 1879), 44.

Katherine appreciated the service and the university's tribute. In a letter sent through Colin Mackenzie, she expressed her gratitude to the vice-chancellor for the respect shown to James' memory, and for the large and representative attendance of members of the university.[24] She died seven years later, on the twelfth of December 1886, also at 11 Scroope Terrace.

James' interment took place in the simple burial ground surrounding Parton Kirk which he had known as a boy. Amongst the few private friends and country gentlemen who attended was George Walker of Castle Douglas.[25] Though there was only a small and invited group, the formalities observed at the grave would echo those described by Sandy Bell in his *Reminiscences*. Attendance at parish funerals was a sacred duty with normally the entire male population turning out unless kept away by works of necessity or mercy. The church elders met the coffin at the churchyard gate. As they escorted it to the grave, they sang the one hundred and third psalm, especially the verse which likens the life of man to grass:

> O thou my soul, bless God the Lord; and all that in me is,
> Be stirred up his holy name, to magnify and bless.
> For he remembers we are dust, and he our frame well knows.
> Frail man, his days are like the grass, as flower in field he grows.

Bell described the psalm's tune, 'Coleshill', as a weird and melancholy melody, yet eminently suitable to the occasion and the surrounding scenery. We do not know if the reduced number who attended James' internment sang the psalm. They may well have. They probably did. Following the words of committal read by the minister, earth was shovelled into the grave by the local red-capped gravedigger. After a short, respectful pause, the chief mourner would say, 'Thank you gentlemen for your attendance' and all would slowly depart.

Appreciation

On Friday the seventh of November 1879, two days after Maxwell's death, George Sturrock addressed the Corsock Mutual Improvement

24 *Cambridge Independent Press*, 22 Nov 1879.

25 Walker, 'James Clerk Maxwell', *Galloway News & Kirkcudbrightshire Advertiser*, 18 May 1883.

Society. His subject was: 'Theology and Science.'[26] Had the topic been chosen beforehand? Was it prompted by Maxwell's death? We do not know. No manuscript record of the lecture survives, but the *Galloway News & Kirkcudbrightshire Advertiser* reported that the talk was listened to throughout with the most marked attention.

Sturrock gave the address in his role as the retiring president of the Society, with Dr Smith of Corsock Free Church occupying the chair, and introducing his brother of the established Church in 'very graceful terms'. Maxwell, with his large-hearted Christian faith, would have delighted in this fellowship across the Scottish ecclesiastical divide. Sturrock repeated the lecture two months later at the Mechanics Institute in Dalbeattie, and newspaper reports indicate he argued that no true science was subversive of the principles of theology: far from geology and kindred sciences being in opposition to divine revelation, they were in unison with it.[27]

Maxwell was fortunate in having had George Sturrock as his minister. Not only was he a capable pastor and preacher but he had a lively interest in the relationship between science and Christian faith. Sturrock's later booklet, *Corsock Parish Church: Its Rise and Progress*, refers to Maxwell's discoveries in electricity, light, and magnetism, especially the connecting of light and electricity. It describes Maxwell as a man who was in many ways childlike, but 'endowed in an eminent degree with a gift of scientific imagination which enabled him to penetrate deeply into regions which had been hitherto regarded as beyond the reach of science.'[28] Sturrock was also aware of Maxwell's work on gases, his imaginary 'demons', and his research on colour which laid the basis for colour photography. He knew the calibre of the man who sat under his ministry Sabbath by Sabbath.

Sturrock became the moving spirit behind a permanent local memorial to his friend's genius, which resulted in the impressive Maxwell window for Corsock Kirk. In 1947 it was this window, along with a plaque in memory of James' father, which was transferred to what had been the Corsock Free Kirk building. By that time the two denominations had reunited locally, and the one-time Free Kirk building was chosen as the one for public worship.

---

26 *Galloway News & Kirkcudbrightshire Advertiser*, 14 Nov 1879.

27 Ibid., 3 Jan 1880.

28 Sturrock, *Corsock Parish Church*, 20.

A month after Maxwell's death, Cambridge's principal academics met in Trinity College to discuss how best to honour their distinguished colleague. All referred to Maxwell in the warmest terms. Professor Stokes, a close friend, spoke eloquently. He reminded his colleagues that their late professor had been a man of the highest scientific eminence, so well known throughout Europe that it was not necessary to dwell upon that aspect of his life.[29] Stokes focussed instead on Maxwell's generous disposition, his retiring habits, his humility, and his simplicity of character. Stokes said he had hardly ever known a man who so completely effaced himself as Maxwell had. He was a man who never pushed himself forward, but took the liveliest interest in the work of others, as the students of Cavendish Laboratory well knew. He alluded to Maxwell's faith, citing: 'the reverence which the late professor retained for divine things, and the retention of his faith therein in spite of the free speculations in which he engaged.'

After more tributes, the meeting agreed to commission a marble bust, to be created by Sir J.E. Boehm, which would be placed in the apparatus room of the Cavendish Laboratory. It was also decided to publish Maxwell's numerous but scattered scientific writings. That task fell to William Niven, Maxwell's associate. A portrait of Maxwell by Lowes Dickinson now also hangs in the hall of Trinity College.

In 1931, J.J. Thomson unveiled memorials to both Faraday and Maxwell in Westminster Abbey. In 1956, a memorial panel to Maxwell was placed in the Mitchell Hall of Marischal College, Aberdeen, with the Aberdeen bronze depicting Maxwell as a young man of under thirty as he had been when a professor in the city. It includes the sign for Saturn, referencing his prize-winning essay on Saturn's rings, plus the formula dp/dt=JCM which was used by friends as a pun for his name.

In 2006, the United Kingdom Parliament in London and the Scottish Parliament in Edinburgh both marked the one hundred and seventy-fifth anniversary of Maxwell's birth. A special session of the

29 *Cambridge Independent Press*, 13 Dec 1879. On the 28th October 1879, Paget, Stokes, and Mackenzie were added by Maxwell as executors of his will, in order to save Katherine distress. She had been sole executrix. Stokes was president of the Victoria Institute in 1886; gave the 1891 Gifford Lectures; and was also vice-president of the British and Foreign Bible Society

Scottish Parliament adopted a motion submitted by Alex Fergusson, Member of the Scottish Parliament for the constituency of Galloway and Upper Nithsdale, which includes Parton and Corsock:

> That the Parliament acknowledges the 175th anniversary of the birth of James Clerk Maxwell on 13 June 2006; recognises his great achievement in discovering the nature of electromagnetic waves which opened the way to the invention of television, radio, radar and the mobile phone; applauds his work on colour perception which enabled the successful development of colour television and colour photography, and believes that he is worthy of greater recognition throughout Scotland, given the acknowledgement of Albert Einstein, who said that 'the special theory of relativity owes its origins to Maxwell's equations of the electromagnetic field', and of Ivan Tolstoy, who wrote 'Maxwell's importance in the history of scientific thought is comparable to Einstein's (whom he inspired) and to Newton's (whose influence he curtailed)'.

In 2008 a statue was finally raised to him in Edinburgh. It is in the city's George Street, next to his boyhood church of St Andrew's, and near the headquarters of the Royal Society of Edinburgh. The location is perfect. For James Clerk Maxwell, his faith, his church and his physics were never far apart.

Reflection and Appraisal

Chapter 30
Science and Faith

Maxwell's theology was mainstream Christian doctrine, set within an evangelical framework. From that context he pursued his science as a natural extension to his faith, seeking to understand God's handiwork. This is expressed in his two prayers which are printed in the opening pages of this book. The same thought is conveyed in his choice of words for the main entrance to the Cavendish Laboratory: '*Magna Opera Domini Exquisita in Omnes Voluntates*' (Psalm 111 verse 2). On the original Laboratory the text was in Latin, but when the new Laboratory was built the text was accompanied by Myles Coverdale's sixteenth-century translation: '*The works of the Lord are great; sought out of all them that have pleasure therein*'.

Christian faith *demanded* science. The image of God in men and women *necessitated* that humanity study the works of God. In return, science *required* religious faith since, for Maxwell, the universe was inexplicable without God. His faith and science were interwoven to such a degree that F.J.A. Hort made the observation that Maxwell was free from the practical dualism between faith and science so often found with academics, and that, for him, every subject had its affinities with the rest of universal truth.[1]

God was the ultimate originator, fashioner, and sustainer of all things. But Maxwell was no deist. The scientist's task was to seek the God-given *natural* cause for each phenomenon, and the natural world had no sense apart from God, with the origins and structure of matter inexplicable without him. The *existence* of the universe was contingent on the *action* of God. And the *logic* of the universe was contingent on the *mind* of God. This philosophical theism intertwined with Maxwell's evangelical faith, prompting his parish minister to describe him as a man of the deepest and the purest piety and of very broad Christian sympathies.[2]

1 Hort, *Letter to Campbell*, 4 Feb 1882, (C&G1, 419).

2 Sturrock, 'Letter to Editor', *Glasgow Herald*, 29 May 1885.

McNatt criticised Maxwell's moderate evangelicalism and Maxwell's weak response to contemporary religious issues on which he should have taken a stronger and more vocal stance.[3] McNatt accepted that Maxwell had an unswerving trust in Christ's atonement and love, but felt that he too readily 'embraced what he judged as fruitful thought by all manner of theologians and sceptics, and included them in his circle of friends.' But McNatt was placing Maxwell in the twenty-first century battlefield between evangelicals and liberals. Maxwell was not of that era. Nor was he of that disposition. He was a man who had to explore every idea, every theory. He had to read what different theologians put forward. In doing so, he followed in his faith the same principle which he followed in his science: of reading every relevant publication possible, good or bad, in order to know what was being thought, and in order to clarify his own position. No ground, holy or otherwise, was to be left unexamined. And that wide reading and extensive dialogue with divergent opinions never undermined the evangelical piety and conservative theology at the heart of his faith, nor did it undercut his trust in Holy Scripture.

At an early age, Maxwell accepted the principles of modern geology, following other Christians such as Thomas Chalmers and Hugh Miller. In adult life, he was unwilling to reject theories of evolution out of hand though he had misgivings, and though he questioned the specific mechanism of natural selection put forward by Darwin. But his faith was such that he approached these issues without fear. George Walker stated that it was Maxwell's 'unqualified testimony' that 'nothing in the very highest results of physical research had, to his mind, militated against the Christian faith.'[4]

New Days, New Science

Maxwell lived in a rapidly changing world. Matthew Stanley's book, *Huxley's Church and Maxwell's Demon,* notes that the Victorian period saw science shifting from a religious context to a naturalistic one, with the origins of the shift much earlier. But was naturalistic science really distinct from, or better than, science carried out by theists? Stanley

3 McNatt, 'James Clerk Maxwell's Refusal to Join the Victoria Institute', 208.

4 George Walker, 'James Clerk Maxwell', *Galloway News & Kirkcudbright Advertiser,* 18 May 1883.

suggests that most of the methodological values underlying scientific practice were in fact identical for theists and non-theists alike. They agreed on the importance of the uniformity of natural laws. They agreed on the use of hypothesis and theory. They agreed on the moral value of science. They agreed on intellectual freedom.

Maxwell was as convinced as a Huxley or a Tyndall that natural phenomena could and should always be explained by natural causes. That was how God had made the world. For Maxwell, the search for natural causes was a *consequence* of belief in a Creator, and not in *opposition* to it. It was precisely because the Creator had given the universe its own logic and structural coherence that each part of nature was linked to every other part. In Maxwell's mind, that dovetailed with his conviction that the whole universe, day by day, was contingent on the power and presence of the living God who, through Christ, had created a physical universe capable of infinite scientific research. There was no foreseeable end to the chain of discoverable natural causes, something which Maxwell embraced, precisely because of his belief that God had given nature an integrity of its own.

When science was understood in this sense, then, for Maxwell, there were reasons aplenty for having no disjunction between the materialist's approach and the theist's approach. They were each simply following the *scientific* approach. On some issues there was scarcely a beam of light between Maxwell on the one hand, and Huxley or Tyndall on the other. Like them, Maxwell held that the integrity and inner consistency of the natural world meant that natural causes could always be found for natural phenomena. Unlike them, he believed this was because God had given the universe its being and its logic. And on the centenary of Maxwell's appointment as a professor at King's College, London, Sir John Randall commented that Maxwell was entranced by the harmony of nature as created by God, that he admired the wonderful coincidence of the laws of nature, and that he marvelled at the human capability of being able to perceive them.[5]

Stanley explores the overlap and shift between theistic and naturalistic science through focussing on Maxwell and Huxley. In Stanley's opinion, when the naturalists gained control of major scientific institutions they were able to reimagine—even rewrite—the

---

5 Randall, 'Aspects of the life and work of James Clerk Maxwell', *Clerk Maxwell and Modern Science*, 25.

history of their discipline, giving the impression that they had liberated science from constraints imposed by religion. But there was no revolution. Ironically, it was the *similarity* between the science conducted by theists and the science pursued by materialists which allowed a smooth transition from the old to the new. The difference was that Maxwell believed that the unity of the created world precluded it being eternal and self-existent, and that it could not be made 'by any of the processes we call natural.'[6] He also insisted that science could not ignore philosophical and theological issues. The nature of the universe posed ultimate questions which had to be addressed.

Unlike positivists, Maxwell saw reality as a unitary whole.[7] In order to get to grips with reality, he sought to understand the *parts* from the *whole,* rather than attempting to integrate the *whole* from the *parts.* Torrance has insisted that the very best enquirers intuitively grasp the whole and understand the parts from it: that is the *actual* methodology of scientific advancement, though, when discoveries are written up and published, the reported methodology can give the impression of the enquirer having progressed from the parts to the whole. In both science and theology, an intuitive grasping of the 'whole' is what enables enquirers to progress beyond simply listing itemised phenomena, to being able to put forward all-embracing theories.

When Peter Tait wrote about his friend for the *Encyclopaedia Britannica*, his observation was: 'In private life Clerk Maxwell was one of the most lovable of men, a sincere and unostentatious Christian.'[8] Yet Tait also wrote of how Maxwell rebuffed any who would shortcut the scientific search: 'Though perfectly free from any trace of envy or ill-will, he yet showed on fit occasion his contempt for that pseudo-science which seeks for the applause of the ignorant by professing to reduce the whole system of the universe to a fortuitous sequence of uncaused events'.

Reality had order. Reality had structure. Reality had a logic. The task of the scientist was to patiently shed light on the nature of reality, aware of its immense depth. It was not scientific to resort to quick

6 JCM, 'Discourse on Molecules', (C&G1, 359).

7 Gaal, *The Faith of a Scientist,* 10.

8 Tait, 'James Clerk Maxwell', *Encyclopaedia Britannica,* (9th ed.) Vol. 15, 647.

spectacular solutions involving the sudden intervention of a *Deus ex Machina,* when that was not how God had done things.

Science never stays still. Maxwell knew that. Sir Edward V. Appleton (1892–1965) one-time principal and vice-chancellor of Edinburgh University, after whom the Appleton Layer in the upper atmosphere is named, has pointed out that Maxwell's electromagnetic theory required considerable revision when dealing with radiations of extremely high frequency, and when in the realm of quantum theory.[9] Since Appleton, the complexities of modern quantum theory have proved the truth of his words even more. But Maxwell's work, developed within the context of his faith, remains foundational for modern science.

The Kindness of Faith

If any of Maxwell's contemporaries had been asked what stood out in their friend's religion, their response may not have focussed on his intellectual assent to credal doctrines—important though these were to him—but on his manner of life, especially his kindliness.[10] Throughout his life, he asked ethical questions: What is the Good? How can I do the Good?

Part of his dissatisfaction with a purely materialist interpretation of life was that it gave no objective basis for a notion of the Good. Materialism seemed to regard the world as little more than the fortuitous arrangement of atoms, viewed humanity as simply the highest form of the evolution of life, and regarded justice as only a social convention. For Maxwell the Good was real: it was God. He sought to live the Good through Christian precepts as he understood them. That living was achieved in union with Christ, through Christ's life within him. On one occasion he stated: 'What is done by what is called myself is, I feel, done by something greater than myself in me.'[11]

In an early letter to Katherine he discussed Galatians 5, reflecting on the Pauline fruit of the Spirit, especially *gentleness*. Their Bible was the Authorised Version which translated the fifth quality of the fruit

9 Appleton, 'Radio and the Ionosphere', *Clerk Maxwell and Modern Science,* 88.

10 C&G1, 170. 175.

11 JCM, as recalled by Hort, *Letter to Campbell,* 4 Feb 1882, (C&G1, 421).

of the Spirit as *gentleness*, whereas some modern translations translate it as *kindness*. Significantly, Maxwell's comment on *gentleness/kindness* interpreted that virtue as an *active* rather than a *passive* quality in the Christian life. As such, it was a positive force which *actively* enabled a Christian to express love, joy, peace, long-suffering, and patience to a high degree. To be gentle or kind was to be initiative-taking and practical in expressing love and care to others. It was in the exercise of this kindness that Maxwell lived out the highest qualities of his own faith. That was what friends noticed, and their homage was not polite convention. It was genuine.

There was kindness to friends and family members when they needed help. There was Charles Robertson, to whom he read textbooks when Robertson's eyes troubled him. There was Robert Pomeroy whom he tended during illness. There was care given to his own father. There was thoughtfulness to Donald Dewar, his brother-in-law. Above all, there was devoted help for Katherine, who was equally solicitous in caring for him. On one occasion, when nursing her during an illness, he did not sleep in a bed for three weeks yet conducted his lectures and other work at the laboratory as usual.[12]

That gentle kindness, issuing from an open-hearted Christian faith, made him the person he was. His mother and father had been like that. And his own experience of receiving warm and affectionate kindness from the Tayler family in Otley when he was critically ill, had prompted him to reflect on how he lived his life. Kindness involved treating others gently, and he lived out that kindness in his scientific life as much as in his church life. He was the opposite of his older Scottish compatriot, the abrasive Sir David Brewster, who was hewn from a quite different rock. No human being is perfect. Maxwell was well capable of expressing annoyance and frustration. But he disliked factionalism, and he disliked confrontation.

Maxwell had an optimistic view of humanity, sidestepping some of the traditional Calvinistic emphasis on the depravity of the human condition. His approach was more in line with the philosophy of Frances Hutcheson who highlighted the goodness of men and women. In an 1882 letter to Lewis Campbell, H.J.A. Hort recalled Maxwell, near his death, expressing such sentiments. Hort remembered him

12 C&G1, 372.

apologising (unnecessarily) for being interested more in things than in people, and saying words to the effect:

> My interest in things has always made me care much more for theology than for anthropology; states of the will only puzzle me. *I cannot ascribe so much to a depraved will as some people do, though I do to a certain extent believe in it.* Much wrong-doing seems to be no more than not doing the right thing; and that finite beings should fail in that does not seem to need the supposition of a depraved will.[13] (Italics ours)

Maxwell's remark was occasioned by Hort saying to him that, though the immediate cause of the miseries of the world is oftener folly rather than wickedness, yet men's folly can frequently be traced back to past misdoing on their part. Maxwell concurred, but added the thought: 'They were foolish because they did not ask for wisdom: not, of course, absolute wisdom, but the wisdom needed for the moment.'

It is a kindly view of humanity which comes through in these words, especially in Maxwell's statement that he could not ascribe so much to a depraved will as some did, though to a certain extent he believed it. His own testimony in the same letter, that he had never had 'a violent shove in all my life', spoke volumes, given the dark days he had experienced. His mother's death affected him deeply. So did his father's. So had Katherine's illnesses. At various times he expressed this openly. Yet, in times of loss he had been helped immeasurably by the care of family and friends. They gave him security, affection, and belonging. As his Aunt Isabella said when he married: 'James has lived hitherto at the gate of heaven.'[14]

Lewis Campbell wrote of his friend as having had a strong sense of humour; a keen relish for witty or jocose repartee expressed not so much by outright laughter as by a twinkle and brightness of the eyes; a serene and placid temper; a patience when others might become vexed or annoyed; and a calmness amidst the vicissitudes of life.[15] Undiluted

---

13 JCM, as recalled by Hort, *Letter to Campbell,* 4 Feb 1882, (C&G1, 421).

14 C&G1, 426.

15 Ibid., 368. Also: 'The leading note of Maxwell's character is a grand simplicity. But in attempting to analyse it we find a complex of qualities which exist separately in smaller men. Extraordinary gentleness is combined with keen penetration, wonderful activity with a no less wonderful repose, personal humility and modesty with intellectual scorn . . . his tenderness

hagiography is always suspicious, yet William Garnett, who observed him at close quarters day by day over several years at the Cavendish Laboratory, shared the following from his direct experience:

> Such complete unselfishness and tender consideration as he exhibited for those around him, and especially for those under his control, are seldom to be met with. During the eight years that he held the chair of Physics in Cambridge, he never spoke a hasty word, even to his attendants. His self-sacrificing devotion to those he loved was the marvel of his friends.[16]

F.J.A. Hort wrote to his eldest daughter, four days after Maxwell's death: 'He was one of the greatest men living, and, I believe it is not too much to say, one of the best; his rare powers being united to a high type of pure and simple Christian character.'[17]

for all living things was deep and instinctive; from earliest childhood he could not hurt a fly. Not less instinctive was the sense of equality amongst all human beings which underlay the plainness of his address. But, on the other hand, his respect for the actual order of the world and for the wisdom of the past, was at least as steadfast as his faith in progress. While fearless in speculation, he was strongly conservative in practice.' (425).

16 Garnett, 'James Clerk Maxwell', *Nature*, Vol. 21/524 (13 Nov 1879), 45.

17 Hort, *Letter to Eldest Daughter*, 9 Nov 1879: Hort, *Life and Letters of Fenton John Anthony Hort*, Vol. 2, 276.

Chapter 31
Physics and Theology

In T.F. Torrance's lecture room, it was his assertion that Maxwell's theology had enabled Maxwell to think radically in his physics which made me sit up. That was a 'Wow!' moment. But was Torrance correct? To answer this question, we need to take a journey.

Neither Maxwell nor Torrance asked if science could prove God.[1] Both were convinced theists. Both took God as a given. Their common aim was to demonstrate that the tenets of Christian faith helped to make sense of the universe, and provided a logic for all scientific thinking. If science was then impelled to look beyond itself and cry out for God, so be it. Science might also aid theology because, even though theological truth has its own root and foundation, science could sometimes give it 'a more graspable formulation through the use of real analogies discovered in nature.'[2] Maxwell and Torrance's primary focus was on how Christian faith could enrich the scientific enterprise. They saw that as the best apologetic.

Sir John Randall wrote that Maxwell's culture was *spiritual*, in contrast to William Thomson's which was *material*.[3] What he had in mind was not Maxwell's theism as such, but Maxwell's readiness to move away from an atomistic view of reality—governed by mechanical interactions between bodies—to something more fluid. Maxwell was open to the possibility that fields of force rather than material points constituted reality, and that such fields might need to be understood in terms of a continuum rather than in

---

1 Torrance, 'The Significance of John Philoponos', 52: 'Philoponos would have nothing to do with any attempt to argue from the nature of the world to the existence of God, for that not only assumes that there is a logical bridge between ideas and reality, but would mean that God is necessarily related to the world. And that, in turn, would imply that the world was not freely created by him out of nothing.'

2 Torrance, 'Christian Faith and Physical Science', 222.

3 Randall, 'Aspects of Maxwell's Life and Work', *Clerk Maxwell and Modern Science*, 25.

terms of objects and particles, all of which Thomson regarded as mysticism. Randall may also have thought that because Maxwell had a belief in non-material reality in one area of life (faith), then he was open to the notion of non-material reality in another area (physics). But to what extent did Maxwell's faith shape his mind-set? Was it a factor in making him receptive to new views of matter, energy, and the universe? Was that how Torrance interpreted him? Was he correct?

Torrance strove to interlock the methodology of natural science and the methodology of theology. Though his award-winning *Theological Science* (1969)—based on his Hewett Lectures of a decade earlier—had no explicit references to Maxwell, his publications after that date, beginning with *Space, Time, and Incarnation* (1969) and *God and Rationality* (1971), cited Maxwell frequently. References to Maxwell increased through the 1970s and 1980s, especially in: *The Ground and Grammar of Theology* (1980); *Christian Theology and Scientific Culture* (1980); *Divine and Contingent Order* (1981); *The Mediation of Christ* (1983) and *The Christian Frame of Mind* (1985).

In Torrance's thinking the ground and grammar of theology was the trinitarian nature of God as expressed in his being and acts, and Torrance wanted the whole of theology to be set within the logic of that trinitarian life of God. As a student he had read Schleiermacher's *The Christian Faith*, but felt something was missing. He decided that the problem with Schleiermacher was that the logic of the triune God was not integrated deep into Schleiermachar's theology; and if, as Torrance believed, proper scientific method in any field of knowledge involves knowing something *according to its nature,* then that lack of a profound integration was a fundamental error. If the Christian Church believed in God as triune, then triunity had to permeate and mould every part of its theology.

Torrance respected Maxwell tremendously. Not only did he see Maxwell as someone who advanced a relational view of space and time, over against the Newtonian container-box model, but he admired Maxwell's ability to break away from some of the crippling dualisms which, in Torrance's view, had afflicted western thinking for centuries. Torrance identified several dualisms which, in his opinion, had infected Christian theology. These included: the

Platonic separation between the *sensible* world and the *intelligible* world; a disjunction between action and reflection, event and idea, becoming and being; and a division between the material and the spiritual, the visible and the invisible, the temporal and the eternal.[4] He saw such dualisms as flowing from the Ptolemaic cosmology which dominated European thought for more than a millennium.

Over against these dualisms, Torrance insisted that entities should be appreciated in their relationship to one another and not as disconnected objects. He regarded relationality as an indispensable quality of true scientific thinking, appealing to Maxwell who had encouraged physics to see reality as an interconnected continuum.

1979 was the centenary of Maxwell's death. Though the Royal Society of Edinburgh had no major commemorative conference on Maxwell that year, it did invite Professor R.V. Jones of Aberdeen to speak on him at its November meeting. Due to unforeseen circumstances Jones was unable to attend in person, but his paper was subsequently published by the RSE.[5] That same year, and coinciding with his retiral from the chair of Christian Dogmatics in the University of Edinburgh, Torrance was elected an FRSE. A year later, in October 1980, Torrance cited Maxwell repeatedly when he delivered the Queen's University 'Lectures in Theology' in Belfast, later published as *Christian Theology and Scientific Culture.*

From 1980 to 1981, Torrance worked on his Warfield Lectures on Nicene theology, which were to mark sixteen-hundred years since the Council of Constantinople in 381. It was that Council which had issued an authoritative version of the Nicene Creed, detailing the trinitarian and incarnational nature of Christian faith. The Warfield Lectures were delivered in 1981 at Princeton Theological Seminary, New Jersey, and later published as *The Trinitarian Faith.*

Torrance was studying Maxwell at the same time as he was preparing the Warfield Lectures, and there is evidence of cross-fertilisation between his two areas of research. For example, in *The Trinitarian Faith* Torrance used the phrase 'action at a distance' when

4 Torrance, *The Trinitarian Faith*, 47.

5 Cf. Jones, 'The Complete Physicist: James Clerk Maxwell, 1831–1879'.

discussing trinitarian relationships.[6] That phrase came straight from the nineteenth-century scientific debates in which Maxwell was involved. Torrance was borrowing terminology normally reserved for physics, and introducing into theology a way of thinking parallel to that which was required in Field Theory.

In December 1981 Torrance reviewed Ivan Tolstoy's *James Clerk Maxwell: A Biography*, for the Church of Scotland's monthly magazine, commenting that every church minister should read this book again and again, as the world was now dominated by the consequences of Maxwell's discoveries.[7] Torrance was referring to the new modes of thought demanded by Maxwell's work, and not only to the technologies which derived from it.

In January 1982, Torrance spoke on Maxwell at a Symposium in Paderborn, West Germany. The title of this major paper was: 'Christian Faith and Physical Science in the Thought of James Clerk Maxwell.'[8] Torrance's Paderborn paper is weighty and detailed, and it is easy to be swept away by the eloquence of his presentation. But did Torrance attribute to Maxwell ideas which Maxwell never had? Did Torrance's determination to make connections between Maxwell's faith and Maxwell's science over-ride the evidence? We consider this below.

Later in 1982, Torrance republished Maxwell's famous 1864/65 paper 'A Dynamical Theory of the Electromagnetic Field', contributing an extensive introduction in which he repeated some of his Paderborn lecture, and in which he explained why—in his view—Maxwell's mode of thinking was critical for both science and theology. In October 1982, Torrance delivered the Didsbury Lectures at the British Isles Nazarene College in Manchester, later published as *The Mediation of Christ* (1983), in which he again made important references to Maxwell.[9]

6 Torrance, *The Trinitarian Faith,* 191.

7 Torrance, 'Review of Ivan Tolstoy's *James Clerk Maxwell: A Biography*', *Life and Work,* (December 1981), 42.

8 Torrance, 'Christian Faith and Physical Science', 215–242.

9 In the 1982 Didsbury Lectures, Torrance built on material delivered in previous lecture series, e.g. at Princeton in 1981.

In 1987, Torrance focussed on Maxwell in a sermon, 'The Theology of Light', preached at St Mary the Virgin Church, Oxford.[10] This complemented the sermons he preached in Corsock in 1979 and in Parton in 1989. In 2000, he published a paper comparing Maxwell with the sixth-century philosopher John Philoponos (c.490–c.570). The paper had the title: 'The Significance and Importance of John Philoponos as a Forerunner of James Clerk Maxwell'.

Other publications could be cited, all of which show that Torrance researched Maxwell in depth. He consulted Campbell and Garnett's *Life*. He worked through Niven's early official edition of Maxwell's collected *Scientific Papers*. He read Maxwell's articles for the ninth edition of the *Encyclopaedia Britannica;* and he accessed papers published for the 1979 centenary. To all of this, he brought the fruit of his vast reading not only in theology but in the philosophy of science, especially in the writings of Albert Einstein and Michael Polanyi.

The Interplay of Science and Theology

Torrance wanted to highlight some of the overlooked instances of science and theology enhancing rather than hindering one another. It has become a cliché that religious ideas stalled the advance of science, with superstition and allegiance to outmoded ways of interpreting the universe acting as a dead weight to progress. The medieval adherence to Ptolemy's geocentric model of the solar system is a case in point. And Torrance was aware of even subtler negative influences at play. He argued that the medieval doctrine of God as Unmoved Mover adulterated the biblical concept of God the free Creator, since it fused the biblical notion of God with aspects of Neo-Platonism and Aristotelianism to produce a static and dualistic relation between God and the universe. In Torrance's view, this defective synthesis was then wedded with Newton's mechanistic and deterministic universe. Over against this, Torrance highlighted the Jewish-Hebraic concept of the living God who continually interacts with the created order, with

10 Torrance, 'The Theology of Light', (8 Feb 1987) published in: Torrance, *The Christian Frame of Mind: Reason, Order, and Openness in Theology and Natural Science*, 147-155.

the universe in a contingent relationship to its Creator. He firmly believed that the new conceptual thinking which Maxwell found he had to adopt when developing Field Theory, enabled both scientists and theologians to start afresh, and make a breach in the 'hard mechanistic structure of Newtonian science.'[11]

However, in contrast to some of the well-known adverse effects of theology on science, Torrance was also keen to cite occasions when, in his view, theological thinking had been a positive creative catalyst in advancing scientific thought. He gave two examples.

First, he maintained that the Judaeo-Christian conviction of God creating the universe out of truly nothing (*creatio ex nihilo*) had enabled science to proceed with confidence, since the doctrine of *creatio ex nihilo* included the assertion that God brought order out of chaos, with belief in the inherent orderliness of the natural world a *sine qua non* for any scientific work. As far as Torrance was concerned, science had been so spectacularly successful because it was built on this conviction, which came from faith in God. Second, Torrance argued that the Church Fathers' trinitarian and incarnational theology had made them receptive to a radically new way of seeing the world in terms of space and time. The doctrines of the Trinity and the Incarnation had required their minds to be open to a relational way of thinking, long before that approach was adopted by moderns such as Maxwell or Einstein.[12]

In Torrance's opinion, the Church Fathers were aided by a Stoic notion of space as something dynamic, and as something which had been forged by an active agency through events.This was in contrast to Greek thought with its static container-box notions of time and space. And, for Torrance, the Church Fathers were open to this different way of seeing things precisely because their thinking had been reshaped by their faith, particularly by the doctrine of the Incarnation which forced them to think through how God created space-time for himself within human flesh. Torrance believed that it was their reflection on the Incarnation, and of how that could be,

11 Torrance, 'Christian Faith and Physical Science', 219.

12 Torrance, *Space, Time, and Incarnation,* 10, 13, 58. Also, Torrance, 'The Significance of John Philoponos', 53f.

which forced them to conceptualise reality in a totally new fashion.[13] On this basis, Torrance claimed that some of the fundamental ideas of modern natural science—which have come to the fore since General Relativity—went back to roots in Christian theology, as theologians, from the fourth to sixth century, thought through the relation between the Incarnation and the creation, and in so doing had to reconstruct their understanding of the foundations of ancient science and culture.[14] Torrance saw parallels to that process in Maxwell's science, and argued that concepts in Maxwell's theology had contributed to Maxwell being able to think differently in his science.

Torrance admired much in Maxwell, especially his conviction that things could be known as they are in themselves, even though the route might be long and circuitous, and even though, at any given time, knowledge is always in need of revision and refinement.[15]

Torrance particularly appreciated Maxwell's conviction that the processes of the mind (logic) could be brought into harmony with the rational structure of nature.[16] Maxwell was never satisfied with mere 'data' but, in Torrance's words, sought a 'theologically and philosophically empowered grasp of things in their natural connectedness and their natural wholeness'.[17] Whereas Maxwell's

---

13 Cf. Torrance, *Incarnation,* 218: '[The Fathers] thought of God in creation as "containing all things by the Word of his power" and as contained by nothing, and of God coming to be with us in the realm of space and time which he created in and with the creation of the universe, yet in such a way that he did not cease to be the creator transcendent over all space and time.'

14 Torrance, *The Ground and Grammar of Theology,* 7. Cf. Torrance, *The Mediation of Christ,* 59: 'When we find a relational way of thinking coming back at us from modern science, it must not be regarded as irrelevant or strange, *for it has long roots in the history of Christian theology to which, under God, it calls us back.'* (Italics ours).

15 Luoma, *Incarnation and Physics,* 108: 'Torrance maintains that Maxwell's view of the universe involved the idea of an inherent intelligibility of its own, which a human inquirer cannot create or project onto reality. This is a manifestation of the contingency of the universe so that it exists under no necessary condition but freely and dynamically.'

16 Torrance, 'Christian Faith and Physical Science', 220. Cf. JCM, 'Analogies in Nature', 1856, (C&G1, 240; Harman, I, 380).

17 Torrance, ibid., 217.

mentor, Sir William Hamilton, saw the mind proceeding from the *parts* to the *whole*, Maxwell's scientific method involved grasping the *whole*, and then understanding the *parts* from it.[18] We recollect Torrance, in his lectures, arguing that whatever scientists claimed to be their *modus operandi*, the history of science threw up numerous instances of researchers intuitively grasping the *whole* before understanding the *parts* and their inter-relation. There is certainly some evidence that was also how Maxwell operated. In his *Treatise on Electricity and Magnetism* (1873), Maxwell wrote approvingly of Faraday:

> Faraday's methods resembled those in which we begin with the *whole* and arrive at the *parts* by analysis, while the ordinary mathematical methods were founded on the principle of beginning with the parts and building up the whole by synthesis. I also found that several of the most fertile methods of research discovered by the mathematicians could be expressed much better in terms of ideas derived from Faraday than in their original form.[19]

Onto-Relations and Personhood

Crucial to Torrance's understanding of Maxwell was the notion of *onto-relations*. Einstein saw Maxwell as initiating a new way of thinking about physical reality, in which reality needed to be thought of in terms of continuous fields rather than in terms of particles and objects. For Maxwell, this field continuum was as much part of an entity as the visible object; and because the field continuum involved an entity's relation to other entities, then entities were not only constituted by the 'stuff' of which they each separately 'consist', but by their *relations*. Nature did not comprise of individual, atomistic, particles: instead, it was object and relation together which constituted an entity.

Torrance termed such relations as *onto-relations*, by which he meant that the *being* (Greek: *ontos*) of a thing is only what it is because of these relations.[20] Relations are constitutive of the *being* of

18 Cf. Gaal, *The Faith of a Scientist*, 17ff.

19 JCM, *Treatise on Electricity and Magnetism*, x, xi. (Italics ours). Cf. Gaal, *The Faith of a Scientist*, 32. Also: Olson, *Scottish Philosophy and British Physics, 1750–1880*, 293f.

20 Cf. Torrance, 'Christian Faith and Physical Science', 230.

a thing. Relations are not *extrinsic* or *accidental* to the fundamental nature of being but are *intrinsic* and *essential*: an entity exists, and only exists, as part of a continuum in a state of dynamic interaction. And this *onto-relational* way of thinking about reality, demanded by the field continuum concept, dovetailed with Torrance's theology of personhood and God.

Torrance was aware that the concept of a field continuum, as a model of the universe, was radically different from what Newton had envisaged.[21] The Newtonian universe had individual objects which either touched one another or had clear gaps between them; when these objects moved about or collided they obeyed Newton's three laws of motion.[22] But light, electricity, and magnetism did not behave like that.

Field Theory offered an alternative solution. It posited a continuum involving lines of force extending out from any and every object, which lines of force interacted with similar entities. Thus, instead of matter being a collection of separate atomistic objects, matter was now envisioned quite differently, with 'objects' regarded as but intense instances of the continuum. Reality was not an accumulation of individual objects complete in themselves—which occasionally interacted with each other—but was a *continuum* in which relationships between 'objects' were part of the being and reality of the 'objects' themselves.[23] In other words, objects only existed in these relationships, and the relationships were 'part' of

21 Cf. ibid., 231, where he describes how Maxwell was presented with a new outlook by Faraday.

22 Maxwell emphasised that Newton deliberately refrained from *explaining* the mechanism by which heavenly bodies act on each other: Newton simply worked out the mathematical laws appropriate to such motions. JCM, 'On Action at a Distance', 48. Cf. JCM, *Lecture on Faraday' Lines of Force,* 1873, (Harman, II/2, 790–811, esp. 798).

23 Torrance, 'Christian Faith and Physical Science', 231f, points out that Maxwell was aware that Faraday never considered bodies as existing with nothing between them but their distance, but conceived of all space as a field of force; and that Faraday even spoke of lines of force belonging to a body as in some sense part of itself. Torrance, Faraday, and Maxwell saw 'bodies' as converging points of force rather than discrete entities in motion in empty space and time.

the object. Torrance wrote: 'The relations between particles in a field of force must be thought of as constituting, in part at least, what particles actually are.'[24] These relations belonged to the nature of the reality itself. Thus they were *onto-relations* or *being-constituting-relations,* with the *being* of a thing having no independent existence apart from such relations. *Being* exists, and only exists, within a dynamic continuum of relations.

Torrance saw Maxwell as a scientist who was increasingly convinced that *relation* was the most important thing to know.[25] For Torrance, this perfectly suited Christian theology which understands God as Father, Son, and Holy Spirit, with each divine person having *being* within, and only within, the continuum of relations between the persons. The persons of the divine Trinity only exist in their relation to one another. Trinitarian doctrine, as much as physics, is governed by *onto-relations* or *person-constituting-relations.*

Torrance saw theology as having made a gigantic error when it attempted to first define the persons of the Trinity and then—after defining each individual person—tried to relate them to one another. This mistake stemmed from a dualistic mind-set in which objects were defined *apart from the continuum* in which they exist.

Torrance argued that Maxwell's new way of seeing the natural world underscored a universal truth—which should be present in any proper theology—that the nature of being can only be understood aright when relations are themselves seen as constitutive of being. Thus, when theologians consider God as Father, Son, and Holy Spirit, they should not start with 'units of being' (the three individual persons of the Trinity) and then try and work out how these can exist in relation to one another. Rather, they should begin by seeing God as intrinsically relational, with relationality constituting the nature of divine being. It was here Torrance thought that Maxwell carried trinitarian-type thinking into his physics.

At the 1982 Paderborn Symposium, Torrance touched on problems introduced by the Christian philosopher, Ancius Boethius (480–524). Using Aristotle's logic of categories, Boethius had

---

24 Torrance, Introduction, *James Clerk Maxwell: The Dynamical Theory of the Electromagnetic Field,* 25.

25 Torrance, 'Christian Faith and Physical Science', 230.

defined 'person' as an *Individual Substance of Rational Nature.*[26] This compact definition was adopted by many (though not all) medieval philosophers and theologians, with personhood viewed as essentially individual.[27] For Torrance, this way of thinking was the theological equivalent of Newtonian mechanics: first, we have objects; second, we try to work out laws governing how these objects inter-relate. But in such a model the relations become accidental and awkward add-ons, rather than *person-constituting-realities.* Unsurprisingly, in his treatise *On the Trinity,* Boethius was forced to argue that triunity should not be predicated *substantially* of God but only *relatively,* hoping that subtle use of Aristotle's Categories might eliminate any logical problems associated with the doctrine of the Trinity.[28] In Torrance's view, this was the wrong way round. Christian theology needed the equivalent of Field Theory.

---

26 Boethius, *A Treatise Against Eutyches and Nestorius,* 3. (*Persona est rationalis naturae individua substantia*). The problem is how, under this definition, how do we relate the persons of the Trinity to one another whilst retaining the oneness and unity of God? The Council of Florence (1439) adopted a solution proposed by Anselm of Canterbury three centuries earlier: *Unity does not lose its consequence unless some opposition of relation stands in the way.* That is: each inner-trinitarian relationship has symmetrical equivalence unless there are specific truths of revelation to the contrary. However, this was predicated on a Neo-Platonist notion of divine perfection and of there being *no unrealised potentialities in God.*

27 Torrance, *Theological Science,* 306, lists honourable exceptions: Richard of St. Victor, Duns Scotus, John Major, and John Calvin. Torrance, 'Christian Faith and Physical Science in the Thought of JCM', 230, maintains that Boethius' view was cemented into western philosophy by Locke and Comte who saw persons as 'separated individuals connected through external relations'. Over against this, Torrance argues that in the Reformed tradition, there was an appreciation that the activity of God in the Incarnation is *person-constituting.* As such, 'relations between persons have ontological force and are part of what persons are as persons.' Torrance argues that this underlay James Clerk Maxwell's concept of union with Christ.

28 Boethius, *On the Trinity,* IV. Boethius made the unity of God the primary reference point. Cf. Fortman, *The Triune God: A Historical Study of the Doctrine of the Trinity,* 164.

As we have seen, Torrance interpreted Maxwell as teaching that 'the relations between particles in a field of force must be thought of as constituting, in part at least, what particles actually are.'[29] In Torrance's mind, this linked with a proper understanding of personhood, as something constituted within a matrix of relations: 'The relations which persons have with one another as persons are *onto-relations*, for they are *person-constituting-relations*.'[30] Relations do not *enhance* persons: instead, without relations personhood does not exist, since personhood exists and only exists within relations:

> That was a concept and a way of thinking developed through the understanding of the Holy Trinity as a Communion of Love in whom Father, Son and Holy Spirit mutually involve and coinhere in the profound onto-relations of that Communion, without any blurring of their hypostatic distinctions or properties as Father, Son and Holy Spirit which would make them no more than modal aspects in an undifferentiated oneness of divine Being.[31]

This was why Maxwell's dynamic view of matter excited Torrance, especially as a theologian for whom the doctrine of the Trinity was the ground and grammar of theology. The three persons of Father, Son, and Holy Spirit, have their being only in dynamic relation to the others, and none of the divine persons can exist in themselves or by themselves as atomistic units of divinity. Onto-relations are constitutive of being within the triune Godhead, which itself can be described as a divine continuum involving Father, Son, and Holy Spirit.

Towards the close of his Paderborn lecture, Torrance went further. He posited not only *person-constituting-relations* but *person-constituting-acts*. This was what he saw Maxwell's ideas leading towards, especially in connection with the Incarnation. Jesus Christ is who he is in his act, and his act is who he is in his being. The two

---

29 Torrance, Introduction, *James Clerk Maxwell: The Dynamical Theory of the Electromagnetic Field*, 25.

30 Torrance, *The Mediation of Christ*, 59.

31 Ibid.

are inextricably conjoined.[32] God is who he is because of his dynamic mode of being involving Father, Son, and Holy Spirit; and this not only involves the life of God within Himself (*ad intra*) but the life of God in relation to the created order (*ad extra*). To put it another way: onto-relations *in dynamic activity*, involving the being and acts of God, must be taken into account when understanding God and his revelation in Jesus Christ.

Torrance believed that this dynamic and relational way of thinking was starting to impact theology. Eleven years before Paderborn, he had expressed satisfaction that modes of thought akin to those employed in Field Theory had at last began to play a role in modern theology despite some false starts having been made: 'What is needed is advance to a higher level of thought in which a unitary theology comprises the two historic approaches to the being of God as discerned in his acts, and to the acts of God as they are discerned issuing from his being.'[33]

The Regulative effect of Faith on Physics

Given the above, was it the case—as Torrance appeared to imply in his class lectures—that Maxwell was so familiar with the notion of onto-relations in the doctrines of the Trinity and the Incarnation that this aided him when he wrestled with the concept of an electromagnetic field?

There seem to be conflicting statements in Torrance's written works. First, in the published version of Torrance's Paderborn paper of January 1982, he declared:

> There is no evidence to show that Clerk Maxwell directly allowed his Christian beliefs in the creation and the Incarnation to influence his understanding of space and time, in the way that these cardinal doctrines affected the mind of the early Greek Fathers who first put

---

32 Cf. Torrance, *The Christian Doctrine of God,* 4: 'The Being of God and the Activity of God in loving may not be separated. God's Being is not some abstract impersonal essence, but dynamic personal Being, for God is who he is in the Act of his revelation, and his Act is what it is in his Being, which Karl Barth used to speak of as "Being in Person".'

33 Torrance, *God and Rationality,* 132. Cf. Torrance, 'The Significance of John Philoponos', 56: '[We need] a recovery of what Karl Barth called the God who acts; the God whose Being and Act are dynamically one and the same.'

> forward the relational notions of space and time, with which we work in science as in theology today.[34]

This restated what Torrance had expressed in his December 1981 review of Ivan Tolstoy's biography of Maxwell.[35] However, nine months after Paderborn, in his Manchester Didsbury Lectures, did Torrance claim more? A *prima facie* reading of the Didsbury Lectures suggests he did. The lectures appear to indicate that Maxwell *consciously* applied theological modes of thought to his physics. This requires examination. To begin with, we note that, at Manchester, Torrance restated his understanding of onto-relations in a theological setting:

> The relations which persons have with one another as persons are onto-relations, for they are person-constituting relations. That was a concept and a way of thinking developed through the understanding of the Holy Trinity as a Communion of Love in whom Father, Son and Holy Spirit mutually involve and coinhere in the profound onto-relations of that Communion, without any blurring of their hypostatic distinctions or properties as Father, Son and Holy Spirit which would make them no more than modal aspects in an undifferentiated oneness of divine Being.[36]

Torrance then linked onto-relational thinking in Maxwell's faith to similar thinking in his physics:

> It is highly significant that it was an onto-relational way of thinking of [a trinitarian] kind *to which James Clerk Maxwell appealed* in the decisive change he initiated in the scientific conception of the structure of the physical universe. *Under the regulative force of his Christian beliefs in the nature of God's creative activity as discerned from the perspective of the Incarnation and in the concept of person,* he made use of being-constituting relations in his development of a dynamical theory of the electromagnetic field, thereby rejecting a mechanical explanation of it and breaking away from the Newtonian tradition in a mechanistic and deterministic conception of nature.[37] (Italics ours)

34 Torrance, 'Christian Faith and Physical Science', 236.

35 Torrance, 'Review of Ivan Tolstoy's *James Clerk Maxwell: A Biography*', 42: 'Clerk Maxwell did not intrude specific theological concepts into his physics, but he brought to his physics a mind profoundly and intuitively geared to God as he has revealed himself in the face of Jesus Christ.'

36 Torrance, *The Mediation of Christ*, 59.

37 Ibid.

In this extract Torrance wrote that Maxwell's faith exercised a *regulative force*. Additionally, Torrance's use of the word *'appealed'* would seem to imply that Maxwell consciously interwove ideas from his faith into his science, since 'appealed' normally carries the sense of an active thought rather than of a subconscious influence.

Was Torrance being more bullish than at Paderborn? Was he revealing his true opinion? Did he believe that the conceptual thinking demanded by Maxwell's faith in the Incarnation and the Trinity—and involving *being-constituting-relations*—made Maxwell conscious of a way of thinking which he could then apply to the natural world in the form of a continuum of being? Was there a modification of Torrance's stance from Paderborn to Manchester? Did his hearers affect his presentation? Paderborn had a relatively secular audience, whereas the Didsbury lectures were given to theology students. Or did Torrance use words and phrases at Manchester which, if challenged, he would have altered?

Torrance's statement in *The Mediation of Christ* gives the strong impression that the conceptual thinking required by Maxwell's faith was *consciously* utilised in Maxwell's physics. This was what I had picked up as a young student in Torrance's lecture room. Yet his comments at Paderborn oblige us to reconsider what he may have meant at Manchester.

Might it be he was applying to Maxwell an interpretative tendency of the 1970s and 1980s: namely, that subconscious concepts from a scientist's culture, philosophy, or faith inevitably inform and mould his or her scientific thinking? Was that the context for his term *'regulative force'*? The word *'regulative'* is crucial to this discussion. Fortunately, we can examine Torrance's use of it, not only at Paderborn (1982), but in *The Mediation of Christ* (1983), and in his John Philoponos paper (publ. 2000).

Paderborn. At one point in his Paderborn paper Torrance discussed various impacts of Maxwell's religious beliefs on his science, including: faith liberating him from subjectivity; faith making him determined that nothing be wilfully left unexamined; faith providing truths which science by itself could not deliver (the origin and order of the cosmos); and faith convincing him that human modes of thought can match physical truths.[38] It was in introducing this list that he wrote of Maxwell's Christian faith exercising 'a *regulative* role in the choice

38 Torrance, 'Christian Faith and Physical Science', 222ff.

and formulation of his leading scientific concepts' and gaving him a cast of mind 'shaped through an intuitive apprehension of the relation of God to his creation.'[39] A few pages later he stated unequivocally: 'There is no evidence to show that Clerk Maxwell *directly* allowed his Christian beliefs in the creation and the Incarnation to influence his understanding of space and time.'[40] Clearly, at Paderborn, Torrance's use of 'regulative' did *not* imply that Maxwell's faith regulated his science *consciously*.

Manchester. Torrance's statements at Manchester appear to say more. His nuanced use of 'regulative' at Paderborn jars with what he expressed there.[41] Did he overstate his case? Did he use unguarded language?

Philoponos. The Paderborn use of 'regulative' then recurs in Torrance's paper on John Philoponos. Torrance wrote that Maxwell's 'Christian faith provided Maxwell with certain analogical truths, root ideas, and fundamental conceptions, for which natural science could not account but which guided him in the scientific task of wedding thought with reality and developing appropriate ideas.'[42]

He further stated—in a sentence closely reflecting his Paderborn position—that Maxwell 'did not intrude theological ideas specifically or directly into his scientific theories, but the Christian faith deeply entrenched in his being exercised a *regulative* role in the choice and formation of his leading scientific concepts.'[43]

In claiming this, Torrance was careful to make clear that this was not a *conscious* step: instead, through union with Christ, Maxwell had 'gained an intuitive apprehension of the relation of God to his creation, which provided Maxwell with what he called "a fiducial point or standard of reference"[44] for discriminating

39 Ibid., 222. (Italics ours).

40 Ibid., 236. (Italics ours).

41 Torrance, *The Mediation of Christ,* 59.

42 Torrance, 'The Significance of John Philoponos', 54.

43 Ibid.

44 Torrance borrows this phrase from Maxwell's 1873 lecture, 'Does the Progress of Physical Science . . . ?' (C&G1, 435; Harman, II/2, 814). A fiducial point is a trustworthy principle or belief which can be assumed as a fixed basis for comparison. Maxwell did not himself relate this phrase to the notion of religious beliefs informing his worldview.

scientific judgements.'[45] This links with Torrance's theory of knowledge, in which the mind of an enquirer—through its harmony with the object of enquiry—develops an intuitive understanding of the object because the object itself has began to mould the very way the observer thinks. In this case Torrance was stating that Maxwell's deep faith in the triune God, through his union with Christ, gave him an intuitive feeling for the notion of onto-relations: it was this which made him open to new forms of structural reality in his science. (We are aware that Torrance's epistemology raises a plethora of questions! We cannot pursue these here!)

What of that 'interpretative tendency' of the 1970s and 1980s to which we referred earlier? Was it that which lay behind Torrance's understanding of 'regulative'?[46] In academic circles the term 'regulative principle' was certainly being used to refer to indirect influences of metaphysical or theological beliefs upon science, and to unconscious or semi-conscious consequences on scientific theorizing from often hidden metaphysical and religious beliefs. Such principles were a topic of discussion. There was an interest in setting scientific theories within their cultural contexts, realising that scientific thinking did not take place in a sealed-off environment, but was impacted by wider movements of thought. Taken in this sense, the phrase 'regulative principle' described the presence of a perhaps hidden or indirect connection between science and extra-scientific ideas.

Ernan McMullin (1924–2011) used the term 'regulative principle' several times in his *Newton on Matter and Activity* (1978).[47] Torrance

---

45 Torrance, 'The Significance of John Philoponos', 54. See also page 49: '[Maxwell] brought his Christian theological convictions to bear upon his mathematics and physics in such a way that, as Einstein declared, "it transformed the logical structure of science."'

46 As a Scottish theologian Torrance would be aware of a quite different use of the phrase 'regulative principle'. In Scottish Presbyterian worship the 'regulative principle' held that only forms and manners of worship explicitly commanded by Scripture should be used in Christian worship, whereas the 'normative principle' allowed any form of worship not explicitly forbidden by Scripture. Maxwell would know the term in that liturgical sense. The Church of Scotland, of which Maxwell was an elder, followed the less stringent 'normative principle', whereas some other Presbyterian denominations adhered to the 'regulative principle'.

47 E.g., McMullin, *Newton on Matter and Activity*, 20, 127.

knew of McMullin's work, and met McMullin at least once, when both presented papers at the 1979 Oxford International Symposium on: 'The Sciences and Theology in the Twentieth Century.'[48] McMullin examined the historiography of science for evidence of non-obvious connections between metaphysics, religion, and science.

Another in the field was E.A. Burtt (1892–1989) who focussed on the metaphysical inputs to scientific theories, and who is cited several times by Torrance.[49] Yet another was John Hedley Brooke (b. 1944), whose article 'Religious Belief and the Natural Sciences: Mapping the Historical Landscape' (1996) has a sub-section specifically entitled 'Religious Beliefs as Regulative Principles.'[50]

It is certainly possible to read Torrance as one of several who held that an individual's subconscious cultural, metaphysical, and religious beliefs can impinge on their work without their conscious knowledge, and that such beliefs 'regulated' their whole manner of thinking and the concepts they employed.

What is definitely the case is that Torrance was convinced that Maxwell's deep Christian faith in the Trinity and the Incarnation opened up his mind to a new way of conceptualising reality, even if Maxwell was himself unaware of transferring such modes of thought from his faith to his physics. Intriguingly, Maxwell had also pondered the non-conscious workings of the mind. In an 1857 letter to Richard Litchfield, he wrote: 'I believe that there is a department of the mind conducted independent of consciousness where things are fermented and decocted, so that when they are run off they become clear.'[51]

On our part, we have found no evidence in Maxwell's extant writings of him acknowledging a direct *conscious* debt to theological concepts in the way which Torrance appears to describe in *The*

48 For the Symposium papers, see: A.R. Peacocke (ed.), *The Sciences and Theology in the Twentieth Century,* (Stocksfield: Oriel, 1981).

49 Burtt, *The Metaphysical Foundations of Modern Physical Science* (Garden City, NY: Doubleday Anchor Books, 1955). For instances of Torrance's use of Burtt, see: *Theological Science*; *Theology in Reconstruction*; and *Transformation and Convergence in the Frame of Knowledge.*

50 Brooke, 'Religious Belief and the Natural Sciences: Mapping the Historical Landscape', [Jitse M. van der Meer, ed. *Facets of Faith and Science: Vol. 1, Historiography and Modes of Interaction,* (New York: University Press of America, 1996)], 1–26.

51 JCM, *Letter to Litchfield,* 29 May 1857 (C&G1, 268; Harman, I, 508f.).

Mediation of Christ. Maxwell may have made an unconscious transference of trinitarian and incarnational onto-relational concepts to his physics, through his faith giving him a cast of mind which influenced his understanding of physical reality. It is also possible that Torrance's immense admiration of Maxwell may have made him make assumptions about the way Maxwell's mind worked![52] Whichever is true, the following fuller extract from Torrance's Philoponos paper expresses his settled view:

> It was not that Clerk Maxwell imported theological conceptions as such into his science, but rather that the pressure of his Christian understanding of God and his creation of the world led him to put forward new ideas and ways of thinking that transformed the basic structure of natural science, and were congenial to the Christian understanding of the universe of space and time. In other words, it was his basic Christian beliefs that prompted his new scientific thinking and exercised a regulative role in the choice and formation of his leading scientific concepts.[53]

Analogies, Gödel, and Levels of Being

When Maxwell presented his 'Analogies in Nature' paper to the Apostles' Club in 1856, his view was that 'analogies of relationship' occured across nature because there were only a finite number of ways in which any two entities could ever relate to each other. Analogies of relationship were important for Torrance since, as we have seen, he was enthusiastic about *onto-relations* as common to both natural science and theological science. But did his assertion that onto-relations exist in the natural world as well as in the spiritual

---

52 Our evaluation echoes Tapio Luoma, *Incarnation and Physics: Natural Science in the Theology of Thomas F. Torrance*, 42f: 'It is useful to emphasize that, in Torrance's opinion, through the Incarnation God does not directly give birth to science but, rather, the doctrine of God's Incarnation is fertile enough to produce a mental attitude that has aided in the development of scientific thinking. Historically, however, Torrance's evaluation of the relevance of Maxwell's religious conviction is not without problems. With regard to Maxwell, this much can be said: in him Torrance easily finds an example to use as a confirmation of the idea of contribution. Maxwell's Christian conviction can be proved without difficulty; in whatever manner its significance is evaluated.'

53 Torrance, 'The Significance of John Philoponos', 55.

world imply that the structural logic of the universe (created-being) reflects the structural logic of God (uncreated-bring)? Does the concept of *onto-relation* become a 'thing' common to the created and the uncreated?

For Maxwell and Torrance, the Judaeo-Christian doctrine that God created the universe *ex nihilo* meant that there are boundaries beyond which science cannot go.[54] Science can investigate causal relationships *within* nature, but cannot explain how things came to be.[55] And *creatio ex nihilo* may also imply that *abstract* entities such as the logic and structure of the universe may be as contingent on the free creative will of God as are matter, energy, space, and time. In other words, the particular form of logic and structure which pervades the universe may not be uniquely necessary, since it is not something which God is subject to, but is itself a free creation by the will of God.[56] Is it necessary for similar forms of logic and structure to apply both to the Creator and to the created order if God is at liberty to give form and order—similar or dissimilar—as he wills?

Torrance held that in the Scottish realist tradition—of which he saw Maxwell as part—all scientific truths were considered to be contingent upon the divine will and had no necessity apart from that will.[57] He cited Duns Scotus (c.1266–1308), interpreting Scotus as understanding contingency in terms of God having the freedom to create the universe in such a way that not only the fact of the universe's existence but also its logic was dependent on the divine volition, with God unhindered by any abstract necessity existing

54 Cf. JCM, *Discourse on Molecules,* (C&G1, 359–360). Also, Torrance, 'Christian Faith and Physical Science in the Thought of JCM', 224, notes the limits which scientific activity could not penetrate, and that science halts when we know (a) X has been made and (b) X has not been made by any known natural process. 'Science is incompetent to reason upon the creation of matter out of nothing,' and 'Science investigates the *form* of matter, not matter in itself.'

55 Cf. Mary Hesse, 'Retrospect', [Peacocke (ed), *The Sciences and Theology in the Twentieth Century*], 283: 'Most scientists would agree that it lies beyond their scope to speak of the beginning or the end of all that is, or to give an ultimate explanation of the existence of regularities in nature.'.

56 Cf. Luoma, *Incarnation and Physics,* 112.

57 Torrance, 'Christian Faith and Physical Science', 223.

independently of, or alongside, him from all eternity.[58] Instead, the laws of nature, and the types of relationships which can exist in the cosmos, derive from the creative will of God. Thomas Chalmers had advanced similar thinking when he referred to 'collocations', i.e. ways in which reality is 'distributed' or 'structured', and we know that Maxwell liked Chalmers' ideas.[59] In this scenario, the logic of the created order has a source beyond the created order itself.

This brings two consequences. On the one hand, God gives the created order a consistent rationality, and in so doing makes it amenable to scientific investigation. On the other hand—and Torrance specifically developed this argument—because the universe has been created by the free and voluntary *will* of God, then the basis for that rationality can never be found within itself, nor can its completeness and consistency ever be demonstrated from within itself.

Maxwell insisted that the natural scientist had to work wholly within a realm of natural causes and effects; he also insisted that there was no line of pure deductive logic which could plot a course from the natural to the supernatural, nor was there any line of causation by which the scientist could link the finite with the infinite. The universe may cry out for God. The existence of truly fundamental entities may demand a Creator. The comprehensibility of the universe may appear to seek an intelligent mind behind it. But these can only be indicated inductively and not deductively. Torrance took a similar position, arguing that it linked with theorems developed by Kurt Gödel in the 1930s.[60]

Put simply, Gödel's theorems propose that any non-trivial mathematical system cannot be proven (within itself) to be both consistent and complete, because all non-trivial mathematical systems contain propositions which are undecidable (i.e. neither provable nor disprovable). Thus, any non-trivial mathematical system requires a further system (outside itself but containing the first) in order to demonstrate its own consistency, and so on *ad infinitum*. In Torrance's view, this implied that mathematics—and indeed nature

58 Torrance, *God and Rationality*, 39; Torrance, *Divine and Contingent Order*, v, x, 109, 153 (fn. 5). See also Torrance's earlier essay: Torrance, 'Divine and Contingent Order', in Peacocke (ed.), *Science and Theology in the Twentieth Century*, 85.

59 JCM, 'Discourse on Molecules', (C&G1, 360).

60 Kurt Gödel, *On Formally Undecidable Propositions of Principia Mathematica and Related Systems*, (Tr. B. Meltzer: New York: Dover, 1962), 1.

as a whole—harbours a disruptive feature: mathematicians can never absolutely prove the logical integrity of their discipline, since any such proof would always require appeal to yet another level of mathematical systems *ad infinitum* in order to establish both completeness and consistency.[61] For Torrance, the ultimate level was that of reality itself.

Gödel's theorems have important corollaries. Because they affect all mathematical logic, then they affect all formal logic. And because they affect all formal logic, then they affect reasoning in any and every field of knowledge, whatever its content or subject-matter. Consequently, human reasoning in any and every branch of knowledge needs reference to a higher/more fundamental level in order to have ultimate validation. For Torrance the ultimate level was *being* itself. The place where the simultaneity of consistency and completeness is resolved is not in a higher level of abstract thought, or indeed in any statement about reality, but in *reality* itself.

This conviction—that the demonstrability of consistency and completeness depends on grasping intuitively the logic of being itself—was central in Torrance's reflections on theological and scientific method. And this was how he interpreted Maxwell's use of analogy, insisting that his own concept of 'layers of knowledge' or 'higher levels' lay behind it since, for Maxwell, a full and comprehensive understanding of nature was impossible from within nature.[62] For clarity, we again emphasise that by 'higher' level Torrance meant a level which was higher in *importance* and therefore more *fundamental*: with the ultimate 'higher' level corresponding to being itself, rather than to statements about being.

Torrance may have pushed some of Maxwell's thinking beyond what Maxwell himself expressed, especially in directly linking Maxwell's reflections on *parallel analogies* with his own emphasis on *strata of knowledge*. Yet, Torrance's insistence that the logic of thought (coherence statements) is only valid when it conforms to the logic of reality (through existence statements) was Maxwellian. Why is logic what it is? It is what it is because it is moulded by the forms of reality as God chose to create reality. This takes Chalmers' 'collocation' argument much further than Chalmers himself went, applying it to laws of logic and not just to physical laws of nature.

61 Torrance, 'Christian Faith and Physical Science', 229.

62 Ibid., 223f.

Chalmers had argued in his Bridgewater essay that it was the *comprehensibility* of nature rather than its *complexity* which pointed to a Creator, as Maxwell acknowledged.[63] Steven Adamson, in his PhD thesis on the apologetics of Thomas Chalmers, pointed out that what Chalmers taught in the 1830s at Edinburgh, was a form of natural theology specifically designed to help his students counter the arguments of David Hume.[64] Chalmers was aware of the impact of Hume's critique of religion on his students, and realised that he had to go deeper than Paley's watchmaker argument. Consequently, he focussed on the overall intelligibility of nature, presaging Einstein's maxim: 'The eternal mystery of the world is its comprehensibility.'[65]

Chalmers directed attention to thc fact that the universe has an intelligible order, and to the apparent impossibility of matter and the laws of nature coming together without a guiding principle. This pivoted on his notion of 'collocation' or 'arrangements' or 'dispositions'. Chalmers maintained that attributing the present order of things to randomness or self-organizing matter was nonsensical, as it confused the *laws* of matter and the *disposition* of matter (i.e. the very tendency of matter to obey laws at all).[66] Moreover, he argued that there should

---

63 JCM, 'Discourse on Molecules', (C&G1, 360).

64 Adamson, *The Apologetics of Thomas Chalmers: The Influences, Methods, and Effects of Chalmers' rebuttals to Objections to Christianity,* (PhD, HTC, 2013), 68.

65 Einstein, *Physics and Reality,* (1936), 33.

66 Chalmers, 'Treatise 1' *Bridgewater Treatises,* 19f, & 21: 'It is in the dispositions of matter more than in the laws of matter, where the main strength of the argument lies, though we hear much more of the wisdom of Nature's laws, than of the wisdom of her collocations.' In a footnote Chalmers adds: 'This distinction between the laws and collocations of matter is overlooked by atheistical writers, as in the following specimen from the *Système dc la Nature*" of Mirabaud. "These prejudiced dreamers", speaking of believers in a God, "are in an ecstasy at the sight of the periodical motion of the planets; at the order of the stars; at the various productions of the earth; at the astonishing harmony in the component parts of animals. In that moment however, they forget the laws of motion; the power of gravitation; the forces of attraction and repulsion; they assign all these striking phenomena to unknown causes, of which they have no one substantive idea." When Professor Robinson felt alarmed by the attempted demonstration of La Place, that the law of gravitation was an essential property of matter, lest the cause of natural theology should be endangered by it—he might have recollected

not only be a differentiating between the *existence* of matter and its dispositions, but also between the *laws* of matter and its dispositions.[67] For Chalmers, the concept of 'dispositions' was offered as the guiding principle of the relationship between matter and the laws of nature.[68]

Maxwell extended Chalmers' reasoning. He pointed out that though nature, left to itself, tends to chaos rather than order, there is the strange but obvious fact that this did not occur with fundamental entities and their unique structure and characteristics.[69] Since Maxwell we have become aware that atoms and molecules *can* disintegrate, but Maxwell would not see that as cancelling his main point. Since there must exist in nature truly fundamental and irreducible entities of *some* type—whatever these may be—then his argument applied to them. These fundamental entities have a *disposition*, which, by definition of 'fundamental', cannot be the product of any other cause. That fundamental and permanent *disposition* exists in itself *sui generis.*

Torrance argued in similar fashion that the logic, rationality, and structure of the universe cannot be found within itself, but came from the creative action of God, *ex nihilo.* The rationality of the universe derives from the divine act. This was why, at Paderborn, he stated that Christian theology and realist philosophy combined to convince Maxwell of the principle of the 'contingent nature of the universe'.[70] The universe has no grounds within itself for either its *being* or its *logic*. Both are contingent upon the will and action of God.[71]

---

that the main evidence for a Divinity lies not in the laws of matter, but in their collocations—because of the utter inadequacy in the existing laws to have originated the existing collocations of the material world. So that if ever a time was, when these collocations were not—there is no virtue in the laws that can account for their commencement, or that supersedes the fiat of a God.' Cf. Adamson, *The Apologetics of Thomas Chalmers,* 163.

67 Adamson, *The Apologetics of Thomas Chalmers,* 164.

68 Chalmers, *Natural Theology,* Vol. 2, 60, 269: i.e. why are there laws of nature, and why does matter obey laws?

69 JCM, 'Discourse on Molecules', (C&G1, 360).

70 Torrance, 'Christian Faith and Physical Science in the Thought of JCM', 223ff.

71 Torrance, 'Divine and Contingent Order', [Peacocke (ed), *Science and Theology in the Twentieth Century*], 85: 'The universe is contingent for it does not exist of necessity: it might not have been at all and might very well have been different from what it is. Yet, in coming to be, the universe is characterised by an open-structured order which partakes of its contingence.'

Contingency and Indeterminacy

Torrance used the word 'contingent' mainly in the sense of 'contingent upon' or 'dependent upon', rather than in the sense of a contingent event being one which is inherently indeterminate or random. When he wrote of the principle of the 'contingent nature of the universe' what he meant was that the structural logic of the universe is not self-complete, nor self-sustaining, nor part of a closed system of physical causality, nor the product of a necessary logic which dictates the form of both divine and created being. But contingency did not mean that chance was at the heart of physical reality.

What Torrance *did* mean by his phrase 'the contingent nature of the universe' was that all reality comes from God, is given sense by God, and is dependent upon God continually sustaining it: the free creative act of God is what gives the universe its rationality, and endows it with dependable laws which humanity can investigate.[72] Torrance saw this as a major difference between a closed-circuit Newtonian system and a Maxwell/Einstein relational system. In his mind, the Newtonian model supposed a mechanistic universe which was *complete*, *consistent*, and *closed* in on itself, whereas the Maxwell/Einstein model required physical reality to refer *beyond* itself for validation of its logic and rationality.

It was the closedness and assumed completeness of the Newtonian system which Torrance described as deterministic. Hence, in defining the Maxwell/Einstein model as non-deterministic, he was not saying it was random, but that it had an 'open' structure rather than a 'closed' one, needing reference to another level for its validity to be established. Torrance was not using Maxwell and Einstein to introduce a 'gap' into causality which might create indeterminacy, and thus allow room for God's action. What he was claiming was that a deistic approach no longer fitted with modern scientific thinking, and for two reasons. First, the freedom of God as personal agent has always meant that God has *never* needed a 'gap' in the structures of created order to *allow* him to interact with it. Second, what Torrance inferred from the Maxwell/Einstein model of the universe, was that—even apart from faith issues—no philosophy which assumes that the causal structure of the universe is closed and complete in itself is valid.

---

72 Torrance, 'Divine and Contingent Order', 84ff.

Thus, for Torrance, contingency did not mean disorder. Rather, contingent being is that which derives its existence, logic, completeness, and consistency, not in itself but from that upon which it is contingent. For both Maxwell and Torrance, the ultimate ground for all being was the free action of God who called the universe into existence, and who sustains it moment by moment in an ongoing dynamic relationship.

What Torrance specifically argued, especially in his use of Gödel, was that any and every level of created existence required reference to another level beyond itself for its own consistency and completeness to be established. This is sometimes termed 'stratification', in which each level is itself contingent order, and with each level open to a higher level which alone provides its 'reason why'. In Torrance's thinking, each 'higher' level was a step *towards* and not *away from* the base level of reality itself. When considering the logic of *statements*, the ultimate 'higher' level was, for Torrance, the *logic of being* itself which the logic of statements sought to represent. Going further, when considering the *logic of being*, the ultimate 'higher' level was the free personal action of God in creation.

Did Torrance sometimes give the impression that the Maxwell/Einstein model indicated indeterminacy (in the sense of randomness) within the structures of the physical world? Tapio Luoma, Mary Hesse, and Ernan McMullin all thought so, and each expressed discomfort with Torrance's position.[73] Luoma was particularly troubled by Torrance's insistence that Maxwell had restored a *non-determinist* contingency to the ontology of the universe through his field theory, over against Newton whose science operated with necessities and the hard causalism. Luoma was concerned that, for Torrance, part of the attraction of Maxwell's science lay in its apparent ability to discard determinism and 'initiate a new era in physics, resulting in the indeterminism of the quantum world'.[74]

73 Cf. Hesse, 'Retrospect', 283, 285f. & McMullin, 'How should Cosmology relate to Theology?' 56, both in Peacocke (ed.), *The Sciences and Theology in the Twentieth Century*. Hesse and McMullin's understanding of how Torrance used words such as 'contingency' and 'determinism' may be questioned, but the fact remains that Torrance gave the *impression* that he believed that the Maxwell/Einstein model allowed ontological indeterminacy.

74 Luoma, *Incarnation and Physics,* 108.

Luoma deconstructed Torrance's argument, focussing on what he regarded as Torrance's error: a conviction 'that the novelty and revolutionary nature of Maxwell's theory of the electromagnetic field lies precisely in its indeterminism.'[75] In Luoma's opinion, for this claim to be correct strict causality should not be applicable to the concept of the field, but in fact it can be, since Maxwell's equations do not allow for indeterminacy or probability—in contrast to those of modern quantum mechanics which do. On this basis, Luoma concluded, 'The notion of determinism cannot be separated from the concept of field as strictly as Torrance would like it to be':

> [Torrance's aim was] to assert that Newtonian science operated with necessities and hard causality or determinism, whereas Maxwell could restore contingency to the ontology of the universe through his field theory. The revolutionary character of Maxwell's science lies, then [for Torrance], in its ability to discard determinism and initiate a new era in physics, resulting in the indeterminism of the quantum world. . . . [But] this requires, however, that strict causality cannot be applied to the concept of field. The counterargument was then put forward that it can indeed be applied. . . . The concept of field is not as free of determinism as Torrance claims.[76]

Having made his critique, Luoma was anxious to emphasise positive features in Torrance's use of Maxwell: 'In spite of the ambiguous and not always convincing way in which Torrance uses both Maxwell and Einstein to support his own position, it should not be overlooked how he wishes to underscore the positive significance of modern physics from the standpoint of ontology.'[77]

But, did Luoma misunderstand Torrance's terminology? Elsewhere, Mary Hesse perceptively remarked that Torrance's arguments were generally directed against *deism* rather than *atheism*.[78] As such, much of Torrance's work on science and theology was focussed on breaking away from a deistical view of science—fostered by a Newtonian universe—which allowed no room for God's activity within the created order since, after its moment of creation,

75 Luoma, *Incarnation and Physics*, 108.

76 Ibid., 155.

77 Ibid., 111.

78 Hesse, 'Retrospect', 285.

the universe trundled along a predetermined path governed by mechanical laws. It was that *deistic* approach which Torrance sought to dispel; and one of the tools which he used was his belief that Gödel's theorem implied that no part of either the physical or the abstract structures of the created order were simultaneously complete and consistent in themselves, but required reference to a higher (more fundamental) level for validation, with the ultimate level being the living God himself in his dynamic interaction with the created order. It was in this sense that Torrance stated that the universe was 'non-determined'. He used terms such as 'non-determined' in the sense of not following a mechanical and deistical model. It was the universe's necessary openness to a higher reality which made Torrance describe the universe as non-deterministic and which he saw in Maxwell.

There is another important question. Did Torrance equate '*what is not demonstrable*' with '*what is not true*'? Gödel did *not* state that any non-trivial mathematical system is either incomplete or inconsistent. What he stated was that it is impossible to *demonstrate* or *prove* the simultaneous consistency and completeness of such a system without reference to another system which envelopes the first. But something *can* be true even if it is impossible to *prove* it to be so. Torrance appears to have assumed that consistency and completeness never exist simultaneously within a particular system, resulting in some outcomes as necessarily indeterminate within that system. His argument was: because the logic of causal sequences derives from the logic of mathematics, then if the mathematical system is incomplete then the causal system must also be incomplete. In Torrance's mind, this dovetailed with his argument that the Newtonian closed container-box view of space and time produced determinism, whereas the Maxwell/Einstein relational and open view of space and time was non-deterministic in the sense explained above.

Notwithstanding these caveats, Torrance's general use of Maxwell is clear. He saw Maxwell as understanding that the universe is not a closed cause-and-effect system complete in itself, but requires reference to a higher level of being and logic for validation. For Torrance, this was implied by Maxwell's approach to analogy and causality.[79] Torrance considered that Maxwell taught that the order which we see in nature has an indeterminate quality, 'even though we must employ the kind of necessary relations we have in mathematics

79 Torrance, 'Christian Faith and Physical Science', 224f.

to enable us to grasp and formulate the kind of laws governing contingent order.' Behind this lay Torrance's conviction that physical laws originally derived from an act of divine freedom, and not from an inherent absolute necessity.[80] As such, he made a strong case for all reality—matter, energy, space, time, *plus* the abstract qualities of logic and structure—being contingent upon the free creative act of God.[81] He wrote: 'The ultimate contingent ground of physical law in the free activity and mind of the Creator must be allowed to determine our understanding of the operation of physical law throughout the chain of physical causes.'[82]

Maxwell would agree. But Maxwell would also be careful to make it clear that this did not establish a non deterministic system of cause and effect within nature. As elsewhere, Torrance's mode of expression can give rise to misunderstanding. He never intended to imply that somewhere within nature there exists randomness. What he was concerned to argue was that the free action of God within the created order is not just *possible* but indeed is *necessary* for the being and logic of the universe to be sustained.

Torrance wanted to allow for the free action of God within the created order, despite that created order being a world of causes and effects. In our view this is best achieved by referencing God's *freedom*, rather than by appealing to *indeterminacy* which is an ambiguous term. God is *'most free'* to act with liberty, though always in accordance with his character and his purposes of covenant love.[83] That was how Maxwell saw it. God is personal being. Personal being is that which has the quality of being able to act freely within an otherwise seemingly deterministic matrix. God is *'most free'*. And it is his gift of freedom to humanity, as the *imago dei,* which rescues humanity from being

---

80 Torrance, 'Christian Faith and Physical Science', 225.

81 Torrance, 'The Significance of John Philoponos', 49: '[The teaching of Scripture] about the creation of the world out of nothing and the redemptive incarnation of the Word of God in space and time in Jesus Christ . . . had the effect of destroying the dualist structure of Greek philosophy and science and of revealing the *contingent* nature of the universe and its God-given *rational* order.'

82 Torrance, 'Christian Faith and Physical Science', 225.

83 For our understanding of God's freedom and will see Chapter 27, 'Automata or Free?'.

mere automata. It is not necessary to posit ontological indeterminism within the structures of physical reality for freedom to exist.

As a moralist as well as a physicist, one of Maxwell's aims was to preserve the notion of free decisions, whose outcomes were not constrained or predetermined by material causes. And for Maxwell, 'freedom' was not derived from physics but from his faith. As Torrance pointed out, when Maxwell spoke of God as the *final* or *ultimate cause,* that was to be understood in terms of the dynamical nature of the living God of Scripture.[84] Maxwell did not try to create 'gaps in nature' or 'gaps of indeterminism' within which free events could occur. What he did was to allow God to be God. God is free to act and to love; humanity is free to act and to be loved.

Embodied Mathematics

For James Clerk Maxwell, forms of thought in the mind had their genesis in the structures of external physical reality; in their most abstract form these forms of thought could be expressed through mathematical symbols and formulae since external physical reality had an 'embodied mathematics'.[85] What he meant by embodied mathematics was the existence in nature of a structure and order which corresponded directly to mathematical models apprehended by the mind.

During his student days in Edinburgh, he was interested in the embodiment of geometrical patterns in physical relations, and Torrance has suggested that it was this which gave him early awareness of real analogies in nature.[86] Because God had given the universe a mathematical order the human mind could explain it through mathematics. This mathematical explanation is what the universe *really* is, and is not just a convenient model imposed by the mind. As such, embodied mathematics was the basis for Maxwell's realist philosophy.

Maxwell also saw mathematics as a creative tool of discovery, so long as mathematicians retained a connection with the actual mathematics of nature. The connection might not always be apparent,

84 Torrance, 'Christian Faith and Physical Science', 225.

85 Ibid., 227. Cf. JCM, 'On Faraday's Lines of Force' (1855/56), *NVN1,* 187.

86 Ibid., 220.

with the history of science throwing up several instances of theories which seemed entirely abstract when first explored but which subsequently proved to be descriptions of something real:[87]

This links with Maxwell's realist approach to science. As a boy he loved to experiment. This disposed him to assume real connections between the empirical and the theoretical, a tendency which continued into his adulthood. Throughout his career he believed that there was a *real* relationship between the empirical (experimental) and the theoretical (mathematical description of reality), and not just a *convenient* relationship. He did not accept that mathematical theories are essentially artificial, and that only by accident do they describe reality. He sought *real* knowledge of what was 'out there', whether in science, theology, or morality.

What about Maxwell's own equations? Were they embedded in nature? Torrance argued they were, and that they only had validity because they described something objectively true; this was because the temporal and spatial structures of the field continuum to which they pointed 'do not depend on the activity of the human observer'.[88] For Torrance, this liberated notions of space and time from the 'massive subjectivity on the Kantian metaphysical idea that they are *a priori* forms of man's sensory perception unaffected by empirical objectivity.'[89]

But did Torrance *prove* that Maxwell's differential equations imply that our concepts of space and time are objective? Maxwell, in his essay on analogy, argued that the idea of number and the notion of spatial coordinates were necessary truths for each and every observer, and hence were objective realities in nature and not subjective impositions of the mind. Torrance argued similarly, but using the qualities of Maxwell's equations rather than the simpler concepts of number and coordinates. It was not only because the equations *worked* pragmatically, but because they provided a new way of seeing reality (as a continuum) which fitted reality like a glove, that they indicated that Maxwell's intuitive leap was not an artificial projection ungoverned by reality itself. That was Torrance's point. A process of step-by-step 'integration', would never have enabled Maxwell to form

87 Torrance, *Divine and Contingent Order*, 48f.

88 Torrance, 'Christian Faith and Physical Science', 236.

89 Ibid.

his equations: instead, he grasped the 'whole' and worked in from there. For Torrance, Maxwell was validated by his results.[90]

The great insight which Torrance took from Maxwell was that objects and forces are not to be regarded as separate objects or separate forces acting at a distance, but are qualities of a field which interpenetrate the space-time continuum of the universe. Reality is *onto-relational.* And Torrance's application of onto-relations to both physics and theology opens up an immensely fruitful way of thinking, especially in relation to understanding personhood and triunity.

Torrance also approved of Maxwell's methodology, with Maxwell prepared to go beyond observations and strict experimental evidence.[91] This fitted with Torrance's insistence that scientific enquiry in any field of knowledge involves an intuitive 'leap of knowledge' in which the enquirer grasps the wholeness of a thing without knowing all the steps leading to it. Torrance believed that this intuitive apprehension was employed by the best minds in both science and theology.

Physics needed time post-Maxwell to free itself from interpreting the field mechanically. Torrance was aware that Maxwell himself worked with two models all his life: one was the old Newtonian system; the other was Faraday's more dynamic and relational concept. Torrance saw the first in Maxwell's paper *On Physical Lines of Force* (1856), with the second in his work *A Dynamical Theory of the Electromagnetic Field* (1865).[92] That may be too simple a division, though it is true that in the interval between the two papers Maxwell became aware that a mechanical interpretation did not explain adequately the real connections in nature. The mechanical interpretation was, at best, a specific limiting case. But well-worn ways of thinking are tenacious, especially when they have worked so well in the past. When physics did make its conceptual leap, Einstein took it into new realms. In Torrance's view, an equivalent shift was needed in theology, where the 'field' is the being and action of God in the person of Jesus Christ.

Torrance admired the ability of physicists to develop new ways of thinking when that was forced upon them by reality and they were forced to think *kata phusin* (in accordance with the nature of things),

90 Cf. ibid., 228f.

91 Torrance, 'Christian Faith and Physical Science', 234.

92 Ibid., 232.

and he was persuaded that a similar transformation of conceptualisation should take place in theology with theologians working within the discipline of God's self-revelation. Torrance wanted to integrate being and act as closely as biblically and theologically possible. He was convinced that an integrated and unified understanding was key to uncovering the fullness of who God is, and what he has done in Jesus Christ.

If a fresh edition of Purnell's *Encyclopaedia of Science* were issued today, would James Clerk Maxwell be more prominent? Assuredly. When Richard Feynman was asked in 1962 by the Californian Institute of Technology to present a course on the fundamentals of physics, he, without hesitation, started with Maxwell's equations.

This book has explored some of the questions raised by Torrance's stimulating lectures to his students, especially his references to Maxwell. What has impressed this author is that the James Clerk Maxwell who emerges from this quest was a man whose humble Christian faith illuminated and advanced his scientific thinking. It gave him a foundation of optimism. It fed his conviction that the natural world could be truly known. And it made him believe that through the scientific enterprise the created order comes to know itself, and in so doing gives glory to its Creator. '*Magna Opera Domini Exquisita in Omnes Voluntates*'.

Bibliography

Abbott, Evelyn & Campbell, Lewis, *Life of Benjamin Jowett,* Vol. 1, (London: Murray, 1897).

Adamson, Steven C., *The Apologetics of Thomas Chalmers: The Influences, Methods, and Effects of Chalmers' rebuttals to Objections to Christianity,* (PhD Thesis, Highland Theological College, 2013), (https://pure.uhi.ac.uk/en/studentTheses/the-apologetics-of-thomas-chalmers)

Anonymous, 'Review of *The Unseen Universe*', *Nature,* Vol. 12/290, (20 May 1875), 41–43.

Arthur, John W., *Brilliant Lives: The Clerk Maxwells and the Scottish Enlightenment,* (Edinburgh: Birlinn, 2016).

Allen, Peter, *The Cambridge Apostles: The Early Years,* (Cambridge: CUP, 1978).

Arianrhod, Robyn, *Einstein's Heroes: Imagining the World through the Language of Mathematics,* (Cambridge: Icon, 2004).

Balfour, Ian L.S., *Revival in Rose Street.* (Edinburgh: Rutherford House, 2007).

Bowler, Peter J., *The Eclipse of Darwinism: Anti-Darwinian Evolution Theories in the Decades around 1900,* (Baltimore: John Hopkins University Press, 1983).

Broadie, Alexander, *The Tradition of Scottish Philosophy,* (Edinburgh: Donald, 2011)

Bronowski, Jacob, *The Ascent of Man,* (London: BBC & Book Club Associates, 1973).

Browne, Thomas, *Religio Medici: The Religion of a Doctor,* (1643).

Burdon-Sanderson, Ghetal, *Sir John Burdon-Sanderson: A Memoir,* (Oxford: Clarendon, 1911).

Burns, Islay, *Memoir of the Rev. William C. Burns, M.A.,* (New York: Carter, 1870)

Calaprice, Alice, *The Ultimate Quotable Einstein,* (Princeton: PUP, 2011).

Campbell, Lewis & Garnett, William, *The Life of James Clerk Maxwell,* (2nd ed.) (Merchant Books, 1884).

Campbell, Neil, & Smellie, R. Martin S, *The Royal Society of Edinburgh (1783-1983)*, (Edinburgh: RSE, 1983).

Carter, M.D; Dempsey, M.W.; etc., (eds.), *The Illustrated Encyclopaedia of Science*, (London: Purnell & Sons, 1960).

Chalmers, Thomas, *Natural Theology*, Vol. 2, (New York: Robert Carter & Brothers, 1857).

Christie, George, *The Story of St. Andrew's Edinburgh*, (Edinburgh: Blackwood, 1934).

Clark, Ronald W., *Einstein: The Life and Times*, (London: Hodder, 1996 ed.).

Cockburn, Lord Henry, *Memorials of His Time*, (New York: Appleton, 1856).

Colebrooke, Andy, *Science and Religion: The Spirituality of James Clerk Maxwell*, (Cambridge: Grove Books, 2020).

Colloms, Brenda, *Victorian Country Parsons*, (London: Book Club Associates, 1977).

Crawford, Thomas J., *The Fatherhood of God*, (Edinburgh: Blackwood, 1868).

Crawford, Thomas J., *The Mysteries of Christianity: Revealed Truths Expounded and Defended*, (Edinburgh: Banner of Truth, 2016 reprint. Baird Lecture for 1874).

Deacon, Richard, *Cambridge Apostles: A history of Cambridge University Elite Intellectual Society*, (London: Royce, 1985).

Dewar, Daniel, *The Work of the Holy Spirit in the Revival of Religion, and the Duty of Christians in regard to It*, (Aberdeen: King, 1839).

Dick, Thomas, *The Christian Philosopher*, (London: Whitaker, 1823).

Dick, Thomas, *The Discoveries of Modern Geology not Inconsistent with Revelation*, (Edinburgh: Dalrymple, 1843)

Domb, C. (ed.), *Clerk Maxwell and Modern Science: Six Commemorative Lectures*, (London: Athlone Press, 1963).

Einstein, Albert, 'Physics and Reality (1936): reprinted: *Daedalus: Journal of the American Academy of Arts & Sciences*, (Fall, 2003), 22–25.

Einstein, Albert, *Autobiographical Notes*, (Illinois: Open Court, 1979).

Elder, Frederick Kingsley, 'The Religious Outlook of James Clerk Maxwell', *History of Physics Group Newsletter* (Spring 2000), 41–47.

Ellicott, C.J., *Modern Unbelief: Its Principles and Characteristics, Six Addresses*, (London: SPCK, 1877).

Ewing, William, *Annals of the Free Church of Scotland*, Vol. 2, (Edinburgh: T&T Clark, 1914).

Francis Everitt, *James Clerk Maxwell: Physicist and Natural Philosopher,* (New York: Charles Scribner's, 1975).

Farrar, Reginald, *The Life of Frederic William Farrar: Sometime Dean of Canterbury,* (London: James Nisbet, 1904).

Fergusson, Alexander, *Chronicles of the Cumming Club and Memories of Old Academy Days,* (Edinburgh: Constable, 1887).

Fergusson, David & Elliot, Mark (eds.), *The History of Scottish Theology, Vol. 2: The Early Enlightenment to the Late Victorian Era,* (Oxford; OUP, 2019).

Flood, McCartney, Whitakers (eds.), *James Clerk Maxwell: Perspectives on his Life and Work,* (Oxford: OUP, 2014).

Forbes, George, *David Gill: Man and Astronomer,* (London: John Murray, 1916)

Forfar, David O., 'Maxwell, Einstein, Newton and Faraday', *NJCMF,* Issue 1 (Spring 2012).

Forsyth, Andrew Russell, 'Old Tripos Days at Cambridge, *The Mathematical Gazette,* Vol. 19, No. 234, July 1935, 162–179.

Fortman, Edmund J., *The Triune God: A Historical Study of the Doctrine of the Trinity,* (London: Hutchinson, 1972).

Gaal, Botond, *The Faith of a Scientist: James Clerk Maxwell,* (Debrecen: Istvan Hatvani Theological Research Centre, 2003).

Garnett, William, 'James Clerk Maxwell', *Nature,* 13 Nov 1879, 43–46.

Garnett, William, *Heroes of Science: Physicists,* (London: SPCK, 1885).

Geike, Archibald, *A Long Life's Work: An Autobiography,* (London: Macmillan, 1924).

Glazebrook, Richard T., *James Clerk Maxwell and Modern Physics,* (New York: Macmillan, 1896).

Goodwin, Harvey, *Education for Working-Men,* (Cambridge: Deighton, Bell & co., 1855).

Goodwin, Harvey, *Memoir of Bishop Mackenzie.* (Cambridge: Deighton, Bell, & Co, 1864).

Goodwin, Harvey, *The Origin of the World according to Revelation and Science,* (London: SPCK, 1880).

Goodwin, Harvey, *The Foundations of the Creed,* (London: Murray, 1890).

Gordon, Margaret, *The Home Life of Sir David Brewster,* (Edinburgh: Douglas, 1881).

Graham, Henry G., *Scottish Men of Letters in the Eighteenth Century,* (London: A&C Black, 1901).

Hamilton, William, *Lectures on Metaphysics and Logic,* Vol. 3, (ed. Mansel & Veitch, Edinburgh: Blackwood, 1866).

Harman, Peter M. (ed.), *The Scientific Letters and Papers of James Clerk Maxwell, Volume I:1846–1862,* (Cambridge: CUP, 1990).

Harman, Peter M. (ed.), *The Scientific Letters and Papers of James Clerk Maxwell, Volume II, Parts 1 & 2:1862–1873,* (Cambridge: CUP, 1995).

Harman, Peter M. (ed.), *The Scientific Letters and Papers of James Clerk Maxwell, Volume III, Parts 1 & 2:1862–1873,* (Cambridge: CUP, 2002).

Harman, Peter M., *The Natural Philosophy of James Clerk Maxwell,* (Cambridge: CUP, 1998).

Hay, David Ramsay, *First Principles of Symmetrical Beauty,* (Edinburgh: Blackwood, 1846).

Heimann, P. M., 'The "Unseen universe": Physics and the Philosophy of Nature in Victorian Britain', *The British Journal for the History of Science,* Jun. 1972, Vol. 6, No. 1 (Jun. 1972), pp. 73-79. (Cambridge: CUP, 1972).

Henry, John, 'Voluntarist Theology at the Origins of Modern Science: A Response to Peter Harrison', *History of Science,* 47/1, (2009), 79–113.

Herschel, John F. W., *Preliminary Discourse on the Study of Natural Philosophy,* (New ed.: London: Longman, 1851).

Hesse, Mary Brenda, *Forces and Fields. The concept of action at a distance in the history of Physics,* (London: Nelson, 1961).

Hertog, Thomas, *On the Origin of Time,* (London: Torva, 2023).

Hort, Arthur Fenton, *Life and Letters of Fenton John Anthony Hort,* Vol. 2, (London: MacMillan, 1896).

Hughes, Thomas, *Memoir of Daniel MacMillan,* (London: MacMillan, 1883).

Huxley, Thomas H., *Collected Essays,* Vol. 1, (London: Macmillan, 1904).

James, Ioan, *Remarkable Physicists from Galileo to Yukawa,* (Cambridge: CUP, 2004).

Jeffrey, Kenneth S., *When the Lord Walked the Land: The 1858–62 Revival in the North-East of Scotland,* (Eugene: Wipf & Stock, 2006).

Jones, Reginald Victor, 'James Clerk Maxwell at Aberdeen, 1856-1860', *Notes and Records of the Royal Society of London,* Vol. 28, No. 1 (Jun. 1973), 57-81.

Jones, Reginald Victor, 'The Complete Physicist: James Clerk Maxwell, 1831–1879, (RSE Yearbook 1980), 5–23.

Keswani, G.H., 'Clerk Maxwell Centenary', *Nature,* Vol. 283 (January 1980), 330.

Knott, Cargill Gilston, *The Life and Scientific Work of Peter Guthrie Tait: Supplementing the two volumes of Scientific Papers published in 1989 and 1900,* (Cambridge: CUP, 1911).

Larmor, Joseph, 'James Clerk Maxwell', *Nature,* (Supplement: Dec 1926), 44–47.

Lindley, David, *Degrees Kelvin: The Genius and Tragedy of William Thomson,* (London: Aurum, 2004).

Longair, Malcolm, *Maxwell's Enduring Legacy: A Scientific History of the Cavendish Laboratory,* (Cambridge: CUP, 2016),

Luoma, Tapio, *Incarnation and Physics: Natural Science in the Theology of Thomas F. Torrance,* (Oxford: OUP, 2002),

Lyell, Charles, 'On the Occurrence of Works of Human Art in post-Pliocene Deposits', *Report of the BAAS,* 1859, (London: Murray, 1860), 93–95.

MacDonald, David K. C., *Faraday, Maxwell, and Kelvin,* (London: Heinemann, 1964).

MacDonald, Ian R., *Aberdeen and the Highland Church: 1785–1900,* (Edinburgh: Saint Andrew Press, 2000).

MacEwen, A. R., *Life and Letters of John Cairns,* (London: Hodder & Stoughton, 1898).

MacIvor, Iain, 'The Story of 14 India Street, Edinburgh', (JCMFL: ED06).

Mackay, Andrew C. & Callander, Sam, *James Clerk Maxwell: Pathfinder of Modern Science,* (Dumfries: Kirkpatrick Durham, Parton & Corsock Community Council,1979).

Mahon, Basil, *The Man Who Changed Everything: The Life of James Clerk Maxwell,* (Chichester: Wiley, 2004).

Maurice, John Frederick Denison, *Theological Essays,* (1853. Reprint, London, James Clarke, 1957).

Maxwell, James Clerk, *Matter and Motion,* (Reprint, London: SPCK, 1920).

Maxwell, James Clerk, *Letter to his Aunt,* 24th May 1858, (manuscript), Basel University Library: Autographen-Sammlung, Geigy-Hagenbach 2168

Maxwell, James Clerk, 'Mathematics and Physics: Presidential Address to the British Association for the Advancement of Science. September 1870' In: *Report of the 40th Meeting of the BAAS, held at Liverpool in September 1870,* (London: Murray, 1871), 1–9.

Maxwell, James Clerk, *Introductory Lecture on Experimental Physics: 25 Oct 1871,* (London & Cambridge: Macmillan, 1871).

Maxwell, James Clerk, *Theory of Heat, (3rd ed.)*, (London: Longmans, 1872).

Maxwell, James Clerk, 'On Action at a Distance', 21 Feb 1873, *Proceedings of the Royal Institution of Great Britain,* Vol. VII (1873-1875), 44–54, (London: Clowes & Sons, 1875).

Maxwell, James Clerk, *'Discourse on Molecules', The London, Edinburgh, and Dublin Philosophical Magazine and Journal of Science,* Vol. 46/308 (1873), *453–469.*

Maxwell, James Clerk, *A Treatise on Electricity and Magnetism,* (Oxford: Clarendon, 1873).

Maxwell, James Clerk, Review of '*Paradoxical Philosophy: A Sequel to the Unseen Universe,*' in *Nature* 19 (1878), 141–143.

Maxwell, James Clerk, 'Atom', *Encyclopaedia Britannica* (9th ed.), Vol. 3, 36–49.

Maxwell, James Clerk, 'The Physical Sciences', *Encyclopaedia Britannica* (9th ed.), Vol. 19, 1-3.

M'Kerly, P. H., *History of the Lands and their Owners in Galloway,* (Edinburgh: Paterson, 1879).

McMullin, Ernan, *Newton on Matter and Activity,* (South Bend: University of Notre Dame Press, 1978).

McNatt, Jerrold L., 'James Clerk Maxwell's Refusal to Join the Victoria Institute', *Perspectives on Science and Christian Faith: Journal of the American Scientific Affiliation,* (Vol. 56, No. 3, Sept 2004), 204–215.

McNaughton, William D, *The Scottish Congregational Ministry: 1794–1993,* (Glasgow: Congregational Union of Scotland, 1993).

Mitchell, Walter (ed.), *The Transactions of the Victoria Institute, or Philosophical Society of Great Britain,* Vol. 1, (London: Hardwicke, 1867).

Moody Stuart, Kenneth, *Alexander Moody Stuart: A Memoir,* (1899. Republished: Edinburgh, Banner of Truth, 2023).

Moorman, J. R. H., *A History of the Church in England.* (London: A&C Black, 1953).

Niven, W.D. (ed.), *The Scientific Papers of James Clerk Maxwell,* Vol. I, (Cambridge: CUP, 1890)

Offord, John, *Faithful Words: Or, The Inner Life,* (London: Shaw, 1867).

Offord, John, *Notes of Sermons by the late J. Offord, taken by a Member of the Congregation (Madame Camilla Soares),* (London: 1870).

Olson, Richard, *Scottish Philosophy and British Physics, 1750–1880: A Study in the Foundations of the Victorian Scientific Style,* (Princeton: PUP, 1975).

Pearson, Karl, 'Old Tripos Days at Cambridge as seen from Another Viewpoint', *The Mathematical Gazette,* Vol. 20, No. 237, 27–36.

Perman, Ray, *James Hutton: The Genius of Time,* (Edinburgh: Birlinn, 2022).

Pesic, Peters, 'The Bell and the Buzzer: On the meaning of Science', *Daedalus: Journal of the American Academy of Arts & Sciences,* (Fall, 2003), 35–44.

Petzold, Charles, 'Maxwell, Molecules, and Evolution', (Internet article accessed 26 Sep 2022: www.charlespetzold.com/etc/MaxwellMoleculesAndEvolution.html)

Ramsay, Edward Bannerman, *A Manual of Catechetical Instruction compiled and arranged for the use of Young Persons,* (Edinburgh: Grant, 1851).

Ramsay, Edward Bannerman, *Pulpit Table-Talk: Remarks and Anecdotes on Preachers and Preaching,* (London: Cassell, Petter & Galpin, 1887 ed.).

Ramsay, Edward Bannerman, *Reminiscences of Scottish Life and Character,* (Edinburgh: Foulis, 23rd ed., 1872).

Raverat, Gwen, *Period Piece: A Cambridge Childhood,* (London: Faber & Faber, 1952).

Rawnsley, H. D., *Harvey Goodwin, Bishop of Carlisle: A Biographical Memoir,* (London: Murray, 1896).

Riddell, J. Oliver, *Clerk Maxwell of Glenlair,* (Castle Douglas: Rae, 1930).

Ritchie, David, *Lectures Explanatory and Practical on the Doctrinal Part of the Epistle to the Romans,* (2 Vols., Edinburgh: Neil & Co, 1831)

Robertson, J. K., *St. Andrews: Home of Golf,* (Edinburgh: Macdonald, revised ed., 1974).

Rodgers, R. E. L., *The Life and Work of Robert Watts,* (Tain: Christian Focus, 1989).

Rogers, Charles, *Memorials and Recollections of the Very Reverend Edward Bannerman Ramsay, LL.D., F.R.S.E., Dean of Edinburgh,* (London: Charles Griffin, 1873).

Sage, Donald, *Memorabilia Domestica: Parish Life in the North of Scotland,* Edinburgh: John Menzies, 1889).

Secord, David, *Victorian Sensation: The Extraordinary Publication, Reception, and Secret Authorship of 'Vestiges of the Natural History of Creation',* (Chicago: The University of Chicago Press, 2000).

Shairp, John; Tait, Peter Guthrie; Adams-Reilly, A, *Life and Letters of James David Forbes, FRS,* (London: Macmillan & Co., 1863).

Shuster, Sir Arthur, *A History of the Cavendish Laboratory, 1871–1910*, (London: Longmans, Green & Co., 1910).

Smith, Charles J., *Historic South Edinburgh*, Vol. 4, (Haddington: Skilton, 1988).

Smith, Robert, *The Quiet Thoughts of a Quiet Thinker: Being Extracts from the Diaries of Robert Smith of Corsock*, (Edinburgh: Oliphant, Anderson & Ferrier, 1896).

Stanley, Matthew, *Huxley's Church & Maxwell's Demon*, (Chicago: UCP, 2015).

Stevenson, W. Scott (ed.), *Famous Edinburgh Students*, (Edinburgh: T. N. Foulis, 1914).

Stewart, Balfour & Tait, Peter G., *The Unseen Universe*, (2nd ed.; London: MacMillan, 1886).

Stewart, Balfour & Tait, Peter G., *Paradoxical Philosophy: A Sequel to the Unseen Universe*, (London: MacMillan, 1879).

Strien, Marij van, 'Vital Instability: Life and Free Will in Physics and Physiology, 1860-1880', *Annals of Science*, 72/3, (2015), 381-400.

Sturrock, George, *Corsock Parish Church: Its Rise and Progress*, (Castle Douglas: Rae, 1899).

Tait, Peter G., 'James Clerk Maxwell', *Encyclopaedia Britannica*, (9th ed.), Vol. 15, 646–647.

Tait, Peter G., *Sketch on Thermodynamics*, (Edinburgh: Edmonston & Douglas, 2nd ed. 1877).

Tait, Peter G. & Thomson, William, *Treatise on Natural Philosophy, Vol. 1, Part 1*, (Oxford: Clarendon, 1867).

Tait, Peter G. & Thomson, William, *Treatise on Natural Philosophy, Vol. 1, Part 1*, (New ed.), (Cambridge: CUP, 1879).

Tayler, Charles B., *A Volume of Sermons*, (London: Hessey, 1828).

Tayler, Charles B., *Facts in a Clergyman's Life*, (New York: Stanford & Swords, 1849).

Theerman, Paul, 'James Clerk Maxwell and Religion', *American Journal of Physics*, Vol. 54/4, 312–317 (1983).

Thompson, Sylvanus P., *The Life of William Thomson: Baron Kelvin of Largs*, Vol. 1, (London: Macmillan, 1910).

Thomson, J. A., *The System of Animate Nature: The Gifford Lectures delivered in the University of St. Andrews in the years 1915 and 1916*, Vol. 1. (London: Williams & Norgate, 1920).

Thomson, J. J. et. al, *James Clerk Maxwell: A Commemoration Volume 1831–1931,* (Cambridge: CUP, 1831., reprint 2011).

Thomson, William, 'The Kinetic Theory of the Dissipation of Energy', *Proceedings of the RSE,* Vol. 8, 325-34, (1874).

Thomson, William, *Mathematical and Physical Papers,* Vol. V, (Cambridge: CUP, 1904).

Thomson, William, *Baltimore Lectures on Molecular Dynamics and the Wave Theory of Light,* (Cambridge: CUP, 1904).

Thomson, William, *Popular Lectures and Addresses,* Vol. II, (London: Macmillan, 1894).

Tolstoy, Ivan, *James Clerk Maxwell: A Biography,* (Edinburgh: Canongate, 1981).

Torrance, T. F., *Theological Science,* (London: OUP, 1969).

Torrance, T. F., *Space, Time, and Incarnation,* (London: OUP, 1969).

Torrance, T. F., *God and Rationality,* (London: OUP, 1971).

Torrance, T. F., *Transformation and Convergence in the Frame of Knowledge,* (Belfast: Christian Journals Ltd, 1980).

Torrance, T. F., *Christian Theology and Scientific Culture,* (Belfast: Christian Journals, 1980).

Torrance, T. F., *Divine and Contingent Order,* (Oxford: OUP, 1981).

Torrance, T. F., 'Review of Ivan Tolstoy's *James Clerk Maxwell: A Biography*', *Life and Work,* (Edinburgh: Church of Scotland, December 1981).

Torrance, T. F., *The Mediation of Christ,* (Exeter: Paternoster, 1983).

Torrance, T. F., *The Christian Frame of Mind: Reason, Order, and Openness in Theology and Natural Science*, (Colorado Springs, Helmers & Howard, new ed. 1989).

Torrance, T. F., *The Trinitarian Faith,* (London: T&T Clark, 2nd ed. 2016).

Torrance, T. F. (ed.), *A Dynamical Theory of the Electromagnetic Field (James Clerk Maxwell),* (Eugene: Wipf & Stock, 1996).

Torrance, T. F., *Scottish Theology: From John Knox to John MacLeod Campbell,* (Edinburgh: T&T Clark, 1996).

Torrance, T. F., *The Christian Doctrine of God: One Being Three Persons,* (London: T&T Clark, 2016).

Torrance, T. F., *The Significance of Philoponos as a Forerunner of Clerk Maxwell,* (History of Physics Group, Newsletter, Spring 2000, No. 13).

Torrance, T. F., *The Ground and Grammar of Theology,* (2nd ed., Edinburgh: T&T Clark, 2001).

Tyndall, John, *Faraday as a Discoverer,* (London: Longmans, Green, & Co., 1868).

Tyndall, John, *Principal Forbes and his Biographers: Strictures on the Life and Letters of John David Forbes by John C. Shairp and Others,* (London: 1873).

Tyndall, John, 'Presidential Address', *Report of the 44th Meeting of the BAAS* (1874), lxvi–xcvii.

Veitch, John, *Memoir of Sir William Hamilton,* (Edinburgh: Blackwood, 1869).

Walker, Robert, 'James Clerk Maxwell', *The Aberdeen University Review,* Vol. 3, No. 9, (June 1916), (Aberdeen: AUP, 1916), 193–213.

Walter, Scott, 'How Maxwell made his mark', *Nature,* Vol. 409, (18 Jan 2001), 283–284.

Walton (ed.), William, *The Mathematical and other Writings of Robert Leslie Ellis, M.A.,* (Cambridge: Deighton, Bell, & Co., 1863).

Watt, Francis, *The Book of Edinburgh Anecdotes,* (Edinburgh: Foulis, 1912).

Whewell, William, *Astronomy and General Physics, considered with References to Natural Theology,* (London: Pickering, 1834)

Whitrow, G.J., *The Nature of Time,* (London: Pelican, 1975).

Williams, John, *Essays on Various Subjects: Philological, Philosophical, Ethnological, and Archaeological,* (London: Russell Smith, 1858).

Wilson, George, *The Life of the Honourable Henry Cavendish,* (London: Cavendish Society, 1851).

Wilson, George, *Researches on Colour Blindness,* (Edinburgh: Sutherland & Knox, 1855).

Wilson, George, *Counsels of an Invalid: Letters on Religious Subjects,* (London: MacMillan, 1862).

Wilson, Jessie Aitken, *Memoir of George Wilson,* (London: MacMillan, 1862).

Index